现代养猪前沿科技与实践应用丛书

现代养猪
应用研究进展

闫之春 ◎ 主编

中国农业出版社
农村读物出版社
北 京

编 写 委 员 会

虽然我国现代养猪业经过数十年的发展取得了很大进步，但是目前现代养猪整体上仍然是一个周期长、困难多、风险大，处于发展期的行业。一方面，养猪企业在实际生产和经营过程中，经常出现一些技术瓶颈、难题，需要尽快解决；另一方面，对企业的经营管理人员来说，在面对多种技术方案并存和未来发展方向的选择方面，也经常需要有专业基础和科学证据才能决定取舍。作为超大型养猪企业，面对这些难题，不可能根据经验臆断来做决策，唯一正确的选择是在企业开展科学试验，为生产经营提供可验证的解决方案。

为了推动基于实际问题的科学研究，促进企业从传统农牧企业向现代企业转型升级，新希望六和股份有限公司于 2018 年 2 月 10 日在青岛正式挂牌成立了新希望六和养猪研究院，成为全公司专门开展现代高效生猪养殖相关技术研发的机构。养猪研究院成立以来，始终遵循"一切从实际问题出发，一切遵循严谨的科学方法，一切着眼于未来发展"的工作方针，在养猪生产技术更新、营养饲料精准优化、猪场配套设施升级，特别是非洲猪瘟预防、控制、净化，以及疫情常态下复产等方面取得了较大的成果，大大提高了生产效率，降低了生产成本。

养猪研究院的科技人员秉承"科技引领，养猪创新"的核心理念，定位为养猪产业新技术、新产品的孵化器，通过与养猪行业内外的合作交流，加速相关技术和产品产出，力争成为企业和行业发展的技术引擎，引领企业快速发展。

高强度、连续开展的应用研究需要不断总结、改进、交流，才能快速提高科研水平。本书收集了 2018—2022 年养猪研究院的部分科研报告、已发表的文章及部分已获得的专利，希望通过交流得到同行校正，以期对将来的养猪应用研究有所裨益，提高我们的科研水平。

放眼未来，现代养猪必将会有更多未知的技术问题，有待科学试验来解答，还有很多艰巨的机制障碍有待突破，但我们相信立足生产实际的科研工作一定能够为养猪事业做出更大的贡献。

编　者

2022 年 8 月

CONTENTS　目录

前言

Chapter

1

猪群健康篇

Assessment and Appliction of Whole-herd-sampling, qPCR-based-testing, and Precision-removal Method in Eliminating ASFv in Four Large Swine Herds in China

（全群抽样、基于 qPCR 检测和精密去除法在中国四个大型猪群中消除 ASFv 的评估和应用）

Xiaowen Li　Weisheng Wu　Peng Li　Junxian Li　Wenchao Gao

Jincheng Yu　Mingyu Fan　Lixiang Kang　Yunzhou Wang　Guiqiang Yuan

Qiannan Yu　Jing Ren　Zhong Yan　Liqiang Zhu　Jintao Li　Xiaoyang Zhang

Zhenwen Shao　Qingyuan Liu　Weiwei Zheng　Shiran Fan　Yakuan Huang　Jing Chen

John Deen　Zhichun Jason Yan

Abstract： Since the firstreport of ASFV in China in 2018，conventional whole herd depopulation method to control ASF has proved unwieldly because of high production intensity and complex trade network. To provide an alternative to conventional methods，we evaluated the feasibility of implementating an extensive sampling method and qPCR tests to determine the status of ASFV in herds，with a rapid response to identified outbreaks.

By assessing and applying these methods without whole herd depopulation，we successfully controlled ASF and eliminated the virus from 4 large swine herds. The time to negative herd was 19，28，14，and 1 days in farms 1-4 respectively. Retention rates of pigs of farm 1 to farm 4 was 69.7%，65%，99.4% and 99.72% respectively.

We anticipated that this innovative method would replace the conventional stamping out one and greatly facilitate the control and eradication of ASFV in China and worldwide.

Keywords： african swine fever virus；quantitative PCR；test removal

The first outbreak of ASF in China was reported on 3rd August, 2018[1] . Without vaccine，the only available tool to prevent ASFV is the implementation of strict bio-security measures at regional and farm level. The response of the veterinary authorities was consistent with the OIE guidelines[2] . This means that the Ministry of Agriculture and Rural Affairs (MARA) of China required depopulation of infected and proximal farms once ASFV was reported[3] . However，it soon became apparent，that these measures were not effective in

China for a number of reasons. These include the high density of pig farms, of which the vast majority were small-scale producers that were connected with each other via highly complex pig and pork trade networks. There may also have been inefficiencies in the response by the industry and government in the early stage of the epidemic. Unlike for other pig diseases such as pseudorabies, there is not any scientific literature that describes the elimination ASFV within a herd without of whole herd depopulation.

Informal reports from affected farmers indicate that the transmission pattern of ASFV in large herds differs from that of other major swine viral diseases. Published information indicates that after infection, time to the manifestation of typical clinical signs varied considerably among pigs, as well as within-pen contact pigs and cross-pen contact pigs. ASFV DNA detection in oral, nasal and rectal swab samples occurred between 0 and 2 days before onset of clinical signs[4,5], suggesting that early detection of viral DNA by qPCR prior to clinical signs may be possible as part of active within-herd ASF surveillance.

It was observed that ASFV spreads relatively slowly within a herd following introduction, with estimated within-pen basic reproduction ratio (R0) of approximately 2.8 and between-pen R0 of approximately 1.4[6]. Moreover, environmental contamination of ASFV can be relatively light, as was shown that introduction of negative pigs into contaminated pens 3, 5, 7 days after infected pigs were removed did not cause infection in this study[5]. In addition, aerosol transmission of ASFV appears to be unlikely to occur as the half-life of ASFV in air was estimated to be 14-19 minutes[7]. Moreover, we observed that most infections within barns were associated with common routes of contact via direct contact, feces or saliva. This published scientific information together with the impression from farms in China indicated that it might be possible to achieve ASFV eradication within a herd by using a test-and-removal method.

Surveillance would be most practical if pooled samples could be used for diagnosis without compromising sensitivity, and one study using pooled EDTA blood to test for ASFV by qPCR indicated that it may be feasible[8].

Our current approach is based on a combination of published research on ASFV together with our own experience with ASFV management in large pig herds. In this study we assessed the feasibility of method in eliminating ASFV Georgia 2007/1 Strain using detailed results from 1 finishing and 3 sow herds.

1　Case presentation

1.1　Validation of maximum number of sample pooling for achieving acceptable sensitivity of ASFV DNA detection

When the initial Ct was approximately 20, even after 2 048 times of dilution, the final

Ct was still around 32 (Fig. 1). However, when the primary Ct was set at around 33, the final Ct values became around 38.5 after 16 folds of dilution. Likewise, as the dilution increased, the variation increased dramatically.

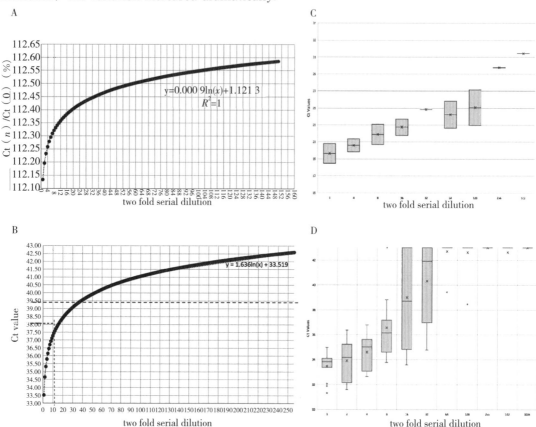

Fig. 1   Validation of maximum number of sample pooling for achieving acceptable sensitivity of ASFV DNA detection. Samples with original Ct value of～20 and～33 respectively were serially diluted and measured by qPCR. Standard curve was established based on qPCR results（A）standard cure of samples with initial Ct of～20（B）standard cure of samples with initial Ct of～33 Variations in Ct values after serial dilution（C）variation of samples with initial Ct of～20（D）variation of samples with initial Ct of～33.

1.2   TTNH (time to negative herd) of a finishing herd and three sow herds

Time to negative herd (TTNH), which is literally defined as time period from the day of first ASFV detection to the day of last ASFV detection in pigs and environment, in Farm 1 was 19 days (Fig. 2A). On farm 1 (a finishing herd), Ct values of ASFV-qPCR in lymph nodes from 4 dead pigs were 20.97（room 1），20.39（room 3），32.67（room 6），29.85 （room 3），respectively on day 0（May 14th，2019）. Considering the low Ct values, the whole pen was depopulated in a bio-secure manner on day 0. On day 2, samples from three dead pigs and two vehicles transporting pigs were found to be qPCR positive. The whole herd sampling，which yielded 241 pig samples，135 enviromental samples，and testing on day 4 showed that only the environmental sample from the pig transfer corridor was qPCR positive with a Ct value of 32.55. After precision removal of pigs and thorough cleaning and disinfection of the pig contact area，the herd remained negative for 14 consecutive days （approximately two incubation periods）until Pen 10 in Room 5 was found to be qPCR positive in oral fluid samples in day 19. The method was carried out again and the herd restored its negative status by removing a total of 1 282 pigs and remained ASFV negative since day 19 until the batches were marketed.

TTNH for Farm 2 was 28 days (Fig. 2B). In Production Line 1 of Farm 2，one out of twenty one samples，ASFV was first detected by qPCR (Ct value of 22. 7) in a Lymph node sample from a dead pig in Stall G17 of Gestation Room 1 in July 11 of 2019 （day 0）. Later in the same day，whole herd sampling and testing of Line 1 were performed. Pooled samples of Row A in Gestation Room 1 and Row K of Gestation Room 2 were found ASFV qPCR positive with a Ct value of 34. 71 and 34. 87，respectively. We observed an intermittent mode of qPCR positive results. The whole surface environmental samples in Gestation Room 1 was found to be qPCR positive almost every day until day 28 when disinfectant sodium hypochlorite was applied. Pigs in Gestation Room 2 restored negative status from day 6，even though the whole surface of environmental samples were found ASFV qPCR positive in day 8 and day 26 and day 27. Samples on 2 working staff were found to be qPCR positive at day 4 and day 23，respectively. The whole herd of Production Line 1 restored negative status from day 28 and remained negative ever since last detection （data not shown）. An intermittent mode of ASFV detection was found probably due to not-fully implemented paired sampling.

TTNH for Farm 3 was 14 days. (Fig. 2C). In Farm 3，a nasal，oral and rectal （NOR）swab sample from an off-feed sow was found ASFV qPCR positive with Ct values of 35. 65 (Stall A-87) in day 0 (January 28th，2020). The Ct value of qPCR results from lymph node sample was 32. 39. All samples from clinically abnormal pigs，whole surface of environment，personnel，and supplies were found qPCR negative after precision removal and thorough cleaning and disinfection. One interesting finding was that the feeder outlet from A87 was shown qPCR positive with a Ct value of 39. 53 on day 14. Farm 3 remained negative since day 14 after the feeder was decontaminated by sodium hypochlorite.

TTNH for Farm 4 was 1 day. (Fig. 2D). In farm 4，the nasal，oral and rectal （NOR）swab samples from an off-feed sow was found ASFV qPCR positive with Ct values of 26. 12 (Stall H26) on day 0 (June 4th，2020). The Ct value of qPCR results from lymph node samples was 32. 39 （data not shown）. After precision removal and thorough cleaning and disinfection，the herd restored its negative status in day 1 and remained negative since then. （data not shown）

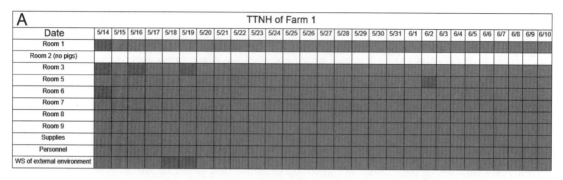

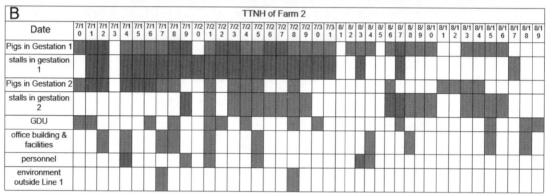

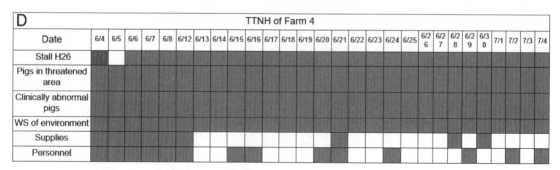

Fig. 2　TTNH (time to negative herds) in Farm 1 (A), 2 (B), 3 (C), 4 (D) respectively. Red box meant positive qPCR result and green box represented negative result. White box meant not detected.

1.3　Retention rate

The retention rate (＝number of retained sows/number of sows prior to ASFV detection) of farm 1 to farm 4 was determined to be 69.7%, 65%, 99.4%, 99.72% respectively (Fig. 3) .

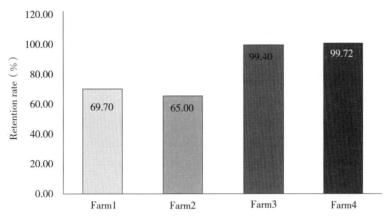

Fig. 3　Retention rate (＝number of retained sows/number of sows prior to detection) in 4 herds.

1.4　Discussions and conclusions

Without vaccines, an innovative and practical way to deal with ASFV was needed due to unsuccessful and expensive application of conventional standard culling measures. By combining scientific knowledge of ASFV with field attempts, we successfully developed a systematic "Whole-herd Sampling, qPCR-based Testing, and Precision Removal" method that successfully eradicated Georgia 2007/1 strain of ASFV in several swine herds.

ASFV was found to be relatively slow in transmission after infection occurs. One study showed that the within pen R0 of ASFV was estimated to be 2.8[3]. Our field practices and observations were consistent with these findings. As demonstrated in the field cases of Farm 1 to Farm 4, if strict measures were taken, ASF can be contained in one area and systematically eliminated until the whole herd regained negative status. For example, in Farm 1, which was most heavily contaminated among four herds, 5 out 10 barns (2, 4, 7, 8, 9) remained negative during the eradication process. This is also consistent with field observations that solid barriers like concrete walls can effectively block the transmission of ASFV between pens. This also supports the idea that ASF is not likely to be an airborne disease[8]. Environmental contamination of ASFV was relatively low, as was reported that introduction of negative pigs into contaminated pens 3, 5, 7 days after ASFV infected pigs were removed did not result in subsequent infection[5]. This seemed contradictory with results of Farm 2 but was consistent with results of Farm 1，3 and 4. In the case of Farm 2，whole surface of environmental samples in Gestation Room 1 found qPCR positive almost every day until day 28 when a new disinfection method using sodium hypochlorite was implemented. We hypothesize that the positive qPCR results were due to non-degraded ASFV DNA but not infectious ASFV particles. Our study showed that sodium hypochlorite was more effective in breaking down DNA of ASFV making sodium hypochlorite more favorable to use by avoiding confusing positive DNA with infectious ASFV particles especially when evaluating the cleaning and disinfection outcome. This is also consistent with findings of a recent report that sodium hypochlorite and chlorine work well in damaging ASFV DNA[1].

These characteristics of ASFV constitute the scientific foundation or our method and vice versa，our results supported these findings. Our method has been constantly upgraded including establishment of quick test labs and using sodium hypochlorite for disinfectants，etc，since the first case of ASFV detection. As showed by the TTNH and retention rate results，the latest removal in Farm 4 holds almost 99% of the herd. This can be attributed to several key points in the application and constant upgrading of the method. The key points are listed below.

（1）Early detection is essential. Early detection means lower levels of contamination and less likelihood of spread of the virus. Compared with the cases of Farm 1 and Farm 2 in 2019，the qPCR results of first ASFV detection in Farm 3 and Farm 4 in 2020 were of higher Ct values，which turned out to be smaller number of pigs infected（only 1 detected pig in Farm 3 and Farm 4 and retention rate was over 99%.）. This early detection contributed to a great extent to the successful implementation of the method.

Early detection can be attributed to several improvement in management. After 2019，each sow farm was equipped with qPCR machine and skilled personnel near the farm，which allows daily monitoring of clinically abnormal pigs，environment，personnel，and incoming supplies. In the cases of Farm 1 and Farm 2 in 2019，samples had to be sent in a qualified lab for analysis，which caused one day delay of results and higher chances of spread. NOR，OF，or PCAS samples were chosen for early detection and lymph node for confirmation. Zhao et al reported that viral DNA appeared 1-3 days earlier in oral fluid than in blood in pigs infected by Chinese strain via contact[9]. Also，NOR，OF，or PCAS samples are collected in a less invasive way and have less chances of contamination of the pigs and environment as compared to blood samples. As was shown in figure 1，average CT value for LN samples was significantly lower than that of NOR or PCAS samples. Since qPCR results with a Ct value of >35 was normally distributed，LN samples with lower Ct values and expected higher concentrations of ASFV DNA were chosen as more reliable samples for ASFV DNA confirmation. Moreover, aggregation of samples made the method cost effective. As was shown in figure 1, aggregation of 2 048 samples and of 16 samples if initial Ct value is of ～20 Ct values and of ～33 respectively would not change the qPCR results.

(2) Precision evaluation of the herd. Upon detecting ASFV in the herd, precision evaluation of the degree of contamination must be carried out promptly. Fourtools are useful in the evaluation: electronic maps, whole herd sampling and paired sampling & testing and qPCR.

Electronic maps and whole herd sampling together generate an accurate picture of ASFV situation in the herd. This helps make decisions in precision removal, tracing sources of contamination, and restarting production. For example, in Farm 3, whole herd sampling suggested that only Stall A87 was positive for one day (day 0), but interestingly, the internal surface of feeder outlet was found qPCR positive with Ct value of 39.53 in day 13. Epidemiological investigation proved that feed plant was contaminated in day-20, so the source of infection might be attributed to the contaminated feed. This example highlights the importance of whole herd sampling, which is not easy to perform, in epidemiological tracing and decisions of resuming normal operations.

The concept of Paired sampling & testing means once again sampling & testing the same grid of the first round of Whole herd sampling, which is marked on the electronic map. At least two rounds of paired sampling & testing were required to identify the ASFV infections. We observed in the field that some farms experienced repeated ASFV infection. Infected pigs via contact varied in time to manifest clinical signs[5], and some studies showed that virus shedding appeared earlier than clinical signs[1], so it is important to conduct paired sampling and testing to avoid misdiagnosing assumed infected animals. As was seen in the case of Farm 2, an intermittent mode of ASFV detection was found probably due to not fully implemented paired sampling.

Similar method was used in the eradication of PRV. In the PRV eradication program, reestablishing QN breeding herd status after PRV infection requires repeated sampling and testing 30 days after the initial test[10]. The underlying mechanism was similar, but the time interval between paired sampling & testing in our practice stems precisely from the latent period of ASFV strain Georgia 2007, the prevalent strain in China[11]. The successful eradication of ASFV from herd proved the credibility of paired sampling & testing.

Although qPCR was widely used in the academic studies, it is uncommon to see the use of qPCR as a tool of swine disease diagnosis or monitoring in China. In less than two years of ASFV outbreak in China, our large production systems were equipped with qPCR machines and reagents. The qPCR test showed a superior advantage over traditional gel-based tests on aspects of sensitivity and promptness. Moreover, since Ct values are inversely proportional to the amount of target nucleic acid in the sample, qPCR results are indicative of level of viral shedding or contamination.

（3）Precision removal was based on precision evaluation. After precision evaluation of level of contamination，individual pigs can be removed in a bio-secure manner that has the least chances of spreading the virus. For example，the number of pigs removed were based on production type（Gestation，Farrowing and Wean to finish or GDU），number of pigs infected in one grid and Ct values. In the gestation，if Ct value was lower than 30，and/or more than two pigs were infected in the grid，the whole grid was depopulated. If Ct value was ＞30，the infected pig and two adjacent pigs were removed. This decision was made based on rough scientific estimation but gives staff clear operation instructions on site and make the method easy to implement，which is essential in eradication of ASFV from herd. Another example of precision removal based on precision evaluation is the design of pig removal routes. Pig removal route was designed based on electronic maps and had least chances of spreading the virus.

In conclusion，we firstly developed an innovative，and systematic method，and successfully implemented it to eradicate ASFV in four farms with most herds retained（nearly 90% in recent two cases）for normal production. The successful eradication of ASFV in herds would greatly facilitate the control and eradication of ASFV in China and worldwide.

2　Methods

When first detection of ASFV occurs in the farm，whole herd sampling and qPCR tests were carried out to evaluate the disease status in the herd. Then infected pigs were accurately removed and environment decontaminated by the precision removal process. After one or more rounds of this method were applied until the whole herd remained negative for 7-14 days.

2.1　Farms

Four different farms were chosen in the time order. First ASFV detection was in February 7th 2019，June 2nd 2019，July 2nd 2019，November 9th 2019 in Farm 1-4，respectively. For each farm，the study period last from the day of first ASFV infection detection to 7-14 days after the day of last positive qPCR result in the herd.

Farm 1 was a wean-finishing site holding 4 231 growing pigs with average weight ranging from 7 kg to 130kg.

Farm 2，a typical commercial sow farm，which had newly developed a herd by introducing 1 484 gilts in the breeding gestation room. The breeding gestation room is part of a typical uniformed production line，which was designed to hold 3 000 sows and include two breeding gestation rooms each with 1 296 stalls and 10 farrowing rooms each with 60 crates. Infection was firstly detected in one breeding gestation room.

Farm 3 was a commercial farrow-wean sow farm with 5 167 in production sows kept in two independent uniformed production lines as described above. Infection was firstly detected in one breeding gestation room.

Farm4 was also a commercial farrow wean sow farm with 3 928 in production sows kept in two independent uniformed production lines as described above. Infection was firstly detected in onebreeding gestation room.

2.2 Whole-herd sampling

Detailed maps showing the precise layout of each farm were produced to inform the sampling protocol. Pigs and the environment were sampled and if one pig was diagnosed as ASFV positive, a whole herd (WH) sampling of pigs and environment was performed, meaning sampling each pig and surfaces in the building within 24 hours. Subsequent whole herd pig samples were collected seven to ten days later.

2.3 Risk based early detection of infected pigs

Two kinds of samples from pigs were collected with a modified method based on those used for PRRSV monitoring[12]. For early detection, all clinically abnormal pigs with signs including off-feed, fever, lethargy, hemorrhagic diarrhea, redness of skin, lameness, and abortion were sampled and tested. Swabs from nasal, oral, rectal (NOR), trough lips, and defecation area surface (pig contact area surface, PCAS) in the sow herd; or oral fluid (OF), trough lips, waterers, and defecation area surface and other floor surface (pig contact area surface, PCAS) swabs in the finishing pig herd were pooled together in a 2ml microtube as one sample. The pooled sample was only valid when at least three out of the five swab samples were successfully taken from the same pig. For confirmation of first ASFV DNA detection, especially when samples showed Ct values of higher than 35, lymph node (LN) samples were collected using an innovative lymph node sample collector. In brief, the pig was held down and the needle-like collector was pierced into an inguinal lymph node. The sample was taken out by the barb of the collector and was then injected into a 2 ml microtube as one lymph node sample.

2.4 Whole-herd sampling of pigs

The NOR samples from each sow in the farm were taken individually based on method described above after ASFV infection was confirmed.

2.5 Whole surface (WS) sampling of supplies, personnel, and environment.

Whole surface sampling can be literally defined as sampling the surface of supplies, personnel, and environment. A 20 cm ×20 cm gauze soaked with 0.9% sodium chloride was used to wipe the surface of supplies, personnel and environment. For environmental sampling, a grid sampling frame was used. A grid represented 20-30 stalls in a gestation room, or a crate in a farrowing room, or a pen with solid walls in a wean-finishing site, or a functional room in the facility such as one dormitory or kitchen. Each pig and grid are clearly marked on the electronic map.

Ground surface, feeder, waterer, slats, and every object in the grid were wiped from top to the bottom. Each grid served as one sample. All samples from hair, face, nasal cavity, glasses, clothes, and boots of individual staff constituted a sample. For supplies, each category of incoming items was swabbed as one sample. The WS samples were put in a valve bag.

2.6 Sample processing and aggregation of samples

NOR or PCAS samples were diluted with 1ml of 0.9% sodium chloride and vortexed and then centrifuged at 4 500 rpm for 30 seconds. Supernatant was collected and storedat −20℃. The WS samples were squeezed for 30 seconds until the dilution were homogenized. The dilutant was then poured into a 1.5ml microtube and centrifuged at 4 500 rpm for 30 seconds. Supernatant was collected and stored at −20℃ for further use. lymph node samples were added with 500 μl of 0.9% sodium chloride and homogenized with a homogenate machine. The homogenate was centrifuged at 4 500 rpm for 30 seconds and supernatant was collected and stored at −20℃ until further use.

No more than 5 NOR and PCAS or WS samples were aggregated as one. Each lymph node sample was tested individually.

3 qPCR testing

3.1 DNA extraction

DNA extraction was performed using a DNA extraction machine Gene Pure Pro 96 from Bioer company (Hangzhou, China) according to the manufacturer's instructions.

3.2 qPCR test procedure

5 μl of extracted DNA was added to 20 μl of qPCR mix from MRD company (Beijing, China) or Thermo Fisher Scientific (Waltham, USA) and qPCR was performed in a Gentier 48E machine from Tianlong company (Xian, China) or Step onePlus from Thermo Fisher Scientific (Waltham, USA) according to the manufacturers' instructions. The procedure for qPCR test was as follows:

50℃ for 2 minutes, 1 cycle,

95℃ for 3 minutes, 1 cycle,

95℃ for 10 seconds, 45 cycles,

60℃ for 20 seconds, 45 cycles.

4　Precision removal

4.1　Materials and equipment used for precision removal

Facial masks, latex gloves, overalls, shoe covers, waterproof polyester clothing, vessels, and carts used for carrying dead pigs were purchased from local markets. Sodium hydroxide and sodium hypochlorite was purchased locally. Virkon was purchased from Lanxess (Cologne, Germany).

4.2　Number of removed pigs

The number of pigs removed was based on production type (gestation, farrowing, wean tofinish or GDU), number of pigs infected in one grid, and Ct values of qPCR test.

In gestation, if the Ct value was lower than 30, and/or more than two pigs were infected in one grid, the whole grid was depopulated. If the Ct value was >30, the infected pig and the two adjacent pigs were removed. In the farrowing room, regardless of Ct values, sows were removed by crate, and suckling piglets in the same crate were also removed. In the finishing site, if one pig was infected, the entire pen of pigs was removed. The two adjacent pens were only removed if the pens were not divided by solid walls.

4.3　Precision removal of the pigs

Pigs were removed in a bio-secure manner. SealedU-shape tunnel were made from waterproof polyester cloth to move the pigs. The pigs were transferred using exclusive carts off the facility. After the pigs were removed, the supplies including gloves, overalls and cloth were incinerated. Afterwards, Virkon was applied in each grid.

4.4　Paired sampling & testing and subsequent daily monitoring and management

Sampling and testing continued afterwards using the same grid as the first round of WH sampling as recorded on the electronic map. At least one round of subsequent sampling & testing was required to eliminate the ASFV. More rounds were needed in the case of heavy contamination. Paired sampling and testing was ceased when previous round found no positive pigs.

After the herd (including pigs and the environment) remained negative for 21 consecutive days, clinically abnormal pigs with signs, such as being off-feed, fever, lethargy, hemorrhagic diarrhea, redness of skin, lameness, and abortion were tested daily. The NOR and PCAS samples were collected for ASFV qPCR test.

4.5 Validation of maximum number of sample aggregation for valid nucleoid acid detection

Verified LN, WS, NOR, and PCAS samples with initial Ct of ~20 and ~33 respectively were serially diluted 2 folds with PBS and quantified by qPCR for 2 replicates.

4.6 Data collection

Data including qPCR result and TTNH (time to negative herd, in which both pigs and environment were negative) were collected from each farm. TTNH was determined by calculating the days from the first ASFV positive qPCR result until the last positive result.

Reference

[1] Lang Gong, et al. African swine fever recovery in China [J] . Vet Med Sci, 2020, 00: 1-4.

[2] https: //www. oie. int/en/animal-health-in-the-world/animal-diseases/african-swine-fever/.

[3] http: //www. moa. gov. cn/nybgb/2017/dsq/201802/t20180201 _ 6136188. htm.

[4] Guinat C, et al. Experimental pig-to-pig transmission dynamics for African swine fever, virus, Georgia 2007/1 strain [J] . Epidemiol Infect, 2016, 144 (1): 25-34.

[5] Olesen A S, et al. Short-time window for transmissibility of African swine fever virus from a contaminated environment [J] . Transbound Emerg Dis, 2018, 65 (4): 1024-1032.

[6] Guinat C, et al. Inferring within-herd transmission parameters for African swine fever virus using mortality data from outbreaks in the Russian Federation [J] . Transbound Emerg Dis, 2018, 65 (2): e264-e271.

[7] De Carvalho H C, Ferreiraet al. Quantification of Airborne African Swine Fever Virus After Experimental Infection [J] . Vet Microbiol, 2013, 165 (3-4): 243-251.

[8] Gallardo C. African swine fever (ASF) diagnosis, an essential tool in the epidemiological investigation [J] . Virus Research, 2019, 271: 197676.

[9] Dongming Zhao, et al. Replication and virulence in pigs of the first African swine fever isolated in China [J] . Emerging Microbes & Infections, 2019, 1.

[10] USDA. Pseudorabies Eradication State-Federal-Industry Program Standards [R], Aphis, 1981.

[11] Jingyue Bao, et al. , Genome comparison of African swine fever virus China/2018/Anhui XCGQ strain and related European p72 Genotype II strains [J] . Transbound Emerg Dis, 2019, 66 (3): 1167-1176.

[12] Linhares, et al. Overview and applications of population-based monitoring protocols for PRRSv detection in breeding herds [R] . Canada, 2019.

集团化养猪公司防控非洲猪瘟实操[*]

李孝文　闫之春

摘要： 集团化的养猪公司在非洲猪瘟防控中，应将养猪场及其投入品的生物安全，以识别风险因素、分区管理、关口控制为框架，构建"生猪产业链闭环生物安全体系"。在生猪供应端、生产端、生产后端和销售端控制可能的病原体载体风险。采用全面科学的采样方案，最大限度地利用新一代实时荧光定量多聚核苷酸链式反应（qPCR）检测及诊断技术，尽早发现可疑个体，及时精准清除，实现全群净化。结合产业链生物安全贯通、种猪场三周净化罗盘法、合同农场分级颜色管理等管理措施，可完全实现非洲猪瘟在养猪产业体系内可防、可控、可净化。

关键词： 非洲猪瘟；生物安全；分级管理；精准剔除；疾病净化

1 闭环生物安全体系理论与实践

1.1 生物安全分级管理理论

按照"长城-关隘守卫"理论，根据进入猪场的风险因素，采用红、橙、黄、绿 4 种颜色分区管理，以达到：①分出等级；②划清界限；③设立关口；④阻断载体。

1.2 闭环生物安全体系

围绕生猪产业链的各环节，构建供应端（饲料原料、饲料加工和运输）、生产端（种猪场场外物资供应、3 km 风险因素、场内各生产单元等）、生产后端（合同农场、多点自育肥场）、销售端及屠宰端的风险因素管理生物安全体系。

2 生猪供应端生物安全

2.1 饲料厂生物安全

整体创新方式：一点生产、多点配送、分区管理、密闭"供、产、消"。

（1）饲料实行全过程管控方法，包括原料选取、运输过程、生产过程，保证各环节安全。

---

* 该文章发表于《中国猪业》第 14 卷第 8 期。

（2）饲料厂参照红橙黄绿分区管理方案，人员、物资、车辆进场采取各级关口管理方法。

（3）考查原料供应商，弃用高风险原料，每批原料采购前进行检测，合格后再选用。

（4）饲料制粒温度保证达到 85℃，并持续 3 min。

（5）所有饲料的运输使用专业运输车，且应按照流程洗消、检查、检测，合格后使用。

2.2　猪场 3 km 风险因素消除

（1）周边环境主要针对 3 km 内养殖散户和关键道路。

（2）针对养殖集团，若周边疫情压力大、且外部生物安全体系不完善，要考虑对周边 3 km 范围内的散养户进行清退并签订补偿协议。

（3）借助卫星地图确定清退范围，分组对各村进行逐户排查。

（4）对于猪场周边主要道路，按照消毒方案制定每周消毒频次，确保有效消毒。

（5）设置安保巡查队，每天定时对猪场周边进行巡查。

2.3　其他风险因素控制

2.3.1　车辆控制

（1）在猪场 5 km 范围内设置三级洗消体系，分别为 5 km，一级洗消；3 km，二级洗消；靠近猪场，三级洗消。

（2）对整体车辆流动制定详细流程，对各级洗消点制定详细洗消流程。

（3）洗消中心、人员均须经过专业培训，合格后上岗。

（4）按照生物安全检查表，生物安全员对洗消后的每辆车进行现场检查。

2.3.2　物资管控

（1）制定详细的物资进场流程。

（2）在距离猪场 3 km 左右设置物资隔离减毒点和食材分洗中心。同时，对进入猪场的食材进行清洗、分切、水煮等预处理。

（3）制定专门物资隔离减毒点和食材分洗中心处理流程，内部进行净污分区管理。

（4）使用专门的物资运输车将物资和食材运至场区门口，卸车过程遵循净污区管理要求。

2.3.3　人员管控

（1）制定详细的人员进出场区流程。

（2）在距离场区 10 km 左右位置设置人员隔离点，在此进行人员的检测、更衣、洗澡、隔离等工作。

（3）人员经过减毒处理且检测合格后，使用专门的人员运输车将其运至场区。

（4）在场区门口再次经过更衣、洗澡等减毒措施后，进入场内。

3　生产端生物安全

本部分主要针对种猪场非洲猪瘟处置环节进行阐述。

3.1　群体与个体诊断

对全场技术员反复培训，场长进行一对一监督，确保栋舍负责人掌握非洲猪瘟四阶的临床症状。每天报告异常猪只临床症状。

3.2　三周净化罗盘法

三周净化罗盘法是一种技术与管理方法的结合，包含了工作目标（3 周内实现群体净化与清除）、工作方法（新一代 qPCR 全群检测——精准清除——全面净化）、群体监测等措施。

3.2.1　全群检测与排查

（1）确定受污染范围，猪群鼻拭子＋肛拭子、环境全群采样。

（2）处置，充分准备后以低污染状态离群。

（3）监测异常猪，持续 3 周。

（4）通常 3 周内可实现猪群、环境等净化，全群监测为阴性。

3.2.2　精准清除阳性猪及受威胁猪

（1）每天检测所有猪，须第一时间将阳性猪清除离群。

（2）赶猪操作：阳性猪赶出猪舍时，所有经过的舍内过道必须铺设地毯和两侧彩条布，并用消毒液（"卫可"1∶100）浸湿。

（3）转猪人员：参与转运的场内人员、赶猪人员、场外人员穿戴一次性隔离服和鞋套，转猪结束后立即进行无害化处理。

3.2.3　网格化精准清除环境病毒

（1）将猪舍按照排布划分若干单元小格，分别为阳性猪单独栏位网格和整体栏舍网格。

（2）单元网格全覆盖检测，阳性栏位采取火焰消毒＋含氯消毒剂清除病原；阴性网格不做处理。

（3）重新检测阳性处理栏位网格和受威胁网格化区域，阳性持续重复处理，阴性则完成环境病原清除。

（4）根据网格化颜色的变化，栋舍全部转为绿色，即实现了全部猪群＋环境的清除和净化。

4　快速实验室检测体系

新一代 qPCR 检测技术是基于全面配套的实验室检测体系，有条件的公司可建立全国性的三级快速检测与监测联动实验室体系，即"三级检测体系"，为"4 h 出采样、6 h 出结果、12 h 有行动"提供有力保障。

5 小结

通过构建与实践闭环生物安全体系的新生物安全策略，识别生猪生产中非洲猪瘟风险因素，并能够有效切断其向群体中传播，大大提高了猪场非洲猪瘟及其他传染病的发生概率。在群体中出现少量阳性猪及风险载体后，通过快速的群体与个体诊断、不断创新的现场采样方法、新一代 qPCR 检测技术等的应用，有效阻断了病原在群体中的传播，快速地实现了疫病的全面净化。目前，构建"生猪产业链闭环生物安全体系"已经成为大多数养猪企业的共识，并在积极实践中不断改进提高。

Study on the Influence of Different Production Factors on PSY and its Correlation[*]
(不同生产因素对 PSY 的影响及其相关性研究)

Ran Guan　Xingdong Zhou　Hongbo Cai　Xiaorui Qian　Xiaoyu Xin　Xiaowen Li

Abstract：Finding out the key reproductive performance factors，affecting piglets weaned per sow per year (PSY) can improve the production efficiency and profitability of pig farms. The objective was to understand the actual distribution of different production factors and PSY of breeding pig farms，analyze the correlation to find the main production factors affecting PSY，and formulating a Production Efficiency Improvement Plan in practice. Data included 603 breeding pig farms from September 28，2020 to September 26，2021. Regression analysis was used to evaluate the relationship between PSY and key production factors，and the characteristics of total pig farms versus high performance (HP) pig farms (the production performance was in the top 10%) or top 5% pig farms were compared. Spearman's rank correlation coefficient was used to analyze the correlation between production factors and find the factors related to PSY. Non - linear support vector regression (NL-SVR) was used to analyze the personalized PSY improvement through a various change of the four key factors.

The median distribution of 15 production factors and PSY in total pig farms were different from those of HP farms. All of data were distributed nonlinearly. Mating rate within 7 days after weaning (MR7DW)，farrowing rate (FR)，number of piglets born alive per litter (PBAL) and number of weaned piglets per litter (WPL) were moderately correlated with PSY，and the correlation coefficients were 0.505 8，0.442 7，0.392 9 and 0.383 9，respectively. When the four factors in NL-SVR changed in medium (0.5 piglet or 5%) or high level (1.0 piglet or 10%)，PSY can be increased by more than 0.5.

NL-SVR model can be used to analyze the impact of changes in key production factors on PSY. By taking measures to improve MR7DW，FR，PBAL and WPL，it may effectively improve the current PSY and fully develop the reproductive potential of sows.

Keywords：PSY；correlation coefficient；mating rate within 7 days after weaning；farrowing rate；number of piglets born alive per litter；number of weaned piglets per litter

---

* 该文章发表于 *Porcine Health Management* 第 8 卷第 1 期。

Piglets weaned per sow per year (PSY) is an important factor to measure the efficiency of pig farms and the reproductive performance of sows. It is closely related to number of weaned piglets per litter, farrowing rate, non-productive days (NPD) and other production factors[1]. By increasing PSY, the purchase cost of gilts and the feeding cost of sows can be shared equally among more weaned piglets, to improve the profits of commercial pig farms[2]. PSY has been used to provide target for the reproductive performance and productivity of breeding herds[3].

Pig farms management based on productive data analysis can help producers and veterinarians maximize the lifetime reproductive potential of sows and improve economic efficiency[4,5]. In China, large-scale pig farms are using data management systems to record production data every day[6]. However, most producers only use basic data for current production arrangements. Researchers mainly use linear models such as mixed effect model[7], general linear model[8] and multiple linear regression model[9] for further analysis of production data, which has certain limitations when making accurate calculation or prediction for nonlinear factors. Therefore, this study intends to carry out personalized PSY prediction of each pig farm through non-linear support vector regression model (NL-SVR) and provide scientific reference for production management to formulate targeted production objectives by counting the change distribution of PSY under three improvement levels of different production factors.

1 Methods

1.1 Farm description

The study did not require approval from the Ethics Committee on Animal Use because no animal was handled. This was a cross-sectional study involving samplesof 603 pig breeding farms from 144 large-scale breeding companies. They fulfilled the following inclusion criteria, which were (1) having a population of 1 000 or more sows, (2) using the internal data management system of the company, (3) complete data records. In addition, all pig farms were two-point breeding farms, and weaned piglets were transferred to special commodity farms for feeding and slaughtering. The replacement gilts in estrus or sows after farrowing were fed in the stalls from checking estrus, breeding to late gestation (generally three days before farrowing).

The farms were from 22 provinces, located in the various regions of the country, namely, the East China (33.0%), North China (17.9%), South China (14.3%), Central China (9.0%), Northwest (9.1%), Northeast (8.3%) and Southwest (8.5%) regions.

All these farms applied automatic feeding system (the feed was transferred from galvanized sheet silo to stainless steel feeders (gilts) or DL6 (a commercial model of feeder, which was suitable for 60 chain-disc feed line, with transparent feed doser and fixed throat band, and the maximum feed storage capacity was 6 L. Manual or electric feed drop can be realized，and the volume in the feed doser can be quantified by adjusting the scale.）feed doser（sows）through auger feed line controlled by feed line controller.）and mechanical ventilation system（climate controller for controlling fans of different size）. At different growth stages，pigs were fed with the corresponding formula of standardized feed（according to the reference feeding amount，gilts and sows were fed the corresponding 12 kinds of feeds in the stages of nursery，growth，fattening，pregnancy and lactation）provided by the company's internal feed factory. All farms used artificial insemination to mate gilts and sows，and two or three inseminations were carried out in each estrus cycle.

1.2　Data collection and manipulation

The production data were uploaded to the internal data management system by each pig farm. All data belonged to the company. The researchers were authorized by the company's production management department and digital technology department to obtain the production data in this study. This study analyzed 603 large-scale（1 000-3 000 sows）pig farms from September 2020 to September 2021. Because this study was based on the statistical analysis of pig farms and the amount of data was relatively small，in order to ensure the basic operation of the algorithm model，there was no excessive processing of the original data.

Firstly，the trend relationship between 15 production factors and normalized PSY in 603 pig farms from management reports was analyzed. The calculation method of normalized PSY was as follow：

$$y_i = \frac{X_i - \min\ \{X_j\}}{\max\ \{X_j\}\ - \min\ \{X_j\}}$$

where：X_i means the actual PSY for i farm；X_j means a vector consisting of all variables of the number j farms，$y_i \in [0\%,\ 100\%]$.

Secondly，spearman's rank correlation coefficient was used to analyze the correlation among 16 production factors，so as to find the factors related to PSY. According to the data distribution trend，the NL-SVR was selected to analyze the personalized impact of the changes of four production factors with the highest correlation on PSY. Mating rate within 7 days after weaning（MR7DW）and farrowing rate（FR）were set at three levels：high（10%），medium（5%）and low（1%），while number of piglets born alive per litter（PBAL）and number of weaned piglets per litter（WPL）were set at three levels：high（1 piglet），medium（0.5 piglet）and low（0.1 piglet）. The distribution of the number of farms corresponding to the change of PSY under different levels of production factors improvement was counted respectively.

1.3　Definitions and categories

Research stage was defined as the stage from September 2020 to September 2021. Total number of piglets was defined as the sum of the total number of piglets sows farrowed during research stage. The NPDs referred to other days except the production days，including mating to pregnancy loss，pregnancy loss to return-service，pregnancy loss to present/culling，weaning – mating，weaning to present/culling. Other definitions were shown in Table 1. The production performance which was in the top 10% referred to high performance (HP) pig farms.

Table 1　Descriptions of productive performance between total pig farms and top 5% pig farms.

	Mean	Median	Minimum	Maximum	SD
	Total pig farms ($n=603$) /Top 5% pig farms ($n=29$)				
Total number of piglets per litter[1]	11.3/12.1	11.4/12.2	9.0/10.0	14.9/14.9	1.0/0.9
Number of piglets born alive per litter[2]	10.4/11.5	10.4/11.5	7.5/9.4	14.1/14.1	1.0/0.9
Number of weaned piglets per litter[3]	9.4/10.5	9.4/10.4	6.8/9.0	12.4/12.4	1.0/0.6
Farrowing rate (%)[4]	78.2/87.4	81.2/88.5	36.6/58.5	99.2/99.2	12.3/6.8
Stillbirth rate (%)[5]	6.4/3.9	5.7/3.6	3.1/0.6	18.4/8.8	2.9/1.7
Mummified piglets rate (%)[6]	2.0/1.4	1.7/1.3	0.6/0.0	9.9/5.2	1.2/0.9
Return-service rate (%)[7]	14.4/6.0	12.9/5.0	2.7/1.2	46.9/13.0	8.3/3.5
Mating rate within 7 days after weaning (%)[8]	55.7/69.2	58.2/69.9	0.0/19.2	85.8/83.8	16.5/12.6
Weaning to breeding interval[9]	17.2/9.6	12.8/9.0	4.7/6.5	23.5/22.0	13.8/3.1
Non-productive days[10]	84.1/50.0	76.8/45.7	27.4/27.7	270.0/113.6	38.2/17.8
Production days[11]	686.5/1 481.0	502.0/763.0	120.0/280.0	5 176.0/5 176.0	768.6/1 368.1
Birth weight of piglets (kg)[12]	1.2/1.2	1.2/1.2	1.0/1.0	1.6/1.5	0.1/0.1
21-day adjusted weight of piglets (kg)[13]	5.7/6.0	5.7/6.2	4.4/4.9	7.3/7.3	0.5/0.5
Longitude	113.8/111.4	113.5/109.0	104.8/103.2	128.8/123.4	5.6/7.1
Latitude	33.9/33.8	35.1/34.1	23.3/23.4	47.7/42.7	6.0/5.3

Note：[1] Total number of piglets per litter：Total number of piglets per litter/Number of litters；

[2] Number of piglets born alive per litter：Number of piglets born alive/Number of litters；

[3] Number of weaned piglets per litter：Number of weaned piglets/Number of litters；

[4] Farrowing rate：Number of farrowed litters/ (Number of farrowed litters ＋ Gestation loss of 115 days pushed forward in the research stage)；

[5] Stillbirth rate：Number of stillbirth/Total number of piglets；

[6] Mummified piglets rate：Number of mummified piglets/Total number of piglets；

[7] Return-service rate：Number of return - service sows/Number of mating sows；

[8] Mating rate within 7 days after weaning：Number of mating sows within 7 days after weaning/ (Number of weaned sows7 days ago - number of culling sows 3 days ago - number of dead sows within 7 days after weaning)；

[9] Weaning to breeding interval：The first mating date of the sow in this breeding cycle - The weaning date of the sow in the same parity；

[10] Non-productive days：(Sum of non - productive days of all sows in the research stage/Days in the research stage) × 365.25/Average number of sows in the research stage；

[11] Production days：Date of the last day of the research stage - Date of production started；

[12] Birth weight of piglets：Sum of birth litter weights in all birth records during the research stage/Number of born alive piglets；

[13] 21-day adjusted weight of piglets：Actual average weaning weight × (2.218－0.081 1 × Average weaning age＋0.001 1 × Average weaning age[2]).

1.4　Statistics analysis

All analyses were conducted with python programming language in PyCharm CE. The farm was considered the experimental unit. In order to reduce the noise in raw data, abnormal data points were deleted, such as the PSY of zero, the farrowing rate of zero, the average number of piglets born alive per litter of zero.

Spearman's rank correlation analysis between 16 factors (including PSY) was performed toconstruct the correlation coefficient matrix. The correlation between variables and PSY, and the collinearity between each variable was analyzed by this analysis.

NL-SVR model in the sklearn algorithm library was used to learn the data of more than 600 pig farms, obtaining the model after fitting and convergence. After a variable in the data was increased by delta, a prediction data set was generated. Followed by the data set prediction with the model, the change of PSY of each pig farm after the variable was increased by delta was obtained. The kernel function is radial basis function (RBF).

2. Results

A total of 16 production factors in 603 pig farms were analyzed. The relationship between 15 production factors and normalized PSY (%) in all farms was shown in Fig. 1. The median of factor versus PSY from all pig farms (green dot) was visually lower thanthat of HP farms (red dot), which was distributed in the left (Fig. 1A B C D G J M), right (Fig. 1E F H I K N O) or middle (Fig. 1L) of the red dot according to different factors. The specific statistical data were shown in Table 1. The mean, median, minimum and maximum of PSY and positive factors (such as total number of piglets per litter (TPL), PBAL, WPL, FR, MR7, production days, and 21-day adjusted weight of piglets in HP farms were higher than that of total pig farms. The mean and median of birth weight of piglets were equal, while the minimum and maximum were still higher. Negatively related factors, such as stillbirth rate (SR), mummified piglets rate, return-service rate, weaning to breeding interval (WBI) and NPDs in HP farms were lower.

The correlation coefficient matrix of 16 production factors in 603 farms were shown in Table 2. PBAL versus TPL represented the highest correlation (0.891 6), followed by PBAL versus WPL (0.848 7). The factor with the highest correlation with PSY was MR7 (0.505 8), followed by FR (0.442 7), PBAL (0.392 9) and WPL (0.383 9), respectively. The top three factors related to MR7DW were PBAL (0.597 8), WPL (0.5780) and WBI (−0.492 2), respectively. The top three factors related to FR were NPD (−0.634 6), WPL (0.405 2) and PBAL (0.374 8), respectively. The another top two factors related to WPL were

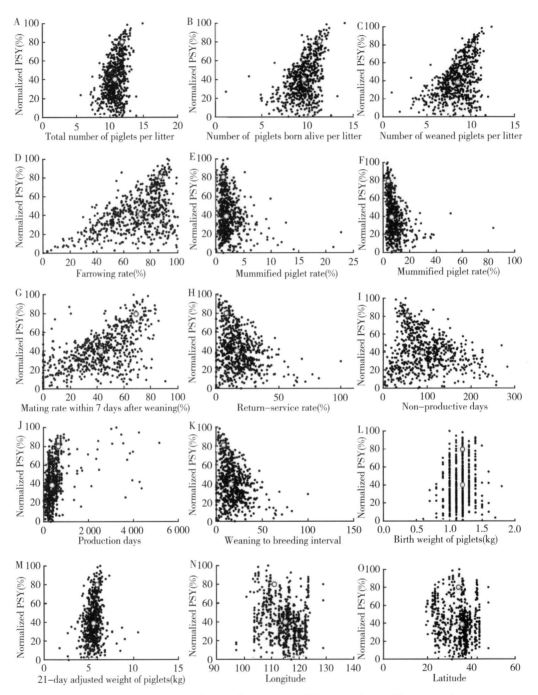

Fig. 1   Relationship between 15 production factors and PSY in 603 farms. The green dot represents
the median of the factor and PSY in all farms ($n=603$), and the red dot represents the median
of the factor and PSY in high performance pig farms (the production performance is in the top
10%, $n=60$).

TPL (0. 711 5) and SR (-0.587 9), respectively.

Table 2　Correlation coefficient matrix of 16 production factors in 603 farms.

	PSY	MR7DW	FR	PBAL	WPL	PD	TPL	21DAWP	BWP	Lat	MPR	NPD	Lon	WBI	RSR	SR
PSY	1.000 0															
MR7	0.505 8	1.000 0														
FR	0.442 7	0.363 4	1.000 0													
PBAL	0.392 9	0.597 8	0.374 8	1.000 0												
WPL	0.383 9	0.578 0	0.405 2	0.848 7	1.000 0											
PD	0.366 9	0.130 8	−0.234 9	0.022 9	−0.081 7	1.000 0										
TPL	0.276 5	0.485 6	0.281 4	0.891 6	0.711 5	−0.000 9	1.000 0									
21DAWP	0.172 1	0.191 9	0.120 3	0.203 1	0.221 8	0.117 2	0.150 8	1.000 0								
BWP	0.102 2	0.076 9	0.058 7	0.004 2	0.029 6	0.138 6	−0.038 5	0.101 3	1.000 0							
Lat	−0.107 0	−0.259 3	−0.367 5	−0.214 7	−0.265 2	0.272 1	−0.169 6	−0.242 2	−0.011 9	1.000 0						
MPR	−0.135 6	−0.183 7	−0.107 0	−0.298 0	−0.315 7	−0.111 9	−0.061 1	0.010 3	−0.072 1	−0.039 8	1.000 0					
NPD	−0.196 5	−0.446 9	−0.634 6	−0.457 3	−0.511 4	0.348 5	−0.369 2	−0.189 9	−0.009 1	0.397 9	0.145 5	1.000 0				
Lon	−0.226 4	−0.246 9	−0.288 4	−0.161 2	−0.152 2	0.206 6	−0.105 2	−0.196 3	−0.009 8	0.657 7	−0.048 7	0.333 1	1.000 0			
WBI	−0.288 9	−0.492 2	−0.151 0	−0.334 5	−0.331 6	0.057 4	−0.305 5	−0.164 8	−0.022 8	0.205 3	0.071 1	0.551 4	0.268 7	1.000 0		
RSR	−0.303 9	−0.113 4	−0.224 7	−0.168 3	−0.150 2	−0.242 5	−0.159 1	−0.226 6	−0.121 0	0.037 9	0.071 0	0.348 7	0.133 4	0.358 9	1.000 0	
SR	−0.366 6	−0.469 9	−0.301 7	−0.566 0	−0.587 9	−0.060 2	−0.227 9	−0.274 9	−0.084 5	0.212 3	0.362 9	0.398 9	0.194 2	0.284 5	0.133 2	1.000 0

Note: Red represents positive correlation and green represents negative correlation. The darker the color, the higher the correlation coefficient. PSY=piglets weaned per sow per year. MR7DW=mating rate within 7 days after weaning. FR=farrowing rate. PBAL=number of piglets born alive per litter. WPL=number of weaned piglets per litter. PD=production days. TPL=total number of piglets per litter. 21DAWP=21-day adjusted weight of piglets. BWP=birth weight of piglets. Lat=latitude. MPR=mummified piglet rate. NPD=non-productive days. WBI=weaning to breeding interval. Lon=longitude. RSR=return-service rate. SR=stillbirth rate.

Fig. 2 showed the distribution of pig farms and corresponding PSY when the four factors were raised at different levels. With the increase of the levels, PSY increased in varying degrees, but the low-level improvement could not increase 0.5 PSY. Among them, PBAL had great promotion potential. When one PBAL was promoted, average PSY can improve nearly 2.5.

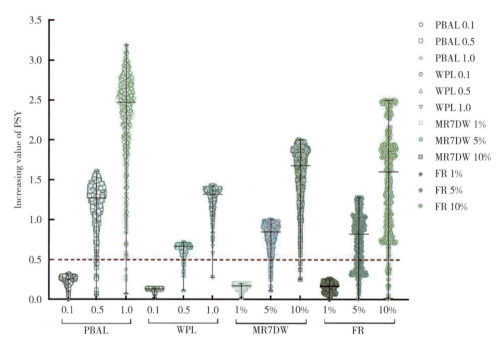

Fig. 2   The improvement of production factors corresponds to the change of PSY and the distribution of pig farms. Each factor is represented by a geometric graph, and the width of graph clustering represents the degree of data concentration. PBAL and WPL were divided into three promotion levels, 0.1 piglet (low level), 0.5 piglets (medium level) and 1.0 piglet (high level), respectively. While MR7DW and FR were divided by 1% (low level), 5% (medium level) and 10% (high level), respectively. The red dotted line represents the distribution of other factors when PSY increases by 0.5. PBAL= number of piglets born alive per litter. WPL = number of weaned piglets per litter. MR7DW=mating rate within 7 days after weaning. FR=farrowing rate.

3　Discussion

Sows have the potential to produce about 60-70 weaned piglets per life[10]. If the annual parities were calculated as 2.27, the average PSY should reach more than 26. However, our data showed that WPL was only 9.4 and the HP farm was 10.5 (Table 1), which indicated that there was a lack of more than two piglets per sow per year, so this was an opportunity to improve farm level productivity. Our results also showed that improving MR7DW, FR, PBAL or WPL can effectively improve the overall PSY of the pig farms (Fig. 2). This was similar with the results of Tani's study[7].

The average of all sows was calculated in each pig farm within one year (September 2020 to September 2021), eliminating the seasonal effect of the observed response[8]. Through scatter plot of data analysis (Fig. 1), we found that the relationship between each variable and PSY was not typical linear but determined the upper boundary of PSY. The higher the dispersion of data, the greater the randomness. Using the general linear models can only get a general trend, but there may be a large deviation from the actual situation. Therefore, instead of linear regression, the NL-SVR model was selected. NL-SVR was a technology widely used in the field of data analysis and prediction. The multivariate nonlinear regression method was used to learn by dividing the vector space, obtaining the differentiated results of PSY changes after each field variable was improved (if linear regression was used, only statistical results can be obtained, and individual results of each field cannot be calculated). The advantage of this approach was on learning the nonlinear relationship between variables and targets and calculated the impact on targets when each variable in the sample changes. When applied to the analysis of the relationship between pig production factors, it can analyze the nonlinear relationship between various factors and PSY, calculate the bottleneck of further improvement of PSY in each field, evaluate the difficulty of improvement, optimize the input-output ratio, and realize the improvement of production efficiency. However, by the great variation in management, facilities, sanitation and feedings, many of which will affect the production performance of piglets and sows, or there were differences in calculation standards, resulted in data fluctuations[8,11].

Among the 15 production factors, the 4 factors with the highest correlation coefficients with PSY were MR7DW, FR, PBAL or WPL, respectively. Their correlations from each other were also very high (Table 2). Reducing NPD can improve productivity and profitability of pig farms. Because with the increase of NPD, sow maintenance cost increased, and profitability decreased[12]. The cost of each NPD of sows ranged from $ 1.60 to $ 2.60[13]. NPD was a comprehensive index, which was significantly affected by management factors, including multiple breedings of sows, MR7DW, parity of culled sows, proportion of return-serviced sows, sow mortality, SR and pig farm scale[12,14]. Increasing MR7DW can shorten NPD, and increasing the lactation period may increase the proportion of estrus in sows within 4-6 days after weaning, which had higher reproductive performance and longer lifetime[15]. However, sows with prolonged lactation will lose a lot of body reserves, which may reduce the farrowing rate[16] and the number of piglets born per sow per year[11], and thus reduce the number of weaned piglets per year[3]. Therefore, it was necessary to optimize feed intake and feeding mode during lactation[16,17].

Among the four key production factors, the change of PBAL had the greatest impact on the improvement of PSY. Koketsu et al. [18] found that the average pre-weaning mortality, number of piglets born alive, number of weaned piglets and PSY of herds increased from 2007 to 2016, which may be related to the genetic improvement of pig industry in the past few decades[19,20]. It was worth noting that the number of weaned piglets didn't increase continuously with the increased number of piglets born alive. When the number of litters increased from 11-12 to 13-16, the pre-weaning mortality of piglets had almost tripled [21,22]. Limited by the reproductive capacity of sow itself, larger litter size can lead to reduced piglet birth weight and increased pre-weaning mortality.

In order to reduce the waste of production costs and economic benefits on pig farms, it is important for managers to maximize the lifetime reproductive performance of all sows. Through our research, we found four production factors with the highest correlation with PSY. Targeted improvement of these factors may improve the productivity of sows (Fig. 3). In addition, it also needs to be combined with appropriate nutrition[23], feeding pattern[24], development of gilts[25], better breeding management (breeding time, semen quality and stockman skills of breeders)[26,27], pig health management (control and prevention of infectious and non-infectious diseases)[28,29], complete buildings (environmental control system, advanced facilities)[30,31], farrowing management (assisted farrowing, colostrum intake and piglet care)[32,33] and trained staff[34].

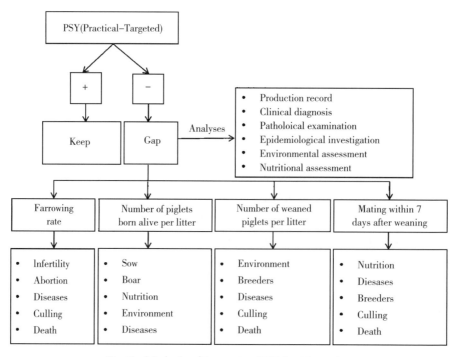

Fig. 3   Methods of improving PSY in this study.

4　Conclusions

Our study revealed the nonlinear distribution of production factors with PSY. Among all the production factors analyzed, we found four key factors associated with PSY, which were MR7DW, FR, PBAL and WPL, respectively. The effects of different factors on PSY of each pig farm were analyzed by NL-SVM model, and the distribution statistics were carried out. If targeted improvements were made to the above four factors, especially PBAL, the PSY of pig farms may be improved effectively.

References

［1］ Bell W, Urioste J I, Barlocco N, et al. Genetic and environmental factors affecting reproductive traits in sows in an outdoor production system ［J］. Livest Sci, 2015, 182: 101-107.

［2］ Abell C E G A H. Evaluation of litters per sow per year as a means to reduce non-productive sow days in commercial swine breeding herds and its association with other economically important traits ［D］. Iowa State University. 2011.

［3］ King V L, Koketsu Y, Reeves D, et al. Management factors associated with swine breeding-herd productivity in the United States ［J］. Prev Vet Med, 1998, 35: 255-264.

［4］ Piñeiro C, Morales J, Rodríguez M, et al. Big (pig) data and the internet of the swine things. A new paradigm in the industry ［J］. Anim Front, 2019, 9: 6-15.

［5］ Paterson J, Foxcroft G. Gilt Management for Fertility and Longevity ［J］. Animals, 2019, 9: 434-448.

［6］ Guan R, Gao W C, Li P, et al. Utilization and reproductive performance of gilts in large-scale pig farming system with different production levels in China: a descriptive study ［J］. Porcine Health Manag, 2021, 7 (1): 1-9.

［7］ Tani S, Piñeiro C, Koketsu Y. High-performing farms exploit reproductive potential of high and low prolific sows better than low-performing farms ［J］. Porcine Health Manag, 2018, 4 (1): 1-12.

［8］ Ek-Mex J E, Segura-Correa J C, Alzina-López A, et al. Lifetime and per year productivity of sows in four pig farms in the tropics of Mexico ［J］. Trop Anim Health Prod, 2015, 47 (3): 503-509.

［9］ Pierozan C R, Callegari M A, Dias C P, et al. Herd-level factors associated with piglet weight at weaning, kilograms of piglets weaned per sow per year and sow feed conversion ［J］. Animal, 2020, 14 (6): 1283-1292.

［10］ Gill P. Nutritional management of the gilt for lifetime productivity-feeding for fitness or fatness? //7th London Swine Conference Proceedings. Today's challenges…tomorrow's opportunities, London, Ontario, Canada, 3-4 April 2007 ［J］. London Swine Conference, 2007: 83-99.

［11］ Pierozan C R, Callegari M A, Dias C P, et al. Herd-level factors associated with non-productive days and farrowing rate in commercial pig farms in two consecutive years ［J］. Livest Sci, 2021, 244: 104312.

［12］ Rix M, Ketchem R. Targeting profit-robbing non-productive days ［J］. National Hog Farmer June, 2009, 7: 2010.

［13］ Koketsu Y. Herd-management factors associated with nonproductive days by breeding-female pigs on commercial farms in Japan ［J］. J Vet Epidemiol, 2005, 9 (2): 79-84.

［14］ Hoshino Y, Koketsu Y. A repeatability assessment of sows mated 4～6 days after weaning in breeding herds ［J］. Anim Reprod Sci, 2008, 108 (1-2): 22-28.

［15］ Koketsu Y, Tani S, Iida R. Factors for improving reproductive performance of sows and herd productivity in commercial breeding herds ［J］. Porcine Health Manag, 2017, 3: 2-10.

［16］ Chantziaras I, Dewulf J, van Limbergen T, et al. Factors associated with specific health, welfare and reproductive performance factors in pig herds from five EU countries ［J］. Prev Vet Med, 2018, 159: 106-114.

［17］ Koketsu Y, Dial G D, Pettigrew J E, et al. Feed intake pattern during lactation and subsequent reproductive performance of sows ［J］. J Anim Sci, 1996, 74: 2875-2884.

［18］ Koketsu Y, Iida R, Piñeiro C. A 10-year trend in piglet pre-weaning mortality in breeding herdsassociated with sow herd size and number of piglets born alive ［J］. Porcine Health Manag, 2021, 7 (1): 1-8.

［19］ Muns R, Nuntapaitoon M, Tummaruk P. Non-infectious causes of preweaning mortality in piglets ［J］. Livest Sci, 2016, 184: 46-57.

［20］ Andersson E, Frössling J, Engblom L, et al. Impact of litter size on sow stayability in Swedish commercial piglet producing herds ［J］. Acta Vet Scand, 2016, 58: 31.

［21］ Weber R, Burla J B, Jossen M, et al. Piglet Losses in Free-Farrowing Pens: Influence of Litter Size ［J］. Agrarforschung Schweiz, 2020: 53-58.

［22］ Nuntapaitoon M, Tummaruk P. Factors influencing piglet pre-weaning mortality in 47 commercial swine herds in Thailand ［J］. Trop Anim Health Prod, 2018, 50 (1): 129-135.

［23］ Kim S W, Weaver A C, Shen Y B, et al. Improving efficiency of sow productivity: nutrition and health ［J］. J Anim Sci Biotechnol, 2013, 4: 26-33.

［24］ Kraeling R R, Webel S K. Current strategies for reproductive management of gilts and sows in North America ［J］. J Anim Sci Biotechnol, 2015, 6 (1): 1-14.

［25］ Patterson J L, Beltranena E, Foxcroft G R. The effect of gilt age at first estrus and breeding on third estrus on sow body weight changes and long-term reproductive performance ［J］. J Anim Sci, 2010, 88: 2500-2513.

［26］ Koketsu Y. Productivity characteristics of high-performing swine farms ［J］. J Am Vet Med Assoc, 2000, 215: 376-379.

［27］ Amaral Filha W S, Bernardi M L, Wentz I, et al. Growth rate and age at boar exposure as factors influencing gilt puberty ［J］. Livest Sci, 2009, 120: 51-57.

［28］ Almond G W, Flowers W L, Batista L, et al. Diseases of the reproductive system ［J］. Dis Swine, 2006, 9: 113-147.

［29］ Iida R, Piñeiro C, Koketsu Y. Removal of sows in Spanish breeding herds due to lameness: incidence, related factors and reproductive performance of removed sows ［J］. Prev Vet Med, 2020,

179: 105002.

[30] Iida R，Koketsu Y. Climatic factors associated with peripartum pigdeaths during hot and humid or cold seasons [J] . Prev Vet Med，2014，115: 166-172.

[31] Cabezon F A，Schinckel A P，Richert B T，et al. Development and application of a model of heat production for lactating sows [J] . J Anim Sci，2017，95（Suppl 2）: 30.

[32] Vanderhaeghe C，Dewulf J，de Kruif A，et al. Non-infectious factors associated with stillbirth in pigs: a review [J] . Anim Reprod Sci，2013，139: 76-88.

[33] Holyoake P K，Dial G D，Trigg T，et al. Reducing pig mortality through supervision during the perinatal period [J] . J Anim Sci，1995，73: 3543-3551.

[34] Knox R. Getting to 30 pigs weaned/sow/year [C] . Proceedings of London Swine Conference，2005: 47-59.

Utilization and Reproductive Performance of Gilts in Large-scale Pig Farming System with Different Production Levels in China: a Descriptive Study[*]

（我国不同生产水平大规模猪场后备母猪利用情况及繁殖性能的描述性研究）

Ran Guan Wenchao Gao Peng Li Xuwei Qiao Jing Ren Jian Song Xiaowen Li

Abstract： This study was to investigate the utilization and reproductive performance of gilts in large-scale pig farms. Data from this descriptive study included 169 013 gilts of 1 540 gilts' batches on 105 large-scale pig farms from April 2020 to March 2021. According to the upper and lower 25th percentiles of piglets weaned per sow per year （PSY） during the research stage，pig farms were divided into three productivity groups：high-performing （HP），intermediate-performing （IP） and low-performing （LP） farms. On the basis of breeds，LP （LP-Total） farms were further divided into LP-breeding pig （LP-BP） and LP-commercial pig （LP-CP） groups. Average utilization，estrus and first mating data were collected from a total of 1 540 gilts' batches. The age-related factors （introduction age，age at first estrus and age at first mating） and litter production （total number of piglets，number of piglets born alive and number of weaned piglets，as well as their proportion distribution） among the HP and LP groups were compared. Litter production in different age groups was also analyzed.

The introduction age，mortality and culling rate of HP farms were lower than those of LP farms. The total number of piglets，number of piglets born alive and number of weaned piglets in HP farms were significantly greater than those of LP groups. The proportion distribution peaks of litter production in HP farms were shifted approximately two more than those in LP groups，and the proportion of low litter production （eight per litter or less） was lower than that in LP groups. The results of different age groups showed that total number of piglets per litter and number of piglets born alive per litter from 220-279 days were the highest，while those from 370 days were the lowest.

---

* 该文章发表于 *Porcine Health Management* 第 7 卷第 1 期。

The overall utilization and reproductive performance of gilts in HP farms was better than those of gilts in LP farms. The difference in utilization was reflected in introduction source, culling rate and mortality. The age at first estrus and first mating, breeds and litter production were the main differences in reproductive performance.

Keywords: gilts; reproductive performance; utilization; litter production

Gilts are the basis for maintaining fertility in large-scale pig farms, representing the largest category in breeding herds, accounting for 18%-20% [1]. When sows are culled from the pig herd due to high parities or low reproductive performance, gilts must be introduced to ensure the reasonable parity structure of the sows in the pig farm and the stability of production objectives. Well-raised gilts are expected to have a good mating rate, farrowing rate, litter production, and even lifetime performance and longevity[2-4].

Piglets weaned per sow per year (PSY) can be used as a benchmark of the productivity and reproductive performance of sows, which varies in different countries[5]. Average PSY is approximately 30.9 in Denmark[6], while in North America is about 25.3[7]. Incredibly, China, as the world's largest consumer and producer of pork, has an average PSY of only about 20[8,9]. The productivity of pig farms can also be divided into high and low levels. A study of high-performing (HP) farms in the United States showed that compared with ordinary farms, their farrowing rate was 9.0% higher, with 0.6 more piglets born alive per litter[10]. The high productivity of HP farms is mainly due to better development of gilts, better breeding management, more advanced productive technology and better piglets care during lactation[11]. Obviously, there are differences between pig farms of different production levels, but no research has been found on HP and low performance (LP) of Chinese pig farms[12].

Since giltsare still in the growth stage, their physical development and reproductive performance are different from those of sows[7,13,14]. To the best of our knowledge, the utilization and reproductive performance of gilts have not been entirely evaluated. Therefore, this study classified all surveyed pig farms into three productivity groups by PSY for a period of time (one year, from April 2020 to March 2021) and analyzed the production and reproductive performance of gilts' batches and individual gilts to provide a database for production and management managers to formulate more tailored policies.

1　Methods

1.1　Farm description

All pig farmsstudied ($n=105$) were from 1 274 pig farms of the same domestic large-scale breeding company in China that fulfilled the following inclusion criteria: (1) continuous and stable production for more than half a year [no major business strategy adjustment or extensive disease epidemic (especially African swine fever)], (2) a population of 1 000 or more productive sows, and (3) the internal data management system of the company. All of these farms applied an automatic feeding system [the feed was transferred from galvanized sheet silo to stainless steel feeders (gilts) or DL6 feed doser (sows) through an auger feed line controlled by a feed line controller], and mechanical ventilation system (climate controller for controlling fans of different sizes). At different growth stages, pigs were fed with the corresponding formula of standardized feed (according to the reference feeding amount, gilts and sows were fed the corresponding 12 kinds of feeds in the stages of nursery, growth, fattening, pregnancy and lactation) provided by the company's internal feed factory. All farms used artificial insemination to mate gilts and sows, and 2-3 inseminations were carried out in each estrus cycle. The average stock of sows was 2 660± 69. 4, while the gilts'stock was 324±24. 9. The average PSY was 19. 9±0. 4.

1.2 Categories and definitions

According to the upper and lower 25th percentiles of PSY (PSY = number of weaned piglets/days during the research stage × 365. 25/average number of sows) ranked by the internal data management system, pig farms were divided into three productivity groups: HP farms (PSY > 23. 5), intermediate-performing (IP) farms (PSY 16. 1-23. 5), and LP farms (PSY<16. 1). LP farms were further divided into three groups by breed: LP-Total (including pure, two-way crossbred and three-way crossbred), LP-breeding pig (LP-BP) groups (including pure and two-way crossbred) and LP-commercial pig (LP-CP) groups (only three-way crossbred). As there were no commercial pigs in HP farms, HP farms were not further classified.

Utilization of gilts was defined assuccessful conception and entering the breeding cycle since the introduction. Nonproductive days (NPDs) referred to days other than production days, including mating to pregnancy loss, pregnancy loss to return service, pregnancy loss to present/departure, weaning mating, and weaning to present/departure. The research stage was defined as the stage from April 2020 to March 2021. Other definitions were shown in Table 1 to Table 3.

1.3 Data collection, study design and exclusion criteria

The production data were uploaded to the internaldata management system by each pig farm. All data belonged to the company. The researchers were authorized by the company's production management department and digital technology department to obtain the production data in this study. This study was a descriptive study that analyzed a total of 169 013 gilts of 1 540 gilts'batches in 105 large-scale (more than 1 000 sows) pig farms from

April 2020 to March 2021. The data analysis of this study was divided into two levels. The utilization and reproductive data of gilts' batches at different production levels were considered batch levels. The age - related factors of gilts and litter production at different production levels were measured at the individual level. To observe the influence of breeds on age-related factors and litter production, the differences among HP, LP-Total, LP-BP and LP-CP farms were compared. The gilts' batch data were complete, without any removal. For gilts used to compare HP and LP farms, records were excluded if they met any of the following exclusion criteria: Total number of piglets per litter was zero (687 gilts); Number of piglets born alive per litter and number of weaned piglets per litter were more than 14 [If these two indexes were ≥15, it exceeded the number of 14 effective nipples of sows (the maximum), the surplus piglets were fostered to other sows with less litter size, resulting in the weaning number of the litter inconsistent with the actual size; or the data was incorrectly entered], (546 gilts) and other incomplete data (865 gilts) . Thus, 35 847 out of 37 045 gilts were used for individual-level studies.

1.4 Statistical analysis

Descriptive statistics were conducted usingWPS Office Excel for Mac version 3.3.0 (Kingsoft Office Corporation, Beijing, China) . The influence of breeds on age-related factors and litter production and the litter production of gilts with different first mating days were analyzed using GraphPad Prism 7.0 (GraphPad Software, Inc. San Diego, CA, US A) . Tukey's multiple comparisons test of ordinary one-way ANOVA was used to study the average utilization, estrus and first mating data of gilts' batches among HP, IP and LP farms. The same method was used for litter production on the HP, LP-Total, LP-BP and LP-CP farms. Normal distribution tests of litter production at different first mating days of gilts were performed by SPSS Statistics software 22.0.0 (IBM Corp. Released 2013. IBM SPSS Statistics for Mac, Version 22.0. Armonk, NY) . $P<0.05$ showed significant difference.

2   Results

A total of 169 013 gilts of 1 540 gilts' batches of 105 large-scale pig farms were analyzed. The source, introduction, mortality and culling of gilts' batches differed in HP, IP and LP farms (Table 1) . The gilts of HP and IP farms were mainly from intracompany farms, most of them were self-sufficient, and less than 10% were from external sources. Nearly 1/5 of gilts in LP farms came from outside the company. The average introduction age of HP farms was the youngest among the three categories of farms. HP farms had the lowest gilt mortality, total mortality, gilt culling rate and total culling rate, although gilt mortality and gilt culling rate were not significantly different from those of IP farms.

Table 1　Average utilization of 169 013 gilts in 1 540 gilts'batches.

		High-performing pig farms ($n=26$)	Intermediate-performing pig farms ($n=53$)	Low-performing pig farms ($n=26$)
		Mean±SEM	Mean±SEM	Mean±SEM
Number	Gilts'batches	235	948	357
Source	Self-breeding[1]	73.0%±8.2%[ab]	80.8%±4.3%[a]	54.6%±6.7%[b]
	Internal introduction[2]	23.8%±7.8%	13.2%±3.3%	27.4%±6.2%
	External introduction[3]	3.2%±1.9%[a]	6.0%±2.4%[a]	18.0%±6.2%[b]
Introduction	Average introduction number of gilts	121±6.94	83±5.89	174±16.21
	Average introduction age	202±3.76[a]	237±1.73[b]	224±2.56[c]
Mortality	Mortality of gilts[4]	1.8%±0.2%[a]	1.7%±0.2%[a]	6.6%±1.0%[b]
	Total mortality[5]	5.6%±0.5%[a]	15.4%±0.9%[b]	19.1%±1.4%[c]
Culling	Culling rate of gilts[6]	9.4%±1.4%[a]	10.5%±0.8%[a]	14.9%±1.4%[b]
	Total culling rate[7]	25.9%±2.0%[a]	32.8%±1.1%[b]	40.4%±1.8%[c]

Note：[1] Self-breeding：Gilts were bred and fed by pig farms themselves；

[2] Internal introduction：Gilts were provided by other pig farms of the internal company；

[3] External introduction：Gilts were provided by the pig farms of the external company；

[4] Mortality of gilts=Deaths from introduction to premating/introduction number of gilts；

[5] Total mortality：Mortality during the research stage, regardless of the production phase (premating, mating, conception, farrowing or feeding) of gilts. Total mortality = Deaths during the research stage/Introduction number of gilts；

[6] Culling rate of gilts=Number of culling gilts from introduction to premating/Introduction number of gilts. The reasons for culling mainly included abnormal estrus, disease or physiological defects；

[7] Total culling rate：Culling rate during the research stage, regardless of the production phase (premating, mating, conception, farrowing or feeding) of gilts. Total culling rate = Number of culling gilts during the research stage/Introduction number of gilts；

[a,b,c] Bars with different letters differ significantly ($P<0.05$).

The estrus of 112 157 gilts of 1 540 gilts'batchesat different production levels showed significant differences (Table 2). The proportion of first estrus of HP farms was significantly lower than that of IP and LP farms, but the proportion of more than second estrus was much higher. The total estrus rate of the HP and IP farms was 77%-78%, which was significantly higher than that of the LP farms. Compared with the IP and LP farms, the HP farms had more estruses and younger average age at first estrus.

Table 2　Average estrus information of 100 811 estrus out of 112 157 gilts in 1 540 gilts'batches.

	High-performing pig farms ($n=26$)	Intermediate-performing pig farms ($n=53$)	Low-performing pig farms ($n=26$)
	Mean±SEM	Mean±SEM	Mean±SEM
Number of gilts'batches	235	948	357
Total number of estrus[1]	21 198	44 917	34 696
Proportion of first estrus[2]	62.4%[a]	92.7%[b]	91.7%[b]

(continued)

	High-performing pig farms (n=26)	Intermediate-performing pig farms (n=53)	Low-performing pig farms (n=26)
	Mean±SEM	Mean±SEM	Mean±SEM
Proportion of second estrus[3]	21.0%[a]	5.1%[b]	4.7%[b]
Proportion ofthird or more estrus[4]	16.6%[a]	2.2%[b]	4.6%[b]
Total estrus rate[5]	77.2%±2.2%[a]	78.1%±1.2%[a]	66.3%±2.1%[b]
Averagetimes of estrus[6]	1.2±0.05[a]	0.9±0.01[b]	0.9±0.03[b]
Average age offirst estrus[7]	209±5.79[a]	224±3.18[b]	213±5.32[b]

Note: [1] Total number of estrus: The total number of gilts with estrus from introduction to premating;

[2] Proportion of first estrus: Average proportion of gilts with one estrus in all estrus gilts of each gilts' batch;

[3] Proportion of second estrus: Average proportion of gilts with two estruses in all estrus gilts of each gilts' batch;

[4] Proportion of third or more estrus: Average proportion of gilts with three times or more estrus in all estrus gilts of each gilts' batch;

[5] Total estrus rate=Number of gilts in estrus/Total number of estrus;

[6] Average times of estrus: Average estrus times of gilts before mating in each gilts' batch;

[7] Average age of first estrus: Average age of first estrus age before mating in each gilts' batch;

[a,b] Bars with different letters differ significantly ($P<0.05$).

The first mating information of 112 157 gilts of 1 540 gilts' batches at different production levels was shown in Table 3. Compared with the IP and LP farms, the HP farms had a lower mating rate at first estrous but a higher mating rate at second or more estrous. For mating weight, there were no differences between 135 kg and 145 kg, but the mating rate under 135 kg in IP farms was significantly higher than that of HP and LP farms. In contrast, HP farms showed a higher mating rate when the mating weight was above 145 kg. For mating age, compared with HP and LP farms, LP farms had higher mating rate under 210 days, lower above 240 days, and the total average mating rate was also lower. Compared with IP and LP farms, the average times of estrus at first mating in HP farms was 1.2-1.5 times more. IP farms had the elder average age at first mating.

Table 3　Average first mating information of 97 998 mating out of 112 157 gilts in 1 540 gilts' batches.

	High-performing pig farms (n=26)	Intermediate-performing pig farms (n=53)	Low-performing pig farms (n=26)
	Mean±SEM	Mean±SEM	Mean±SEM
Number of gilt's batches	235	948	357
Total number ofmating	19 819	43 867	34 312
Mating rate at first estrus[1]	54.9%±2.6%[a]	74.6%±1.2%[b]	63.9%±2.1%[c]
Mating rate at second or more estrus[2]	18.4%±1.9%[a]	2.6%±0.4%[b]	1.9%±0.5%[b]
Mating rate under 135 kg[3]	12.8%±1.7%[a]	20.0%±1.2%[b]	7.3%±1.2%[ac]
Mating rate between 135 kg and 145 kg[4]	47.9%±2.5%	50.6%±1.5%	50.5%±2.2%
Mating rate above 145 kg[5]	12.6%±1.6%[a]	6.6%±0.7%[b]	7.9%±1.3%[b]
Mating rate under 210 days[6]	7.2%±1.2%[a]	8.9%±0.8%[a]	13.6%±1.6%[b]
Mating rate between 210 days and 240 days	24.6%±1.9%	23.5%±1.2%	23.3%±2.0%

(continued)

	High-performing pig farms ($n=26$)	Intermediate-performing pig farms ($n=53$)	Low-performing pig farms ($n=26$)
	Mean±SEM	Mean±SEM	Mean±SEM
Mating rate above240 days[8]	41.5%±2.3%[a]	44.8%±1.4%[a]	28.9%±1.9%[b]
Totalaverage mating rate[9]	73.3%±2.2%[a]	77.2%±1.2%[a]	65.8%±2.1%[b]
Average times of estrus at first mating [10]	2.4±0.96[a]	0.9±0.03[b]	1.2±0.26[b]
Average age at firstmating[11]	216±5.99[a]	224±3.21[b]	213±5.33[ab]

Note：[1]Mating rate at first estrus＝Number of mating at first estrus/Total number of mating；

[2]Mating rate at second or more estrus＝Number of mating at second or more estrus/Total number of mating；

[3]Mating rate under 135 kg＝Number of mating at weight under 135 kg/Total number of mating；

[4]Mating rate between 135 kg and 145 kg＝Number of mating at weight between 135 kg and 145 kg/Total number of mating；

[5]Mating rate above 145 kg＝Number of mating at weight above 145 kg/Total number of mating；

[6]Mating rate under 210 days＝Number of mating at age under 210 days/Total number of mating；

[7]Mating rate between 210 days and 240 days＝Number of mating at ages between 210 days and 240 days/Total number of mating；

[8]Mating rate above 240 days＝Number of mating at age above 240 days/Total number of mating；

[9]Total average mating rate＝Total number of mating/Total mating/Total number of introductions；

[10]Average times of estrus at first mating：Average times of estrus at first mating in each gilts'batch；

[11]Average age at first mating：Average age of gilts at first mating in each gilts'batch；

[a,b,c] Bars with different letters differ significantly ($P<0.05$).

Table 4 showed the differences among the HP，LP-Total，LP-BP and LP-CP groups. The LP-CP groups had a unique breed (three-way crossbred pigs) among all groups. The LP-CP groups had an elder introduction age and elder age at first estrus than the other groups. Although the age at first mating in the HP farms and LP-Total groups differed significantly，there was no significant difference between the LP-CP groups and HP farms.

Table 4   Average age-related and litter production of 35 847 gilts.

	High-performing pig farms ($n=26$)	Low-performing pig farms-Total pigs ($n=26$)	Low-performing pig farms-Breeding pigs ($n=22$)	Low-performing pig farms-Commercial pigs ($n=18$)
	Mean±SEM	Mean±SEM	Mean±SEM	Mean±SEM
Number ofgilts	11 833	24 014	10 952	13 062
Introduction age	201±0.46[a]	210±0.25[b]	208±0.38[c]	211±0.34[d]
Age at first estrus	242±0.42[a]	252±0.28[b]	249±0.34[c]	254±0.42[d]
Age at first mating	255±0.38[a]	253±0.28[b]	250±0.35[c]	255±0.42[a]
Total number of piglets born per litter	11.8±0.02[a]	9.7±0.02[b]	9.9±0.03[c]	9.6±0.02[d]
Number of piglets born alive per litter	11.0±0.02[a]	8.7±0.02[b]	8.9±0.03[c]	8.6±0.03[d]
Number of weaned piglets per litter	9.3±0.04[a]	7.1±0.03[b]	7.0±0.03[c]	7.3±0.03[d]

Note：[a,b,c,d] Bars with different letters differ significantly ($P<0.05$).

For litter production, the total number of piglets per litter, number of piglets born alive per litter and number of weaned piglets per litter in the HP farms were significantly higher than those of the LP groups (Table 4). As shown in Fig. 1, the proportion distribution peaks of litter production in HP farms were shifted approximately two more than those in LP groups, and the proportion of low litter production (eight per litter or less) was lower than that in LP groups.

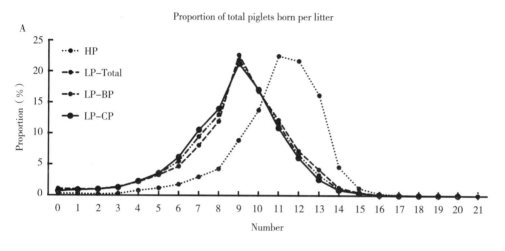

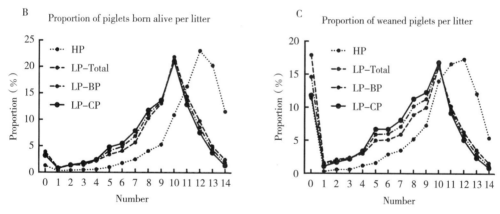

Fig. 1   Litter proportion distribution of gilts at different production levels. The proportion distribution peaks of litter production in HP farms were shifted approximately two more than those in LP groups, and the proportion of low litter production (eight per litter or less) was lower than that in LP groups. Values represent Mean $\pm$ SEM. (HP farms, $n = 11\ 833$; LP-Total groups, $n = 24\ 014$; LP-BP groups, $n = 10\ 952$; LP-CP groups, $n = 13\ 062$).[a,b,c,d] Bars with different letters differ significantly ($P < 0.05$).

Table 5 showed the litter production of gilts with different first mating days. The data of the total number of piglets born and the number of piglets born alive on different first mating days of gilts were close to a normal distribution. The Skewness were -0.619 and -0.794, respectively. The Kurtosis were -0.739 and -0.361, respectively. The total number of piglets born per litter and number of piglets born alive in 250-279 days and 340-369 days was the highest, followed by 220-249 days. The age above 370 days was the lowest under these two litter production parameters. Differences in the number of weaned piglets were only seen from 190-219 days and 280-309 days, and no significant difference was found among the other days.

Table 5　Litter production at different first mating days of 35 847 gilts.

Age at first mating days	Number of gilts	Litter production		
		Total number ofpiglets born	Number of piglets born alive	Number of weaned piglets
		Mean±SEM	Mean±SEM	Mean±SEM
160-189	898	9.9±0.10[a]	9.1±0.11[a]	7.9±0.13[ac]
190-219	7 082	10.1±0.03[a]	9.3±0.04[ac]	7.9±0.04[a]
220-249	10 575	10.5±0.03[e]	9.6±0.03[d]	7.8±0.04[ac]
250-279	8 261	10.7±0.09[d]	9.8±0.03[e]	7.9±0.05[ac]
280-309	5 522	10.3±0.03[a]	9.2±0.05[ac]	7.7±0.06[bc]
310-339	2 259	10.3±0.06[a]	9.3±0.07[ac]	7.9±0.08[ac]
340-369	651	10.8±0.11[de]	9.6±0.14[acde]	8.1±0.17[ac]
370-399	347	9.3±0.16[bc]	8.4±0.18[b]	7.3±0.21[ac]
≥400	252	9.5±0.13[bc]	8.4±0.18[b]	7.7±0.21[ac]

Note: [a,b,c,d,e]Bars with different letters differ significantly ($P<0.05$).

3　Discussion

This study clearly presented the multiple effects of pig farms with different production levels on the utilization and reproductive performance of gilts through statistical analysis of large amounts of data. The effects of utilization were mainly reflected in source, culling rate and mortality. The age at first estrus, age at first mating, breeds and litter production were the main factors affecting reproductive performance. Among them, the negative effects of breeds (three-way crossbred) on conception rate, farrowing rate and litter production cannot be ignored.

In our study, more than 90% of the gilts in HP pig farms and IP pig farms came from internal companies (self-breeding and internal introduction). For farrow-to-finish pig farms, self-breeding was a common method of introduction. It can not only cut the costs of gilt purchase but also reduce the risk of new pathogens brought by external introduction[11,12]. By comparison, 82% of the gilts in LP farms came from internal companies. This source structure may increase the risk of introducing unknown diseases into LP farms.

Our study showed that the culling rate ofgilts on the LP farm was 14.9%, which was significantly higher than that on the HP (9.4%) and IP pig farms (10.4%). The main reasons were undesirable limb configuration, repeated mating infertility and anestrus[14,15,16]. Tani et al. [17] suggested that when the age at first mating increased from 220 days to 300 days, the risk of culling gilts due to reproductive failure increased by 2.1%. Because the average introduction age in LP farms was 223 days, the records of first induced estrus and subsequent estrus were missed. Therefore, it was easy to delay the best time of breeding, resulting in the culling of these gilts due to reproductive problems. In general, the annual renewal rate of gilts was 45% - 60%, which was a very important production cost of pig farms[18]. From an economic point of view, gilts must serve more than three parities in the breeding herd to reduce substantial renewal costs[19]. On the premise of ensuring the parity structure of sows, the lower renewal rate of gilts, the higher profit of pig farms[20]. However, during the entire one-year research stage, the total culling rate of LP farms was as high as 40.4%. Together with the total mortality of 19.1% (Table 1), the loss of gilts was nearly 60%, indicating that the overall utilization of gilts in LP farms was very low.

Estruswas the basis for the reproductive performance of gilts. Gilts were bred in subsequent estrus cycles rather than their first cycle[21]. The breeding technical standard of gilts in our company was to introduce gilts (age at 130 days), raise them in a pen with more than a dozen gilts, induce estrus at the age of 160 days, record the first estrus, and transfer the estrus gilts to a single column. Afterwards, when the gilts were aged at 210 d and weighed more than 135 kg, they could be bred for the first time. However, due to the shortage of sow sources, the age of introduction/first mating in all groups was generally older, especially in the LP-CP group. This not only caused the gilts to miss the optimal mating period (second or third estrus), increased the NPDs and feed consumption due to excessive feeding time but also further reduced the estrus rate due to excessive weight gain[1,22]. Gilts that missed the mating period will occupy a corresponding number of individual columns while waiting for the next estrus period. As a result, other breeding gilts that need to be transferred to individual columns can only continue to be raised and bred in pens. Once they fought with each other and caused severe acute stress, they continued to affect their reproductive performance, which not only reduced animal welfare but also shortened the sow's production life[23]. This study found that the total estrus rate and total mating rate of LP farms were lower than those of HP and IP farms, which may be one of the reasons for poor production performance.

An interesting finding of this studywas that in all classified pig farms, only LP pig farms had three-way crossbred gilts, and the proportion was as high as 54%. It was reported that breeds had an important impact on the reproductive performance of sows[24]. To study the effect of breeds on the reproductive performance of gilts, LP farms were further divided into breeding pig groups (LP-BP) and commercial pig groups (LP-CP). The results showed that the average total number of piglets born per litter and average number of piglets born alive per litter of the LP-CP groups were the lowest in all groups (Table 4). The proportion distribution peaks of litter production in the LP groups were shifted approximately two times less than those in the HP farms, and the proportion of low litter production (eight per litter or less) was higher than that in the HP farms (Fig. 1). This may be the characteristic of three-way crossbred pigs, that is, high growth performance and superior meat quality but poor reproductive performance[25]. In March 2019, the price of pork in China continued to rise[26]. Coupled with the influence of African swine fever, there was an extreme shortage of breeding sows. In this case, a large number of three-way crossbred sows originally used as commercial pigs were used as breeding sows[27]. With the recovery of pig production, three-way crossbred sows, which were used as emergency supplements during special periods, were gradually culled.

We also found that there was a certain relationship between litter production and the age at first mating. When the age at first mating was 160-279 days, the average total number of piglets born per litter and average number of piglets born alive per litter increased with increasing age and decreased after 280 days, but there was a small peak at 340-369 days. This was inconsistent with the report of Saito[4] because he calculated the annualized lifetime number of pigs born alive. Nevertheless, the overall trend was similar.

This study had some limitations. Litter production was only statistically analyzed at the first parity. If the reproductive performance of all parities throughout the lifetime can be recorded, the impact of farm performance on the reproductive performance of gilts will be more accurately assessed. The current data represented many different management conditions and levels, ignoring the possible interaction between these factors. However, this study was worthwhile to describe the overall trend of the production level of large-scale pig farms. The gap between HP farms and LP farms in terms of the comprehensive utilization and reproductive performance of gilts was emphasized, which provided reference indicators for LP farms to be improved.

4　Conclusion

There were differences in the overall utilization and reproductive performance of gilts in pig farms of different production levels. The production level of HP farmswas significantly higher than that of LP farms. The differences in the overall utilization of gilts were mainly reflected in the introduction source, culling rate and mortality in the gilt stage and the research stage. The age at first estrus, age at first mating, breed and litter production were the main factors that affected reproductive performance.

References

[1] Faccin J E G, Laskoski F, Lesskiu P E, et al. Reproductive performance, retention rate, and age at the third parity according to growth rate and age at first mating in the gilts with a modern genotype [J]. Acta Sci Vet, 2017, 45: 1-6.

[2] Houška L. The relationship between culling rate, herd structure and production efficiency in a pig nucleus herd [J]. Czech J Anim Sci, 2009, 54 (8): 365-375.

[3] Safranski T. Management of ReplacementGilts [J]. A Platform For Success, 2016: 95.

[4] Patterson J, Foxcroft G. Gilt management for fertility and longevity [J]. Animals, 2019, 9 (7): 434.

[5] Saito H, Sasaki Y, Koketsu Y. Associations between age of gilts at first mating and lifetime performance or culling risk in commercial herds [J]. J Vet Med Sci, 2010: 1012030403.

[6] Lopes R, Kruse A B, Nielsen L R, et al. Additive Bayesian Network analysis of associations between antimicrobial consumption, biosecurity, vaccination and productivity in Danish sow herds [J]. PrevVet Med, 2019, 169: 104702.

[7] Kraeling R R, Webel S K. Current strategies for reproductive management of gilts and sows in North America [J]. J Anim Sci Biotechno, 2015, 6 (1): 1-14.

[8] Luo Y, Li S, Sun Y, et al. Classical swine fever in China: a minireview [J]. Vet Microbiol, 2014, 172 (1-2): 1-6.

[9] Lei K, Teng G H, Zong C, et al. The study of urine hormone index based on estrus mechanism of sows. 2020 ASABE Annual International Virtual Meeting [C]. American Society of Agricultural and Biological Engineers, 2020: 1.

[10] Koketsu Y. Productivity characteristics of high-performing swine farms [J]. J Am Vet Med Assc, 2000, 215: 376-379.

[11] Tani S, Piñeiro C, Koketsu Y. High-performing farms exploit reproductive potential of high and low prolific sows better than low-performing farms [J]. Porcine Health Manag, 2018, 4 (1): 1-12.

[12] Lambert M È, Denicourt M, Poljak Z, et al. Gilt replacement strategies used in two swine production areas in Quebec in regard to porcine reproductive and respiratory syndrome virus [J]. J Swine Health Prod, 2012, 20 (5): 223-230.

[13] Forner R, Bombassaro G, Bellaver FV, et al. Distribution difference of colostrum-derived B and T cells subsets in gilts and sows [J]. PloS One, 2021, 16 (5): e0249366.

[14] Chitakasempornkul K, Meneget M B, Rosa G J M, et al. Investigating causal biological relationships between reproductive performance traits in high-performing gilts and sows [J]. J Anim Sci, 2019, 97 (6): 2385-2401.

[15] Sasaki Y, Koketsu Y. Reproductive profile and lifetime efficiency of female pigs by culling reason in high-performing commercial breeding herds [J]. J Swine Health Prod, 2011, 19 (5): 284-291.

[16] Wang C, Wu Y, Shu D, et al. An Analysis of Culling Patterns during the Breeding Cycle and Lifetime

Production from the Aspect of Culling Reasons for Gilts and Sows in Southwest China [J] . Animals, 2019, 9 (4): 160.

[17] Tani S, Koketsu Y. Factors for Culling Risk due to Pregnancy Failure in Breeding-Female Pigs [J] . J Agric Sci, 2016, 9: 109-117.

[18] Davies P R, Funk J A, Morrow W E M. Fecal shedding of Salmonella by gilts before and after introduction to a swine breeding farm [J] . J Swine Health Prod, 2000, 8 (1): 25-29.

[19] Casey T, Harlow K L, Ferreira C R, et al. The potential of identifying replacement gilts by screening for lipid biomarkers in reproductive tract swabs taken at weaning [J] . J Appl Anim Res, 2018, 46 (1): 667-676.

[20] Stalder K, Knauer M, Baas T, et al. Sow longevity [J] . Pig News Inf, 2004, 25 (2): 53N-74N.

[21] Green M L, Diekman M A, Malayer J R, et al. Effect of prepubertal consumption of zearalenone on puberty and subsequent reproduction of gilts [J] . J Anim Sci, 1990, 68 (1): 171-178.

[22] Cottney P D, Magowan E, Ball M E E, et al. Effect of oestrus number of nulliparous sows at first service on first litter and lifetime performance [J] . Livest Sci, 2012, 146 (1): 5-12.

[23] Lagoda M E, Boyle L A, Marchewka J, et al. Mixing aggression intensity is associated with age at first service and floor type during gestation, with implications for sow reproductive performance [J]. Anim, 2021: 100158.

[24] Knecht D, srodoń S, Duziński K. The impact of season, parity and breed on selected reproductive performance parameters of sows [J] . Arch Anim Breed, 2015, 58 (1): 49-56.

[25] Kriauziene J, Rekštys V. Compatibility of various pig breeds at common and complex crossbreeding [J] . Veterinarija ir zootechnika, 2003, 21 (43): 85-89.

[26] Liu Y, He L, Li D, et al. Correlation Analysis of Chinese Pork Concept Stocks Based on Big Data [C]. International Conference on Artificial Intelligence and Security, Springer, Cham, 2020: 475-486.

[27] Woonwong Y, Do Tien D, Thanawongnuwech R. The future of the pig industry after the introduction of African swine fever into Asia [J] . Anim Front, 2020, 10 (4): 30-37.

The Construction of Recombinant *Lactobacillus casei* Vaccine of PEDV and its Immune Responses in Mice[*]
(PEDV 重组乳酸杆菌疫苗的构建和小鼠免疫应答)

Xiaowen Li　Bingzhou Zhang　Dasheng Zhang　Sidang Liu　Jing Ren

Abstract： Porcine epidemic diarrhea (PED) is a contagious intestinal disease caused by porcine epidemic diarrhea virus (PEDV) characterized by vomiting, diarrhea, anorexia, and dehydration, which have caused huge economic losses around the world. At present, vaccine immunity is still the most effective method to control the spread of PED. In this study, we have constructed a novel recombinant *L. casei*-OMP16-PEDVS strain expressing PEDVS protein of PEDV and OMP16 protein of *Brucella abortus* strain. To know the immunogenicity of the recombinant *L. casei*-OMP16-PEDVS candidate vaccine, it was compared with BL21-OMP16-PEDVS-F, BL21-OMP16-PEDVS, and BL21-PEDVS recombinant protein.

The results showed that we could detect higher levels of IgG, neutralizing antibody, IL-4, IL-10, and INF-γ in serum and IgA in feces of *L. casei*-OMP16-PEDVS immunized mice, which indicated that *L. casei*-OMP16-PEDVS candidate vaccine could induce higher levels of humoral immunity, cellular immunity, and mucosal immunity.

Therefore, *L. casei*-OMP16-PEDVS is a promising candidate vaccine for prophylaxis of PEDV infection.

Keywords： PEDV; *Lactobacillus casei* vaccine; PEDVS protein; OMP16; immune responses

1　Introduction

Porcine epidemic diarrhea (PED) is caused byporcine epidemic diarrhea virus (PEDV) with symptoms including diarrhea, vomiting, anorexia, dehydration, and weight loss in piglets[1,2]. Pigs of all ages can be infected with different symptoms and the mortality rate in piglets is up to 100%[3], which have led to huge economic losses all around the world. To control the spread of PEDV, most kinds of vaccines are constructed, such as aluminum-hydroxide-adjuvanted inactivated vaccine, bivalent inactivated Transmissible Gastroenteritis

＊　该文章发表于 *BMC Veterinary Research* 第 17 卷第 1 期。

Virus（TGEV）and PEDV vaccine，and attenuated PEDV vaccine[4]．Although these vaccines play an important role in controlling PED，they all have their defects．Inactivated vaccines cannot activate cellular immune responses，attenuated vaccine is not very safe，and they all cannot induce sufficient production of virus-specific IgA antibodies of mucosal immune responses．Therefore，it is necessary and urgent to develop a new vaccine to control PED．

Lactobacillus casei is often considered to be a kind of safe vector system for targeted delivery of antigens in oral immunization，with beneficial effects on the health of humans and animals[5]．Meanwhile，it can be used as a delivery system to regulate the T-helper cell response and stimulate the secretion of specific IgAs for mucosal immunity[6]．On the other hand，*Lactobacillus casei* recombinant vaccine is easier administration，lesser chance of hypersensitivity reaction，and more cost-effective compared with traditional vaccines．Based on the reports，*Lactobacillus casei* recombinant vaccines have been successfully used in the prevention and control of human papillomavirus，*Streptococcus pneumonia*，and *Escherichia coli*[7-9]．There are also some similar attempts in designing of PED vaccines．The researches find that a recombinant *Lactococcus lactis* strain expressing a variant porcine epidemic diarrhea virus S1 gene could induce high levels of IL-4 and IFN-γ in immunized mice[10]．*Lactobacillus casei*-based anti-PEDV vaccine expressing microfold cell-targeting peptide Co1 fused with the COE antigen of PEDV could also induce effective immune response[11]．To improve the effectiveness of PEDV vaccine．In this study，we construct a new *Lactobacillus casei* recombinant vaccine of PED，which can stimulate stronger mucosal，humoural and cellular immune responses against PEDV infection via oral administration．

PEDV，a member of thecoronaviridae family，consisted by four structural proteins which contain the 150-220 ku glycosylated spike（S）protein，the 20-30 ku membrane（M）protein，the 7 ku envelope（E）protein，and 58 ku nucleocapsid（N）protein[12]．Thereinto，the S protein can be divided into S1（1-735 amino acid）and S2（736-last amino acid）domains[13]·and S1 protein includes the receptor-binding region and the main neutralizing epitopes[14]．Vaccine adjuvant acts as an immunomodulator can induce and enhance immune responses against co-delivered antigens．OMP16 protein of *Brucella abortus* strain was verified that could activate dendritic cells in vivo，induces a th1 immune response，and was a promising self-adjuvanting vaccine[15-17]．Therefore，OMP16 protein was inserted into the *Lactobacillus casei* recombinant vaccine in our study．

So far，few studies about the *Lactobacillus casei* recombinant vaccine of PEDV are reported．Therefore，this study is aimed to construct a novel *Lactobacillus casei* candidate oral vaccine，which can supply better humoral immunity，cellular immunity，and mucosal immunity to prevent the spread of PED．

2　Materials and methods

2.1　Bacterial strains, viruses, culture conditions, plasmids, and primers

The bacterial strains, plasmids, and primers used in this study are listed in Table 1. The standard reference strain of *Lactobacillus casei* ATCC 393 was cultured in de Man Rogosa and Sharpe (MRS) broth at 37℃[18]. The BL21 (DE3) and DH5α were cultured in Luria-Bertani (LB) medium at 37℃[19]. The recombinant *Lactobacillus casei*, BL21 (DE3), and DH5α strains were cultured in the corresponding medium with proper antibiotics, respectively. *Brucella abortus* was grown in Tryptic Soy Broth (TSB) or Tryptic Soy Agar (TSA) medium (Difco Laboratories, Detroit, MI, USA) at 37℃. The Vero cells infected with PEDV strains were cultured in DMEM (Gibco, Langley, VA, USA) supplemented with 10 μg/mL trypsin (Gibco, Langley, VA, USA)[20]. The pVE5523 and pET28a (+) plasmids were expression vectors of *Lactobacillus casei* and *Escherichia coli*, respectively.

Table 1　Characteristics of bacterial strains, plasmids, and primers used in this study.

Strain/plasmid/primer	Characteristics and/or sequences	Source/reference
Strain		
DH5α	Genotype: supE44ΔlacU169 (φ80lacZ△M15) hsdR17 recA1 endA1 gyrA96 thi-1 relA1	TaKaRa (Otsu, Japan)
BL21	Genotype: F-ompT hsdS (rB-mB) gal dcm (DE3)	TaKaRa (Otsu, Japan)
Lactobacillus casei	ATCC393, used as a vector system for targeted delivery of antigens in oral immunization.	Preserved in our lab
Brucella abortus	*Brucella abortus* S19, supply gene sequence of *omp*16 gene	Preserved in our lab
PEDV	AJ1102, variant strain, isolated in China.	Preserved in our lab
Plasmids		
pET28a (+)	The expression vector of *Escherichia coli*, Kanr	Preserved in our lab
pVE5523	The expression vector of *Lactobacillus casei*, Ampr	BioVector NTCC
pET28a-PEDVS	The recombinant vector of pET28a and S protein of PEDV, Kanr	This study
pET28a-OMP16-PEDVS	The recombinant vector of pET28a and fusion protein (S protein of PEDV and OMP16 protein of *Brucella abortus*), Kanr	This study
pVE5523-OMP16-PEDVS	The recombinant vector of pVE5523 and fusion protein (S protein of PEDV and OMP16 protein of *Brucella abortus*), Ampr	This study
Primers		
PEDVS-F1/R1	F1: CCGGAATTCATGCTGAGTCATGAACAGCC R1: TGCTCTAGATTAATATGCAGCCTGCTCTG	This study

（continued）

Strain/plasmid/primer	Characteristics and/or sequences	Source/reference
PEDVS-F2/R2	F2：GGTGGTGGCGGTAGCGGCGGTGGTGGCTCTGGT GGCGGCGGTTCTTTCTTTTGTTACTTTGCCAT R2：CGCGATATCTTAATATGCAGCCTGCTCTG	This study
OMP16-F/R	F：CGCGTCGACATGGCGTCAAAGAAGAACCTTCCG R：AGAACCGCCGCCACCAGAGCCACCACCGCCGC TACCGCCACCACCGGTACCCCGTCCGGCCCCGT	This study

2.2 The construction of recombinant *Lactobacillus casei* and BL21 strains

The recombinant expression plasmids were constructed based on the plasmids and primers in Table 1. At first, the partial sequence of PEDV S gene (493-708 amino acid), the partial sequence of PEDV S ` gene (493-708 amino acid), and the partial sequence of *Brucella abortus* OMP16 gene (26-168 amino acid) were amplified using primer pairs PEDVS-F1/R1, PEDVS-F2/R2, and OMP16-F/R, respectively. Subsequently, the overlap extension method was used in OMP16 and PEDVS` fragments to construct a new fragment OMP16-PEDVS with linker polypeptide (GGGGSGGGGSGGGGS) stuck in the middle. Then, the fragments PEDVS was inserted into pET28a plasmid with EcoRI/XbaI restriction enzymes to generate recombinant plasmid pET28a-PEDVS, and the new fragments OMP16-PEDVS was inserted into pET28a and pVE5523 plasmids with EcoRI/XbaI and SalI/EcoRV restriction enzymes to generate recombinant plasmid pET28a-OMP16-PEDVS and pVE5523-OMP16-PEDVS, respectively. The recombinant plasmids (pET28a-PEDVS, pET28a-OMP16-PEDVS, and pVE5523-OMP16-PEDVS) were transformed into BL21 (DE3) or *Lactobacillus casei* ATCC393 by transformation or electroporation based on the reported paper[21].

2.3 Analysis of protein expression by western blot

The protein expression of BL21-pET28a-PEDVS, BL21-pET28a-OMP16-PEDVS, and L. casei-pVE5523-OMP16-PEDVS strains were detected based on the reported method with somemodification[21-23]. The recombinant strains BL21-pET28a-PEDVS and BL21-pET28a-OMP16-PEDVS were cultured in LB broth and IPTG was used to harvest to pET28a-PEDVS and pET28a-OMP16-PEDVS proteins. The blank vector was used as a negative control. After cell lysis and centrifugation, the supernatant and sediment were collected. Then, the target proteins were purified using the Nickel affinity chromatography column based on the previous study[24]. Meanwhile, the cultural supernatant of the recombinant L. casei-pVE5523-OMP16-PEDVS strain was also harvested by centrifugation at 9 000×g for 10 mins at 4℃. Whereafter, the samples with sodium dodecyl sulfate (SDS) loading buffer were boiled 10 mins. The proteins were separated by 12% sodium dodecyl sulfate-polyacrylamide gel electrophoresis (SDS-PAGE) and then transferred into PVDF membranes (Millipore, Mississauga, ON, Canada). Membranes were blocked with 5% skimmed milk for 2 h at 37℃ and then incubated with murine monoclonal antibody of S protein overnight at 4℃ and HRP conjugated goat anti-mouse IgG (ABclonal, Wuhan, China) for 2 h at 37℃. The protein bands were visualized using the Clarity™ Western ECL Blotting Substrate (Bio-Rad, Hercules, CA, USA).

2.4 Immunization and sample collection

The immunogenicity of recombinant *Lactobacillus casei* vaccine was evaluated using six-week-old female specific pathogen-free (SPF) BALB/c mice[10,18], which were purchased from Shandong Agricultural University animal center. A total of 50 mice were randomly divided into 5 groups with 10 mice in each group. The mice were immuned with recombinant *Lactobacillus casei* vaccine and purified protein (pET28a-PEDVS, pET28a-OMP16-PEDVS, pET28a-OMP16-PEDVS + Freund's complete adjuvant), respectively. The immunization protocol was performed based on Table 2 and a booster immunization was given after 13 days.

For the serological study, serum was collected on 0, 14, and 28 days post-immunization (dpi) via tail vein punching and stored at −20℃ until use. Feces were collected one day before vaccination and every boosting time. For IgA detection, feces were diluted (w/v) with 0.05 mol/L sodium EDTA at 1 ∶ 4 ratios just after collection and incubated for 14 h at 4℃ following proper mixing. The supernatant was collected by 12 000× g centrifugation and preserved at −20℃ until use. At 28th dpi; three mice from each group were sacrificed and the intestine was processed for IgA detection according to the author described previously[10,18,23].

Table 2　The immune protocol of BALB/c mice.

Grouping	Characteristics	Immune methods	Immunizing dose
L. casei-OMP16-PEDVS	Recombinant strain *L. casei* contains pVE5523-OMP16-PEDVS vector	PO	200 μL
BL21-PEDVS	pET28a-PEDVS protein	IM	200 μg
BL21-OMP16-PEDVS	pET28a-OMP16-PEDVS protein	IM	200 μg
BL21-OMP16-PEDVS-F	pET28a-OMP16-PEDVS protein with Freund's complete adjuvant	IM	200 μg
PBS	Negative control	IM	200 μL

2.5　Determination analysis of antibody levels

The levels of IgG in the sera and IgA in the feces were measured bythe ELISA methods with some modification[18,23]. The methods were as follows: Polystyrene microliter plates were coated overnight at 4℃ with 100 μL 10 μg/mL PEDVS protein, OMP16-PEDVS protein, or 100 μL recombinant L. casei-OMP16-PEDVS strain. After blocking with 5% skimmed milk, the collected samples were serially diluted in PBS, added in triplicate, and incubated at 37℃ for 1 h. Then, an HRP conjugated goat anti-mouse IgG or IgA antibody (Invitrogen, USA) was added to each well (1 : 5 000) and incubated for 1 h at 37 ℃. The polystyrene microtiter plates were washed 5 times during each step. At last, 100 μL of TMB substrate (tetramethylbenzidine and H_2O_2) was added to each well and 50 μL of stop solution was added after 10 mins. The OD values at 630 nm were measured using a multimode plate reader (EnVision).

2.6　Virus neutralization assays

The neutralizing antibody titers of PEDV in sera were examined according to the methods with some modifications[25,26]. Briefly, the murine serum was heat-inactivated (56℃ for 30 mins) and then serially two-fold diluted in 96 well plates (Corning, USA) with triplicates of each sample. Then, an equal volume of 200 $TCID_{50}/50\mu$L PEDV strains were added to 96 well plates and incubated for 1 h at 37℃. The mixture was added to new 96 well plates coated with Vero cell monolayers and incubated for 1 h at 37℃. Cells were then washed and incubated in the maintenance medium at 37℃ in 5% CO_2. After 2 days, the cytopathic effect (CPE) was observed using an inverted microscope and the neutralizing concentration was defined as the lowest concentration of antibodies in the serum.

2.7　Cytokine detection

To detect the secretion of cytokines, supernatants were obtained from the laboratory mice (0, 14, and 28 days). Levels of secreted IL-4, IL-10, and IFN-γ were determined using commercial ELISA kits (Elabscience Biotechnology Co., Ltd, Wuhan) according to the manufacturer's recommendations, respectively. Cytokine was quantified from the different standard curves prepared from standard reagents provided by the kits respectively and optical density (OD) value was detected at 450 nm from each plate using a multimode plate reader (EnVision)[23,27].

2.8　Statistical analysis

All data were obtained from at least three independent experiments, and results were presented as the means±standard deviation (SD). The statistical analysis was performed using two-tailed t-tests and one-way analysis in Graph Pad Prism 7. 0 (GraphPad Software Inc. , USA). The significant difference was defined as * $P < 0.05$, and the various degrees of significant difference were designated as **$P < 0.01$, ***$P < 0.001$, respectively.

3　Results

3.1　The verification of recombinant *Lactobacillus casei* and BL21 strains

　　To verify the constructed recombinant plasmids, the recombinant expression plasmids pET28a-PEDVS, pET28a-OMP16-PEDVS, and pVE5523-OMP16-PEDVS were digested using NcoI/XhoI restriction enzymes, and the enzyme digestion results were shown in Fig. 1. The sizes of target bands in electrophoretograms were the same as the expected results and sequencing results indicated that the recombinant expression plasmids exhibited no mutation. These results indicated the successful construction of pET28a-PEDVS, pET28a-OMP16-PEDVS, and pVE5523-OMP16-PEDVS recombinant plasmids.

　　The results of western blot showed that the recombinant proteins pET28a-OMP16-PEDVS and pET28a-PEDVS were expressed inthe supernatant of BL21-pET28a-OMP16-PEDVS and BL21-pET28a-PEDVS strains, respectively. The pVE5523-OMP16-PEDVS protein was also verified in the cultural supernatant of L. casei-pVE5523-OMP16-PEDVS strain. The specific bands from Fig. 2 showed that the recombinant proteins pET28a-OMP16-PEDVS, pVE5523-OMP16-PEDVS, and pET28a-PEDVS were all harvested successfully (Fig. 2).

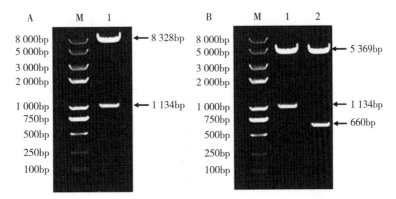

Fig. 1　The enzyme digestion results of recombinant plasmids. A: The enzyme digestion results of the pVE5523-OMP16-PEDVS plasmid; M: D8000 DNA Ladder Marker; 1: pVE5523-OMP16-PEDVS plasmid; B: The enzyme digestion results of pet28a-OMP16-PEDVS and pET28a-PEDVS plasmid; 2: pET28a-PEDVS plasmid.

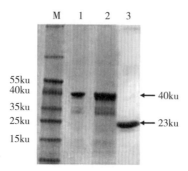

Fig. 2   The verification of recombinant expression proteins. M: protein marker; 1: the secretory protein form recombinant L. casei-pVE5523-OMP16-PEDVS strain; 2: the purified protein from BL21-pET28a-OMP16-PEDVS strain; 3: the purified protein from BL21-pET28a-PEDVS strain.

3.2 The IgG antibody levels in serum of mice immunized with candidate vaccines

To evaluate the specific immunogenicity of generated vaccine candidates, BALB/c mice were selected and divided into 5 groups. Then, the levels of IgG in the serum and IgA in the feces were measured with commercial ELISA kits. The results revealed that there were no substantial differences for IgG levels among the vaccinated groups and almost no IgG antibody was found in all mice before immunization. However, substantial differences were subsequently found after the first vaccination and IgG antibody levels in serum of 28 days were obviously higher than that of in 14 days. Among 5 group mice, the mice immunized with L. casei-OMP16-PEDVS and BL21-OMP16-PEDVS-F showed similar and highest immunogenicity. Therefore, L. casei-OMP16-PEDVS and BL21-OMP16-PEDVS-F could produce highest immunogenicity, followed by BL21-OMP16-PEDVS and BL21-PEDVS (Fig. 3).

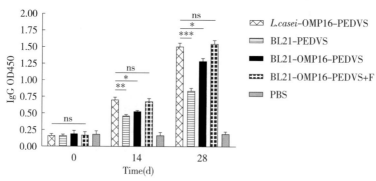

Fig. 3   The IgG antibody levels of candidate vaccines in the serum of immunized mice. Serum was collected on days 0, 14, and 28 days before or after immunization and examined via commercial ELISA kits and measured at an absorbance of 450 nm. Bars represent the mean±standard deviation of three independent experiments. * $P<0.05$,** $P<0.01$, and *** $P<0.0001$ represent increasing degrees of significant differences, respectively, and ns means no significant difference.

3.3 The IgA antibody levels in feces of mice immunized with candidate vaccines

To evaluate the specific immunogenicity of generated vaccine candidates, the levels of IgA antibody in feces of mice were also evaluated. The results showed that there was no special anti-PEDVS IgA antibody existed before immunization. However, large amounts of IgA antibody in feces of *L. casei*-OMP16-PEDVS immunized mice were detected and it was obviously higher than that of in BL21-OMP16-PEDVS-F, BL21-OMP16-PEDVS, and BL21-PEDVS group mice at 14 days after immunization. At 28 days after immunization, the IgA antibody levels of *L.* casei-OMP16-PEDVS immunized mice reached its highest maximum. Meanwhile, the IgA antibody levels in the other three groups did not present an obvious increase. Therefore, the candidate vaccine *L. casei*-OMP16-PEDVS could stimulate higher levels of antibody in immunized mice compared with BL21-OMP16-PEDVS-F, BL21-OMP16-PEDVS, and BL21-PEDVS immunized mice (Fig. 4).

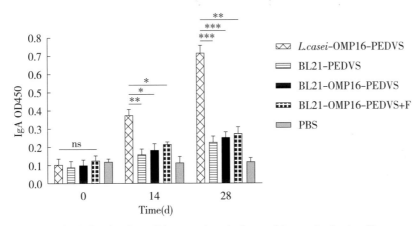

Fig. 4　The IgA antibody levels of candidate vaccines in feces of immunized mice. Feces were collected on day 0, 14, and 28 days before or after immunization and examined via commercial ELISA kits and measured at an absorbance of 450 nm. Bars represent the mean±standard deviation of three independent experiments. * $P < 0.05$, ** $P < 0.01$, and *** $P < 0.000\,1$ represent increasing degrees of significant differences, respectively, and ns means no significant difference.

3.4 The neutralizing antibody levels of serum in immunized mice

To evaluate the protective effect of candidate vaccines of *L. casei*-OMP16-PEDVS, BL21-OMP16-PEDVS-F, BL21-OMP16-PEDVS, and BL21-PEDVS in mice, the neutralizing antibody levels were measured. Results showed that no neutralizing antibody was detected before immunization. Neutralizing antibody was detected at 14 days after immunization and it increased at 14 days after booster immunization. The antibody response in mice that received *L. casei*-

OMP16-PEDVS possessed a stronger anti-PEDV neutralizing activity than that in mice orally administered with BL21-OMP16-PEDVS-F，BL21-OMP16-PEDVS，and BL21-PEDVS. Therefore，the candidate vaccine *L. casei*-OMP16-PEDVS could stimulate highest neutralizing antibody level，followed by BL21-OMP16-PEDVS-F，BL21-OMP16-PEDVS，and BL21-PEDVS（Fig. 5）.

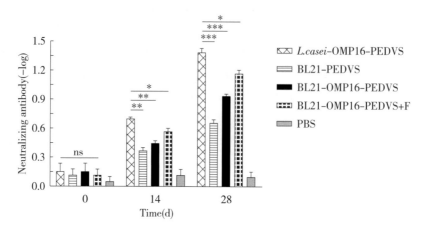

Fig. 5   The neutralizing antibody levels of candidate vaccines in the serum of immunized mice. Serum was collected on days 0，14，and 28 days before or after immunization and examined via neutralization test. Bars represent the mean ± standard deviation of three independent experiments. * $P < 0.05$,** $P < 0.01$，and *** $P < 0.000\ 1$ represent increasing degrees of significant differences，respectively，and ns means no significant difference.

3.5 Cytokine levels

To compare the cellular immune response level of *L. casei*-OMP16-PEDVS，BL21-OMP16-PEDVS-F，BL21-OMP16-PEDVS，and BL21-PEDVS immunized mice，IL-4，IL-10，and IFN-γ were determined，respectively. The results showed that the levels of cytokines IL-4，IL-10，and IFN-γ in the sera of mice were all very low and have no significant difference before immunization. Whereas，similar changes were observed in the results of IL-4，IL-10，and IFN-γ. At 14 days after immunization，the level of IL-4，IL-10，and IFN-γ in *L. casei*-OMP16-PEDVS immunized mice were higher than BL21-OMP16-PEDVS-F，BL21-OMP16-PEDVS，and BL21-PEDVS immunized mice. At 14 days after the booster immunization，a higher IL-4，IL-10，and IFN-γ level in *L. casei*-OMP16-PEDVS immunized mice were detected compared with that of in other three groups. Therefore，the candidate vaccine *L. casei*-OMP16-PEDVS could stimulate highest IL-4，IL-10，and IFN-γ level，followed by BL21-OMP16-PEDVS-F，BL21-OMP16-PEDVS，and BL21-PEDVS（Fig. 6）.

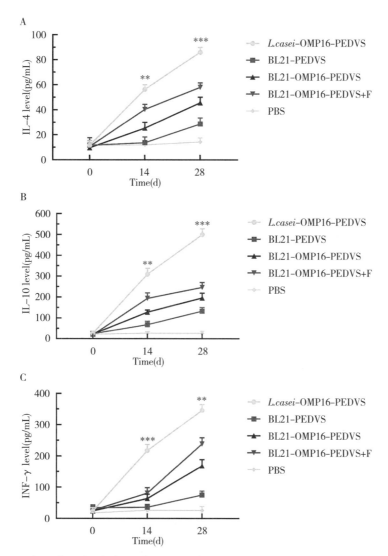

Fig. 6　Detection of cytokine levels from the serum of immunized mice. Serum was collected on days 0， 14， and 28 days before or after immunization and examined via commercial ELISA kits. The absorbance value was measured at an absorbance of 450 nm for IL-4 （A）， IL-10 （B）， and IFN-γ （C）， respectively. Bars represent the mean±standard deviation of three independent experiments. * $P<$ 0. 05，** $P<$ 0. 01， and *** $P<$ 0. 000 1 represent increasing degrees of significant differences， respectively.

4　Discussion

Since a large-scale outbreak of PED that caused by PEDV variants occurred in October 2010， which has resulted intremendous economic losses in China and all around the world[1]. However， traditional vaccines are all designed based on CV777 classical strain which cannot supply sufficient protection to PEDV variant[4]. To control the spread of PEDV and reduce the economic losses, novel vaccines of PEDV variant strains are also designed. At present, PEDV inactivated vaccine and attenuated vaccine of PEDV variant strains are all approved by Chinese government and there are all exhibiting promising prospects in controlling PED. But defects are also existed in the two kinds of novel vaccines. PEDV infects swine through the digestive tract and has intestinal tissue tropism. Therefore, Mucosal immunity is more effective than systemic immunity in preventing PEDV entry into intestinal epithelial cells, and vaccines must provide mucosal protection effectively in the intestinal tract. So, In this study, we construct a new kind of vaccine which can stimulate stronger anti-PEDV-specific IgG and sIgA antibodies.

Lactobacillus casei has potential immune-modulatory properties as a vaccine delivery vehicle and the expression of bioactive compounds on the cell wall of this bacterium can stimulate appropriate immune responses[28,29]. It is widely used for expressing several heterologous antigens of human papillomavirus, *Streptococcus pneumonia*, and *Escherichia coli* as vaccines in animal models, which all showed excellent immunogenicity[7-9]. Compared with the inactivated vaccine and attenuated vaccine, the *Lactobacillus casei* vector vaccine can also stimulate higher IgA level and cellular immune response. Studies have shown that IgA's first line of defense in the intestine would be better than IgG in protecting piglets from PEDV infection[10,30]. Therefore, it is promising to develop a kind of *Lactobacillus casei* vector vaccine of PED.

Based on the reports, the S protein of PEDV can bedivided into S1 (1-735 amino acid) and S2 (736-last amino acid) domains[13]· and S1 protein includes the receptor-binding region and the main neutralizing epitopes[14]. The core neutralizing epitope (COE) can induce strong neutralizing antibodies against PEDV[31,32]. Combining with the antigenicity analysis, a partial sequence of the S1 gene (1 477-2 124bp) was selected to construct the recombinant plasmid. The selected small fragment was proved that not only had good immunogenicity but also contributed to the secreted expression of *Lactobacillus casei*. On the other hand, *Pasquevich* found that *Brucella abortus* outer membrane protein 16 could activate dendritic cells in vivo, induces a th1 immune response, and was a promising self-adjuvanting vaccine against systemic and oral acquired brucellosis[15]. Similar research that unlipidated outer membrane protein omp16 from Brucella spp. as nasal adjuvant could induce a Th1 immune response and modulates the Th2 allergic response to cow's milk proteins was also proved[16]. Therefore, a partial sequence of the *omp*16 gene was chosen to construct recombinant plasmid to enhance to immune function in our study.

To know whether the novel*Lactobacillus casei* recombinant vaccine could induce humoral immune responses, the IgG, IgA, and neutralizing antibody levels were measured.

The IgG antibody level of *L. casei*-OMP16-PEDVS recombinant vaccine immunized mice had no significant difference with BL21-OMP16-PEDVS-F recombinant vaccine immunized mice. But the IgA and neutralizing antibody levels were obviously higher than that of in BL21-OMP16-PEDVS-F, BL21-OMP16-PEDVS, and BL21-PEDVS recombinant vaccine immunized mice. The results showed that *L. casei*-OMP16-PEDVS could induce stronger humoral immune responses, especially IgA antibody level. Studies have shown that IgA was the first line of defense in the intestine and would be better than IgG in protecting piglets from PEDV infection[30]. The research also verified the result that *L. casei*-OMP16-PEDVS recombinant vaccine could supply better immunological protection to PEDV.

To explore the type of immune response induced by recombinant *L. casei*-OMP16-PEDVS, the levels of IL-4, IL-10, and IFN-γ were detected to evaluate the activity of T lymphocytes. Based on the report, IFN-γ plays an important role in cellular immune response caused when pathogens invade the body, IL-4 plays an important role in the humoral immune response and promoting immune tolerance and mucosal immunity[33]. IL-10 plays essential roles in fighting against mucosal microbial infection and maintaining mucosal barrier integrity within the intestine[34]. Meanwhile, results of cytokine detection showed that the mice immunized with *L. casei*-OMP16-PEDVS recombinant vaccine could induce stronger expression of IL-4, IL-10, and IFN-γ, which supported the results that *L. casei*-OMP16-PEDVS recombinant vaccine could induce stronger humoral immune response, IgA antibody level, and cellular immune response, respectively.

5　Conclusion

In summary, *L. casei*-OMP16-PEDVS, BL21-OMP16-PEDVS-F, BL21-OMP16-PEDVS, and BL21-PEDVS candidate vaccines were constructed in this study. Meanwhile, the humoral immune response and cellular immune response levels of these candidate vaccines in mice were evaluated. The results showed that the mice immunized with *L. casei*-OMP16-PEDVS could produce higher levels of IgG, IgA, neutralizing antibody, IL-4, IL-10, and INF-γ compared with the mice immunized with BL21-OMP16-PEDVS-F, BL21-OMP16-PEDVS, and BL21-PEDVS. Therefore, the recombinant *L. casei*-OMP16-PEDVS candidate vaccine may establish the ground for the development of a safe, effective, and convenient recombinant mucosal vaccine for prophylaxis of PEDV infection.

References

[1] Wang D, Fang L, Xiao S. Porcine epidemic diarrhea in China [J]. Virus Research: An International Journal of Molecular and Cellular Virology, 2016: 7-13.

［2］ Sueyoshi M，Tsuda T，Yamazaki K，et al. An immunohistochemical investigation of porcine epidemic diarrhoea ［J］. Journal of Comparative Pathology，1995，113（1）：59-67.

［3］ Sun R Q，Cai R J，Chen Y Q，et al. Outbreak of Porcine Epidemic Diarrhea in Suckling Piglets，China ［J］. Emerging Infectious Diseases，2012，18（1）：161-163.

［4］ Tong Y E，Feng L，Li W J. Development of bi-combined attenuated vaccine against transmissible gastroenteritis virus and porcine epidemic diarrhea virus ［J］. Chinese Journal of Preventive Veterinary Medicine，1999，21（6）：35-39.

［5］ Pouwels P H，Leer R J，Boersma W. The potential of Lactobacillus as a carrier for oral immunization： development and preliminary characterization of vector systems for targeted delivery of antigens ［J］. Journal of Biotechnology，1996，44（1-3）：183.

［6］ Tsai Y T，Cheng P C，Pan T M. The immunomodulatory effects of lactic acid bacteria for improving immune functions and benefits ［J］. Applied Microbiology and Biotechnology，2012，96（4）：853-862.

［7］ Adachi K，Kawana K，Yokoyama T，et al. Oral immunization with a Lactobacillus casei vaccine expressing human papillomavirus（HPV）type 16 E7 is an effective strategy to induce mucosal cytotoxic lymphocytes against HPV16 E7 ［J］. Vaccine，2010，28（16）：2810-2817.

［8］ Campos I B，Darrieux M，Ferreira D M，et al. Nasal immunization of mice with Lactobacillus casei expressing the Pneumococcal Surface Protein A：induction of antibodies，complement deposition and partial protection against Streptococcus pneumoniae challenge ［J］. Microbes & Infection，2008，10（5）：481-488.

［9］ Wen L J，Hou X L，Wang G H，et al. Immunization with recombinant Lactobacillus casei strains producing K99，K88 fimbrial protein protects mice against enterotoxigenic Escherichia coli ［J］. Vaccine，2012，30（22）：3339-3349.

［10］ Guo M，Yi S，Guo Y，et al. Construction of a Recombinant Lactococcus lactis Strain Expressing a Variant Porcine Epidemic Diarrhea Virus S1 Gene and Its Immunogenicity Analysis in Mice ［J］. Viral Immunology，2019，32（3）：144-150.

［11］ Wang X，wang L，Zheng D，et al. Oral immunization witha Lactobacillus casei-based anti-porcine epidemic diarrhea virus（PEDV）vaccine expressing microfold cell-targeting peptide Co1 fused with the COE antigen of PEDV ［J］. Journal of Applied Microbiology，2017，124（2）：368-378.

［12］ Duarte M，Tobler K，Bridgen A，et al. Sequence analysis of the porcine epidemic diarrhea virus genome between the nucleocapsid and spike protein genes reveals a polymorphic ORF ［J］. Virology，1994，198（2）：466.

［13］ Lee D K，Park C K，Kim S H，et al. Heterogeneity in spike protein genes of porcineepidemic diarrhea viruses isolated in Korea ［J］. Virus Research，2010，149（2）：175-182.

［14］ Sun D B，Feng L，Shi H Y，et al. Spike protein region（aa 636789）of porcine epidemic diarrhea virus is essential for induction of neutralizing antibodies ［J］. Acta Virologica，2007，51（3）：149-156..

［15］ Pasquevich K A，Samartino C G，Coria L M，et al. The protein moiety of Brucella abortus outer membrane protein 16 is a new bacterial pathogen-associated molecular pattern that activates dendritic cells in vivo，induces a Th1 immune response，and is a promising self-adjuvanting vaccine against systemic an ［J］. Journal of Immunology，2010，184（9）：5200.

[16] Ibanez A E, Smaldini P, Coria L M, et al. Unlipidated Outer Membrane Protein Omp16 (U-Omp16) from Brucella spp. as Nasal Adjuvant Inducesa Th1 Immune Response and Modulates the Th2 Allergic Response to Cow's Milk Proteins [J]. Plos One, 2013, 8 (7): e69438.

[17] Pasquevich K A, Estein S M, Samartino C G, et al. Immunization with recombinant Brucella species outer membrane protein Omp16 or Omp19 in adjuvant induces specific CD4$^+$ and CD^8+ T cells as well as systemic and oral protection against Brucella abortus infection [J]. Infection and Immunity, 2009, 77 (1): 436-445.

[18] Ma S, Wang L, Huang X, et al. Oral recombinant Lactobacillus vaccine targeting the intestinal microfold cells and dendritic cells for delivering the core neutralizing epitope of porcine epidemic diarrhea virus [J]. Microbial Cell Factories, 2018, 17 (1): 20.

[19] Chu S, Zhang D, Wang D, et al. Heterologous expression and biochemical characterization of assimilatory nitrate and nitrite reductase reveals adaption and potential of Bacillus megaterium NCT-2 in secondary salinization soil [J]. International Journal of Biological Macromolecules, 2017, 101: 1019.

[20] Guo N, Zhang B, Hu H, et al. Caerin1. 1 Suppresses the Growth of Porcine Epidemic Diarrhea Virus In Vitro via Direct Binding to the Virus [J]. Viruses, 2018, 10 (9).

[21] Chen Z, Lin J, Ma C, et al. Characterization of pMC11, a plasmid with dual origins of replication isolated from Lactobacillus casei MCJ and construction of shuttle vectors with each replicon [J]. Applied Microbiology & Biotechnology, 2014, 98 (13): 5977.

[22] Wang X, Wang L, Huang X, et al. Oral Delivery of Probiotics Expressing Dendritic Cell-Targeting Peptide Fused with Porcine Epidemic Diarrhea Virus COE Antigen: A Promising Vaccine Strategy against PEDV [J]. Viruses, 2017, 9 (11): 312.

[23] Bhuyan A A, Memon A M, Bhuiyan A A, et al. The construction of recombinant Lactobacillus casei expressing BVDV E2 protein and its immune response in mice [J]. Journalof Biotechnology, 2018, 270: 51-60.

[24] Noi N V, Chung Y C. Optimization of expression and purification of recombinant S1 domain of the porcine epidemic diarrhea virus spike (PEDV-S1) protein in Escherichia coli [J]. Biotechnology & Biotechnological Equipment, 2017, 31 (2): 1-11.

[25] Li C, Li W, Esesarte E, et al. Cell Attachment Domains of the Porcine Epidemic Diarrhea Virus Spike Protein Are Key Targets of Neutralizing Antibodies [J]. 2017, 91 (12): 1-16.

[26] Li F Q. Oral administration of coated PEDV-loaded microspheres elicited PEDV-specific immunity in weaned piglets [J]. Chinese Journal of Public Health, 2019, 09 (014): 161-166.

[27] Qi G, Zhao S, Tao Q, et al. Effects of porcine epidemic diarrhea virus on porcine monocyte-derived dendritic cells and intestinal dendritic cells [J]. Research in Veterinary Science, 2016, 106: 149-158.

[28] Bonet M, Chaves A S, O Mesón, et al. Immunomodulatory and Anti-Inflammatory Activity Induced by Oral Administration of a Probiotic Strain of Lactobacillus Casei [J]. European Journal of Inflammation, 2006, 4 (1): 31-41.

[29] Grangette C, H Müller-Alouf, Geoffroy M C, et al. Protection against tetanus toxin after intragastric administration of two recombinant lactic acid bacteria: impact of strain viability and in vivo persistence [J]. Vaccine, 2002, 20 (27-28): 3304-3309.

［30］ Song D S，Oh J S，Kang B K，et al. Oral efficacy of Vero cell attenuated porcine epidemic diarrhea virus DR13 strain ［J］. Research in Veterinary Science，2007，82（1）：134-140.

［31］ Makadiya N，Brownlie R，Jan V，et al. S1 domain of the porcine epidemic diarrheavirus spike protein as a vaccine antigen ［J］. Virology Journal，2016，13（1）.

［32］ Chang S H，Bae J L，Kang T J，et al. Identification of the epitope region capable of inducing neutralizing antibodies against the porcine epidemic diarrhea virus ［J］. Molecules & Cells，2002，14（2）：295-299.

［33］ Sumi T，Fukushima A，Fukuda K，et al. Differential contributions of B7-1 and B7-2 to the development of murine experimental allergic conjunctivitis ［J］. Immunology Letters，2007，108（1）：62-67.

［34］ Xue M，Zhao J，Ying L，et al. IL-22suppresses the infection of porcine enteric coronaviruses and rotavirus by activating STAT3 signal pathway ［J］. Antiviral Research，2017，142：68-75.

病毒性腹泻的分子流行病学调查

李　洋　乔梦丽

摘要：本试验旨在对体系内主要的腹泻病毒进行分子流行病学调查，掌握病毒的流行情况，为病毒性腹泻的防控提供数据支持。将采自不同场线腹泻仔猪的粪便、肠管；成年猪的粪便、口腔液处理后进行 PEDV S1 基因和 PoRV VP7 基因测序。结果：①成功测得 24 个场线 45 个样品的 PoRV VP7 序列，其中，G9 型占比 51.6%，是流行最多的毒株亚型，其次为 G5、G1、G4；24 个场线中，有 7 个场线为不同 G 型 PoRV 混合感染，混合感染比例达到 29.2%。②共测得 67 个场线的 PEDV S 基因序列，其中与 NH-TA2020 位于同一进化分支的毒株占比最高，达 61.2%，有 4 个场线流行 S-INDEL 毒株，并且出现一批以饶阳毒株为代表的新毒株。流行病学调查结果表明，目前 G9 型为优势基因型，并且很多场线发生不同 G 型混合感染；PEDV 体系内流行毒株与 NH-TA2020 同源性较高，有多个场线出现了新的流行毒株。

关键词：分子流行病学调查；PEDV；PoRV

临床上引起猪腹泻的病毒性病原有很多，比较常见的有 3 种：猪传染性胃肠炎病毒（swine transmissible gastroenteritis virus，TGEV）、猪流行性腹泻病毒（porcine epidemic diarrhea virus，PEDV）和猪轮状病毒（porcine rotavirus，PoRV）。各年龄段的猪对这 3 种病毒都易感，尤其会给幼龄仔猪带来很大影响，造成极高的病死率[1]。自 2010 年 10 月猪流行性腹泻在我国南方大面积暴发，迅速蔓延至全国以来，对 PEDV 的流行情况进行持续的监测十分必要。PEDV 根据全基因组和 S 基因可以分为 1 型基因群（G1）、2 型基因群（G2）以及 3 型基因群（G3），G1 型和 G2 型又可分为 G1a、G1b 亚型及 G2a、G2b 亚型。近年来 PEDV 在我国流行，分离获得的 PEDV 毒株大多数是 G2 型的成员，其基因组发生明显变异，对我国防制该疫病增加了困难[2,3]。S1 基因的插入和缺失突变模式是更好地理解 PEDV 在亚洲、欧洲和北美洲出现和重复出现的一个重要靶点。尽管在中国猪群中广泛进行了基因分型[4-6]，但关于 PEDV 大流行毒株的信息相当有限。因此，PEDV 的分离与鉴定对掌握该病的流行病学和防控方法具有重要意义。

A 群轮状病毒（rotavirus A，RVA）为呼肠孤病毒科轮状病毒属成员，是引起婴幼儿及其他多种幼龄动物病毒性腹泻的主要病原体之一。RVA 基因组分为 11 个基因节段，分别编码 6 种结构蛋白（VP1、VP2、VP3、VP4、VP6 和 VP7），以及 5 种或 6 种非结构蛋白（NSP1～NSP5/6），RVA 的 11 个基因节段之间均易发生重配事件。其中结构蛋白 VP4 和

VP7 形成了外层，VP7（G 型）和 VP4（P 型）基因的序列分析是一种被广泛接受的确定 RV 基因型的方法[7]。根据 VP6 蛋白的抗原性可将 RV 分类为 10 个血清群（A～J），A、C、E 和 H 群 RV 已经被证实可感染猪，其中感染人类和动物最常见的为 A 群轮状病毒[8]。RVA 感染仔猪呈世界流行，各国猪场中 RVA 的阳性率为 3.3%～67.3%，给世界养猪业造成了严重的经济损失。研究表明，猪 RVA 是人源 RVA 感染的潜在来源[9]，因此了解猪 RVA 的流行情况和分子特征对 RVA 的防控以及防止猪 RVA 传播给人类至关重要。

1　试验材料与方法

1.1　病料采集

对不同生产线腹泻仔猪的粪便、肠管；成年猪的粪便、口腔液等样本进行主动收集。

1.2　主要试剂

HiScript Ⅱ One Step RT-PCR Kit（Dye Plus）、DL2000 Marker。

1.3　引物

根据美国国家生物信息中心（National Center for Biotechnology Information，NCBI）数据库中找到的 PEDV 19 个参考毒株的序列，用 DNAMAN 软件比对其保守区，根据 S1 基因保守区设计 PEDV S1 基因测序引物，全长 2 367 bp；在 NCBI 中查找下载 PoRV VP7 基因各个 G 型代表性毒株序列，根据保守区设计 VP7 基因测序引物，全长 1 000 bp。引物交由生工生物工程（上海）股份有限公司合成。

1.4　方法

（1）将采集到的粪便加入生理盐水重悬离心后取上清液，肠管研磨后加入生理盐水重悬离心后取上清液，用 DNA\RNA 共提核酸提取试剂盒对样本进行核酸提取，将检测结果为 PEDV 阳性的样品进行 S1 基因扩增，检测结果为 RV 阳性的样品进行 VP7 基因的扩增。反应体系：2×One Step Mix 25 μL，One Step Enzyme Mix 2.5 μL，上、下游引物各 2 μL，模板 9 μL，RNase free ddH$_2$O 9.5 μL。反应程序：45℃ 25 min，95℃预变性 5 min；95℃变性 30 s，55℃退火 30 s，72℃延伸 1 min（VP7 1 min，S1 2.5 min），共 32 个循环；再于 72℃延伸 8 min，4℃终止反应。

（2）将测序结果用 DNAstar 软件进行拼接后与 PEDV、RV 各参考序列进行同源性比对分析，用 MEGA7 进行遗传进化分析。

2　试验结果

2.1　PEDV 流行病学调查结果

共测得 67 个场线的 PEDV S 基因序列，与 NH-TA2020 位于同一进化分支的毒株占比最高，达 61.2%，出现一批以饶阳毒株为代表的新毒株（图 1）。

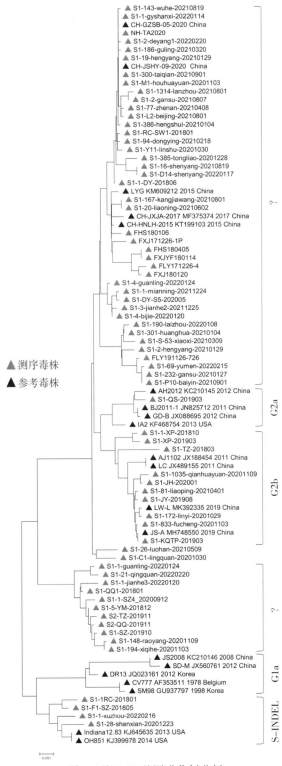

图 1　基于 S1 基因进化树分析

2.2 RV 流行病学调查结果

成功测得 24 个场线的 45 个样品的 PoRV VP7 序列，其中 G9 型占比 51.6%，是流行最多的毒株亚型，其次为 G5、G1、G4（图 2、图 3）；24 个场线中，有 7 个场线为不同 G 型 PoRV 混合感染，混合感染比例达到 29.2%（表 1）。

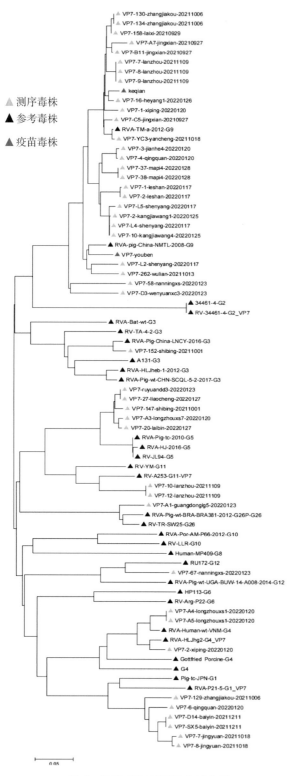

▲ 测序毒株
▲ 参考毒株
▲ 疫苗毒株

图 2 基于 VP7 基因进化树分析

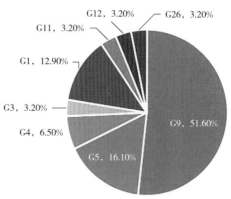

图 3 VP7 基因各型占比情况

表 1 VP7 基因混合感染情况

混合感染场线	毒株类型
施秉县	G3+G5
兰州市	G9+G11
三台县清泉村	G1+G9
张家口市	G1+G9
西屏镇	G4+G9
龙州县响水镇	G4+G5
南宁市西盛村	G9+G12
混合感染占比	29.2%
混合感染含 G9 型占比	62.5%

3　讨论

3.1　PEDV 流行病学调查

PED 是威胁全球养猪业的最严重的腹泻疾病之一。2010 年，一种新的变异 PEDV（G2 基因型）首次在中国中部地区出现，并在全国迅速蔓延，造成了巨大的经济损失[10]。随后，这种类型的病毒传播到美国、亚洲和欧洲的许多国家[11]。2014 年，另一个 PEDV 基因型，命名为 S-INDEL，首次在美国检测到[12]。自此，世界范围内发现了越来越多的变异 PEDV 毒株。PEDV 流行病学调查结果表明，体系内大多数场线流行 G2 型，有些场线开始出现新的毒株，PEDV 的毒力决定性 S1 基因正在发生快速突变。PEDV 是 RNA 病毒，在其复制的过程中容易发生突变，导致疫苗接种失败，这使控制该类疫病增加了困难。对当前流行毒株的分离、鉴定和遗传进化分析，可以为探究该病毒致病机制和开发新型高效疫苗提供帮助。

3.2　PoRV 流行病学调查

猪 A 型轮状病毒的所有 G 型中，G5 是全球最主要的流行基因型，占所有毒株的 45.8%，G3 占 11.2%，G4 占 9.6%。G5 是流行最广泛的毒株。G 型有 G9、G11、G2、G4、G5、G26、G3 基因型的报道[13,14]。对全球重要的猪 RV 进行回顾，发现 G3、G4、G5、G9 和 G11 广泛传播，对猪 RV 流行病学研究很重要。在本试验的样本中也发现了这些 G 型。此外，G5、G3 和 G4 被列为全球和欧洲排名最多的亚型，流行病学调查结果表明，G9 型为目前流行的优势基因型，2008—2015 年均有中国检测出 G9 基因型的报道，显示 G9 基因型可能是中国乃至亚洲最新流行的毒株，一段时间内优势基因型的变化也可能发挥非常重要的作用[15]。来自这类地区的样本较少，可能是缺乏检测优势的一个原因。

4　结论

PEDV S1 基因流行病学调查结果表明，PEDV 毒株与 NH-TA2020 毒株同源性较高，多个场线开始流行 PEDV 新毒株；PoRV VP7 基因流行病学调查结果表明，G9 型为优势基因型，多个场线存在不同 G 型混合感染的情况。

参考文献

[1] 许梦怡，王彦红，薛峰. 猪传染性胃肠炎病毒猪流行性腹泻病毒猪轮状病毒多重荧光 RT-PCR 检测方法的建立 [J]. 中国兽医科学，2020，50（12）：1500-1508.

[2] 隋灵. 猪流行性腹泻病毒 CH/HLJ/18 株的分离与鉴定 [D]. 黑龙江：东北农业大学，2021.

[3] 翟新国，周琼琼，郑培培，等. 河北省一株猪流行性腹泻病毒 S1 基因的遗传变异与重组分析 [J]. 黑龙江畜牧兽医，2021，9（12）：56-62.

［4］ Chen P，Wang K，Hou Y，et al.Genetic evolution analysis and pathogenicity assessment of porcine epidemic diarrhea virus strains circulating in part of China during 2011-2017 ［J］.Infect Genet Evol，2019，69：153-165.

［5］ Wang E，Guo D，Li C，et al.Molecular Characterization of the ORF3 and S1 Genes of Porcine Epidemic Diarrhea Virus Non S-INDEL Strains in Seven Regions of China，2015 ［J］.PLoS One，2016，11 (8)：e0160561.

［6］ Wen Z，Li J，Zhang Y，et al.Genetic epidemiology of porcine epidemic diarrhoea virus circulating in China in 2012-2017 based on spike gene ［J］.Transbound Emerg Dis，2018，65 (3)：883-889.

［7］ Wenske O，Rückner A，Piehler D，et al.Epidemiological analysis of porcine rotavirus A genotypes in Germany ［J］.Vet Microbiol，2018，214：93-98.

［8］ VlasovaA N，Amimo J O，Saif L J.Porcine Rotaviruses：Epidemiology，Immune Responses and Control Strategies ［J］.Viruses，2017，9 (3)：48.

［9］ Malasao R，Khamrin P，Kumthip K，et al.Complete genome sequence analysis of rare G4P ［6］ rotavirus strains from human and pig reveals the evidence for interspecies transmission ［J］.Infect Genet Evol，2018，65：357-368.

［10］ Li W，Li H，Liu Y，et al.New variants of porcine epidemic diarrhea virus，China，2011 ［J］.Emerg Infect Dis，2012，18 (8)：1350-1353.

［11］ Lee C.Porcine epidemic diarrhea virus：An emerging and re-emerging epizootic swine virus ［J］.Virol J，2015，22 (12)：193.

［12］ Wang L，Byrum B，Zhang Y.New variant of porcine epidemic diarrhea virus，United States，2014 ［J］.Emerg Infect Dis，2014，20 (5)：917-919.

［13］ Zhang H，Zhang Z，Wang Y，et al.Isolation，molecular characterization and evaluation of the pathogenicity of a porcine rotavirus isolated from Jiangsu Province，China ［J］.Arch Virol，2015，160 (5)：1333-1338.

［14］ 原霖.猪轮状病毒 G4P ［6］ 株的分离鉴定及新型检测方法的建立与应用 ［D］.北京：中国农业大学，2018.

［15］ Papp H，László B，Jakab F，et al.Review of group A rotavirus strains reported in swine and cattle ［J］.Vet Microbiol，2013，165 (3-4)：190-199.

非洲猪瘟弱毒尾根血拭子试验

李 鹏 高文超 蒋晓雪 田晓刚 石运通 王 青 李孝文

摘要：为寻找一种可替代前腔静脉采血分离血清进行 qPCR 诊断的采样方法。本试验分别在黄骅市、高密市、夏庄镇等弱毒场线采集栏位微环境、咽拭子、尾根血，分析其与前腔静脉采血的敏感性和特异性。旨在寻找新式采样方法，实现降本增效、优化人员操作。同时，为提高血拭子检测的灵敏性，本试验比对了常规 qPCR 与一体机之间的差异。结果表明：尾根血相对于前腔静脉血清检测敏感性为 62.41%（95%CI，56.8%~68.1%）；一体机（1∶10 稀释）比常规 qPCR 敏感性高约 3 倍；尾根血拭子＋一体机相对于前腔静脉血检测的敏感性为 85.71%（95%CI，79.2%~92.2%），阳性检出率差异不显著（$P=0.608$）。本试验为采集尾根血拭子应用一体机检测这种新方法奠定了基础。

关键词：尾根血；一体机；检测敏感性

非洲猪瘟（African swine fever，ASF）是发生在家猪和野猪身上的一种病毒性出血性疾病。ASF 临床症状广泛，其中以高热、厌食、嗜睡和全身出血为特征。非洲猪瘟防控目前暂无安全有效的疫苗可用，因此快速、可靠的检测方法对其防控至关重要。目前应用的非洲猪瘟诊断技术主要表现在病原、核酸、抗原与抗体四个方面[1]。每种诊断技术都有其特点，从而为不同情况下的非洲猪瘟病毒（ASFV）诊断提供了多种选择。通过无接触采集养殖场样品进行隔离监测[2]。唾液是感染非洲猪瘟弱毒株动物的主要排毒途径，病毒血症是间歇性的[3]。同时要考虑感染猪排泄物的病毒水平，对环境的污染情况。环境污染在动物反复感染周期中起很重要的作用[4]。防控非洲猪瘟的主要方法是早期预警和主动监测。然而，取样困难，因此使用快干棉签的采样方法可以促进采样。有研究者使用血液、骨骼和器官样本评估使用干血拭子检测 ASFV 抗体和基因组[5]。因此，进一步研究栏位微环境、咽拭子、尾根血的潜在价值很重要。

1 试验材料与方法

1.1 试验材料

采样纱布、采血针、普通采血管、咽拭子采样器、植绒拭子、针头、自封袋、手套、隔离服、核酸提取试剂盒、检测试剂盒等。

1.2 采样方式

（1）栏位微环境采样

①擦拭母猪嘴能碰到的栏杆。

②擦拭靠猪只面的料槽边缘。

③擦拭猪只能接触到的下料口。

④擦拭栏位前部猪只嘴能接触到的地方约 30cm 处（包括料槽下面部分），务必采集到唾液。

⑤擦拭小水槽边缘、底部及栏板。

⑥擦拭母猪水碗边缘及底部。

⑦擦拭前栏门、栏杆，料道人员将纱布传递给粪道人员。

⑧擦拭猪只后栏杆、栏门内侧面、栏板。

⑨擦拭猪只栏位地板后约 60 cm 的部分，采集到尿液后将采样纱布放入自封袋内并编号。

（2）前腔静脉血采样[6]

①前腔静脉血采集需将猪只保定，做好防护。

②在两侧第 1 肋骨与胸骨结合处，倾斜 30°插入采血针，采集血液。

（3）尾根血拭子采样

①做好防护，在粪道进行采样。准备采样针、植绒拭子、止血棉签、1.5 mL 离心管。

②对准采血部位扎针使其出血。

③出血适量后，使用植绒拭子蘸取血液。

④将蘸满血样的植绒拭子放入 0.5 mL 洗脱液的离心管中洗脱。

⑤采集好的样品放入自封袋，样品采集完成。

⑥对猪只出血点止血。

⑦手套反装，包裹植绒拭子。

⑧对猪只后部栏位进行消毒。

（4）咽拭子采样

①戴好长臂手套，将咽拭子采样器从头部褪去防护膜。

②平行插入猪只口腔，稍用力使咽拭子采样器进入咽喉处。

③来回抽拉几次，缓慢取出咽拭子采样器。

④将防护膜归位，防止污染。

⑤将采集好的样品放入自封袋内并编号，取下长臂手套，重复上述步骤采集下一头猪。

1.3 样品检测

所有采集的样品冷藏送实验室检测 ASFV。

2　试验结果

2.1　各类样品与血清检测结果比较

从组别 1 检测结果来看，弱毒场线栏位微环境检测敏感性为 25.52%（95% CI，19.4%～31.7%），特异性为 95.56%（95% CI，92.5%～98.6%）；血清阳性率为 51.22%，栏位微环境为 15.18%。从组别 2 检测结果来看，弱毒场线栏位咽拭子检测敏感性为 71.57%（95% CI，62.8%～80.3%），特异性为 85.11%（95% CI，74.9%～95.3%）；血清阳性率为 68.46%，咽拭子阳性率为 53.69%。从组别 3 检测结果来看，弱毒场线尾根血拭子检测敏感性为 62.41%（95% CI，56.8%～68.1%），特异性为 92.95%（95% CI，89.6%～96.3%）；血清阳性率为 55.40%，尾根血拭子阳性率为 37.72%。由此可知，弱毒场线中咽拭子及尾根血拭子可作为下一步全群检测的备选方法（表 1、表 2）。

表 1　各类样品与血清的检测敏感性、特异性比较（%）

组别	样品		血清检测结果（份）		敏感性 (95% CI)	特异性 (95% CI)	Cohen's kappa (95% CI)
			阳性	阴性			
1	栏位微环境	阳性	48	8	25.52 (19.4～31.7)	95.56 (92.5～98.6)	0.206 (0.136～0.276)
		阴性	141	172			
2	咽拭子	阳性	73	7	71.57 (62.8～80.3)	85.11 (74.9～95.3)	0.503 (0.369～0.637)
		阴性	29	40			
3	尾根血拭子	阳性	176	16	62.41 (56.8～68.1)	92.95 (89.6～96.3)	0.533 (0.465～0.601)
		阴性	106	211			

表 2　各类样品与血清检测结果比较

组别	检测方法	样品类型	检测总数（份）	阳性（份）	阴性（份）	阳性率（%）	χ^2（P 值）
1	qPCR	血清	369	189	180	51.22	0
		栏位微环境		56	313	15.18	
2	qPCR	血清	149	102	47	68.46	0.009
		咽拭子		80	69	53.69	
3	qPCR	血清	509	282	227	55.40	0
		尾根血拭子		192	317	37.72	

注：组别 1 样品来源为黄骅市猪场、夏庄镇猪场、高密市猪场；组别 2 样品来源为黄骅市猪场、夏庄镇猪场；组别 3 样品来源为高密市猪场。

2.2　常规 qPCR 与一体机检测比较

从 171 份样品检测结果来看，常规 qPCR 的敏感性相对于一体机为 33.06%（95% CI，24.7%～41.4%），即一体机（1∶10 稀释）比常规 qPCR 的敏感性高约 3 倍，特异性为 98.03%（95% CI，94.2%～100%）；一体机检测阳性率为 70.35%，常规 qPCR 检测阳性率为 23.84%（表 3、表 4）。

表 3　弱毒场线常规 qPCR 与一体机检测敏感性、特异性比较（%）

检测方法		一体机检测结果（份）		敏感性（95% CI）	特异性（95% CI）	Cohen's kappa（95% CI）
		阳性	阴性			
qPCR	阳性	40	1	33.06（24.7～41.4）	98.03（94.2～100）	0.214（0.134～0.294）
	阴性	81	50			

表 4　弱毒场线常规 qPCR 与一体机检测敏感性、特异性比较

检测方法	检测总数（份）	阳性（份）	阴性（份）	阳性率（%）	χ^2（P 值）
一体机	172	121	51	70.35	0
qPCR		41	131	23.84	

注：样品来源为高密市猪场。

2.3　弱毒场线血清＋常规 qPCR 检测与尾根血拭子＋一体机检测比较

据 152 份样品检测结果可知，尾根血拭子＋一体机检测的敏感性为 85.71%（95% CI，79.2%～92.2%），特异性为 70%（95% CI，55.8%～84.2%）。血清＋常规 qPCR 检测的阳性率为 73.68%，尾根血拭子＋一体机检测的阳性率为 71.05%，差异不显著（卡方检验：$P＝0.608＞0.05$）。故尾根血拭子＋一体机检测可替代血清在普检中的使用（表 5、表 6）。

表 5　血清＋常规 qPCR 检测与尾根血拭子＋一体机检测的敏感性、特异性比较（%）

检测方法		血清＋常规 qPCR 检测结果（份）		敏感性（95% CI）	特异性（95% CI）	Cohen's kappa（95% CI）
		阳性	阴性			
尾根血拭子＋一体机	阳性	96	12	85.71（79.2～92.2）	70.00（55.8～84.2）	0.54（0.390～0.689）
	阴性	16	28			

表 6　血清＋常规 qPCR 检测与尾根血拭子＋一体机检测的结果比较

检测方法	检测总数（份）	阳性（份）	阴性（份）	阳性率（%）	χ^2（P 值）
血清＋常规 qPCR	152	112	40	73.68	0.608
尾根血拭子＋一体机		108	44	71.05	

注：样品来源为高密市猪场。

3　讨论

在弱毒株感染时，唾液检出的病毒量要稍高于其他排毒途径。而且，唾液中病毒留存时间要长于血液和粪便。这表明经口感染弱毒株的动物，会在扁桃体和淋巴结入口处控制其局部感染，而不会通过病毒血症造成全身传播[6-8]。本试验中弱毒场线栏位咽拭子相对于血清检测其敏感性为 71.57%（62.8%～80.3%），可能是一种有效的以无创方式检测个体或环境病毒的方案。

被感染动物血液内的病毒载高[9]。由于全群采集前腔静脉血而耗费大量人力、物力，所以选择微创采血来代替全群采血在实际操作中显得尤为重要。本试验中弱毒场线栏位尾根血检测敏感性为 62.41%（56.8%～68.1%），有可能成为下一步全群检测的替代方法。

一体机检测理论上比 qPCR 检测方法敏感性高 9～100 倍，在本试验中一体机检测（1：10 稀释）比常规 qPCR 检测敏感性高约 3 倍，一体机＋尾根血拭子检测相对于血清检测其敏感性提高到了 85.71%（79.2%～92.2%），阳性检出率差异不显著（$P=0.608$）。优化采样条件后，结合尾根血采血有望成为弱毒场线全群检测的替代方法。

参考文献

[1] 苗春，李伟，杨思成，等．非洲猪瘟诊断技术研究进展［J］．中国动物检疫，2022，39（1）：82-89.

[2] 许崇友，韩勇，司西波，等．一种适用于规模猪场的非洲猪瘟病毒监测与清除方法［P］. CN113916401A，2022-01-11.

[3] Aleksandra K，Estefanía C F，Sandra B，et al. Distinct African Swine Fever Virus Shedding in Wild Boar Infected with Virulent and Attenuated Isolates［J］. Vaccines，2020，8（4）.

[4] Olesen A S，Lohse L，Boklund A，et al. Short Time Window for Transmissibility of African Swine Fever Virus from a Contaminated Environment［J］. Transbound Emerg Dis，2018，65：1024-1032.

[5] Carlson J，Zani L，Schwaiger T，et al. Simplifying sampling for African swine fever surveillance：Assessment of antibody and pathogen detection from blood swabs［J］. Transbound Emerg Dis，2018，65（1）：e165-e172.

[6] 张文志，刘金，Aleksandra Kosowska，等．感染非洲猪瘟强毒株和弱毒株的野猪排毒方式的差异［J］. 猪业科学，2021，38（8）：20-24.

[7] McVicar J W. Quantitative aspects of the transmission of African swine fever［J］. Am. J. Vet. Res，1984，45：1535-1541.

[8] Oura C A L，Eards L，Batten C A. Virological Diagnosis of African Swine Fever-Comparative Study of Available Tests［J］. VirusRes，2013，173：150-158.

[9] Arias M，Jurado C，Gallardo C，et al. Gaps in African Swine Fever：Analysis and Priorities［J］. Transbound Emerg Dis，2018，65：235-247.

猪 α 干扰素对 RV 腹泻仔猪预防和治疗的试验

卿 杰 王泽炜

摘要： 本试验旨在探究两种猪 α 干扰素对仔猪轮状病毒（RV）腹泻的预防及治疗效果。通过对猪场 RV 分子流行病学调查，选取甘肃 BYJY，RV 腹泻仔猪作为试验对象。在产房仔猪感染 RV 疫情期间，分别对 1 日龄仔猪和 7 日龄以下 RV 腹泻仔猪注射猪 α 干扰素进行预防和治疗试验。结果：①注射干扰素对 1 日龄仔猪进行 RV 感染预防，对照组发病率为 24.46%，注射猪 α 干扰素预防组仔猪 RV 发病率为 12.24%～13.75%，有效降低了低日龄仔猪发病率，且预防组感染仔猪症状减轻，两种猪 α 干扰素效果基本一致；②对感染 RV 的 7 日龄以下的仔猪注射猪 α 干扰素进行治疗，治疗效果与常规使用抗生素进行治疗无明显差异。

关键词： 干扰素；仔猪腹泻；猪轮状病毒

1 研究背景及意义

轮状病毒（Rotavirus，RV）为双股 RNA（dsRNA）病毒，属于呼肠孤病毒科（Reoviridae）轮状病毒属（*Rotavirus*）。轮状病毒是引起人类新生儿和猪、牛、羊等多种幼龄动物的急性胃肠道传染病的主要病原之一，临床表现以呕吐、腹泻、脱水为主要特征[1-3]。RV 最早是由 Mebus 等在 1969 年从犊牛粪便中分离得到。在我国，庞其方等在 1979 年首次从儿童腹泻粪便中检测出 RV，此后，我国兽医研究者又陆续从仔猪、犊牛和羔羊等多种动物的粪便中检测到 RV[4]。成年动物一般呈隐性感染。轮状病毒感染对动物的健康有很大危害，常导致养殖业巨大的经济损失，在全球已成为一个重要的公共卫生问题。

猪轮状病毒是引起哺乳仔猪病毒性腹泻的重要病原之一[5,6]。仔猪感染轮状病毒引起的腹泻在我国和世界许多养猪国家普遍存在[7-12]。轮状病毒在猪群中呈地方流行性，猪轮状病毒主要存在于患病猪或带毒猪的消化道内，随粪便排出体外，经粪-口途径引起传播和流行，可在不同年龄个体的成年猪和哺乳猪之间相互感染并长期根植在养猪场，而且易感宿主只需很少的病毒粒子就可以引起疾病[13]。虽然较年长的猪，如断奶猪和母猪的临床症状较轻，但会严重影响出生仔猪的生长和成活率[14,15]。猪轮状病毒主要感染小肠绒毛顶端的小肠成熟肠上皮细胞并在其中繁殖，其次也可在盲肠、结肠的成熟上皮细胞中复制[16,17]。轮状病毒感染引起的肠道病变主要限于消化道，胃壁弛缓，内充满凝乳块和乳汁，小肠黏膜呈条状或弥散性出血[18,19]。本病多发生在冬春季节，在许多受感染地区由于多数的成年猪都已经获得了免疫能力，因此，发病猪多是哺乳期的幼小仔猪，死亡率 7%～20%。中国自 20 世纪 80 年代从腹泻的猪中分离得到了猪轮状病毒以来，目前猪轮状病毒在我国也大面积存在。

20 世纪 30 年代，病毒学家在研究两种病毒感染同一种细胞时发现了病毒干扰现象，科学家一直在寻找引起这种现象的活性物质。1957 年，病毒学家 lsaacs 和研究员 Lindenman[20] 在研究流感病毒时，首次发现细胞能产生一类可溶性的分泌物来干扰和抑制病毒的复制，并将其命名为干扰素（interferon，IFN）。1980 年，国际干扰素命名委员会正式定义干扰素。干扰素是一类活性蛋白，有广谱、高效的抗病毒效果，具有影响细胞活性分化和调节机体免疫等功能[21]。α 干扰素（IFN-α）主要由 Namalwa 细胞、B 淋巴细胞、白细胞、KG-1 细胞、慢性髓性白血病细胞、B 淋巴细胞、水疱口炎病毒（VSV）或鸡新城疫病毒（NDV）诱导后的成纤维细胞产生。IFN 并不直接灭活病毒，而是通过作用于靶细胞，激活靶细胞内的抗病毒基因，合成抗病毒蛋白（antiviral protein，AVP），使细胞产生一种抗病毒状态而发挥作用。已知的具有酶活性的 AVP 主要包括双链 RNA 依赖性蛋白激酶（PKR）、2′-5′寡腺苷酸合成酶和 Mx 蛋白。这些酶可以使病毒蛋白合成受到抑制，使病毒降解。IFN-α 主要是通过对淋巴细胞和巨噬细胞的调节或诱导上调细胞表面 MHC-Ⅰ类分子的表达起免疫调节作用[22]。IFN-α 不仅能上调细胞表面 MHC-Ⅰ类分子的表达，增强 CD8$^+$ 细胞毒性 T 细胞反应，而且能启动细胞毒性 T 细胞（CTL）和 NK 细胞的分化。IFN-α 还能分别诱导巨噬细胞提高 Fc 受体的表达和淋巴细胞释放肿瘤坏死因子（TNF）。某些 IFN-α 亚型可能增加 MHC-Ⅲ类分子的表达。病毒感染细胞或宿主时因病毒种类的差异，在一定的感染时期均能诱导机体产生干扰素。干扰素防御系统属于非特异性免疫反应，在动物机体抵抗外源病原感染时出现较早，是已知机体防御系统中防御反应出现最早的，其作用迅速，具有广谱性，但持续时间较短。当干扰素系统被激活后，动物可在 1～3 周内对其他病毒的重复感染具有抵抗力。干扰素能抑制猪轮状病毒、伪狂犬病毒、传染性胃肠炎病毒、猪流行性腹泻病毒、猪繁殖与呼吸综合征病毒、新城疫病毒和口蹄疫病毒等多种病毒的复制[23]。

猪轮状病毒 RV 是引起仔猪病毒性腹泻、体重下降甚至死亡的关键诱因。冬春季节气温低，温度波动大，RV 对仔猪的威胁越来越大。干扰素作为抗病毒产品已应用于病毒感染治疗。干扰素具有抗病毒、免疫调节的作用，通过诱导细胞产生一类蛋白质而发挥作用。这些蛋白质可以使病毒蛋白合成受到抑制，使病毒降解。干扰素具有以下特点：①干扰素抗病毒不是杀灭病毒而是抑制病毒；②不同的细胞种类、不同的病毒感染类型对干扰素敏感程度不同，所以临床效果也有所不同。因此，本试验分别将猪 α 干扰素应用于 RV 感染发病场线进行预防和治疗试验，以探明实际生产中干扰素对 RV 的防控效果，对目前日益严重的 RV 感染提供一种防控方案。

2　试验材料与方法

2.1　基本情况

试验开展时间：2022 年 1 月 5 日—26 日。

试验地点：甘肃 BYJY 猪场 N 线，试验期间仔猪只发生 RV 感染。

发病场线情况：RV 确诊时间为 2021 年 8 月 21 日；发病日龄为 1～10 日龄；发病率为 30%。

2.2 试验材料

本次试验材料如表 1 所示。

表 1 猪 α 干扰素应用于 RV 感染试验材料

商品名	主要成分	使用剂量
干扰素 A	猪 α 干扰素、黄芪多糖提取物	1∶1 溶解，3 mL/头
干扰素 B	猪 α 干扰素	2 mL/头

2.3 试验方法

（1）预防方案 选取反复发病的场线，选择刚分娩产房单元进行试验。将 1 日龄仔猪分为 3 组：A 组 170 头仔猪注射干扰素 A；B 组 156 头仔猪注射干扰素 B；C、D 组共 341 头仔猪作为对照组，对猪只不做处理。实验猪只每窝注射后更换针头与手套。如 2 日龄的仔猪发生腹泻则补注射一针。统计试验组与对照组的仔猪总头数、腹泻发病日龄、腹泻头数、腹泻程度以及死亡头数。

（2）治疗方案 对刚发生腹泻且 7 日龄以下的仔猪，采集肛拭子检测病原 Ct 值（每窝猪样品混合检测），再使用干扰素治疗。腹泻仔猪分为 3 组：A 组 33 头腹泻仔猪，注射干扰素 A；B 组 31 头腹泻仔猪，注射干扰素 B；C 组 31 头腹泻仔猪，使用场线内常规治疗方案（庆大霉素饮水/黏杆菌素饮水）。实验猪只每窝注射后更换针头与手套。连续注射治疗 2 d。治疗后每天对各个组的猪只采集肛拭子进行检测，记录猪只腹泻头数、腹泻指数。

腹泻程度：0 为正常，粪便固态成型；1 为轻度腹泻，粪便稀软成型；2 为中度腹泻，粪便呈黄色水样；3 为重度腹泻，粪便呈水样喷射。

3 试验结果及讨论

3.1 干扰素对 RV 病毒的预防效果

3.1.1 干扰素对 RV 发病率的影响 注射干扰素对 1 日龄仔猪进行 RV 感染预防，试验组与对照组每天的腹泻率如图 1 所示。对照组发病率为 24.46%，试验组仔猪 RV 的发病率为 12.24%～13.75%，有效降低低日龄仔猪对 RV 的感染率 50.0%～56.2%，干扰素 A 和干扰素 B 对 1 日龄仔猪的 RV 预防效果无明显差异。

3.1.2 干扰素预防对仔猪腹泻指数的影响 与对照组相比，使用干扰素 A 进行预防，可降低腹泻仔猪的临床症状，水样腹泻占腹泻仔猪最大比例为 5.4%，低于对照组（水样腹泻最大比例为 11.9%）（图 2）。

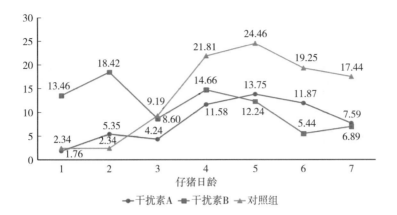

图 1 干扰素预防后仔猪腹泻率随日龄的变化情况

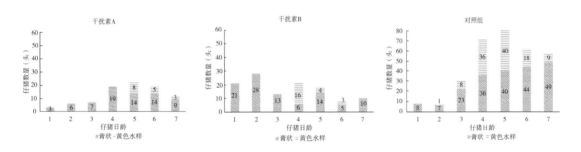

图 2 干扰素预防后仔猪腹泻程度随日龄的变化情况

3.2 干扰素对 RV 腹泻仔猪的治疗效果

3.2.1 干扰素治疗对仔猪排毒的影响 仔猪每天采样检测 RV Ct 值情况如图 3 所示。结果显示，注射干扰素治疗组及抗生素治疗组随仔猪日龄增大，仔猪排毒均会减少。

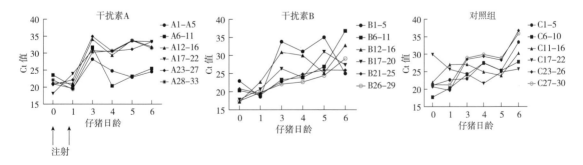

图 3 干扰素注射后仔猪 RV Ct 值随时间的变化情况

3.2.2 干扰素治疗对仔猪腹泻指数的影响 仔猪每天腹泻指数变化情况如图 4 所示。结果显示，注射干扰素会使仔猪腹泻症状减轻，但不能停止腹泻。

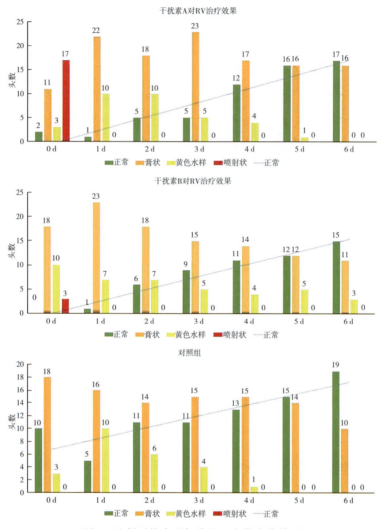

图4   注射干扰素后仔猪腹程度数变化情况

3.2.3   干扰素治疗对 RV 腹泻率的影响   试验 6 d 后，与对照组（治疗率：49.39%）相比，试验组（干扰素 A：48.49%；干扰素 B：51.72%）的腹泻率无明显下降（图 5）。

4   结论

猪 α 干扰素可在仔猪 1 日龄时进行注射，可有效预防 RV，降低仔猪对 RV 的感染发病率，并能降低感染后的感染症状。

对感染 RV 的 7 日龄以下的仔猪使用干扰素进行治疗，治疗效果与常规使用抗生素进行治疗无明显差异，不建议使用干扰素对腹泻仔猪进行治疗。

还需在不同腹泻场线进行验证，明确猪 α 干扰素的使用效果及使用场景。

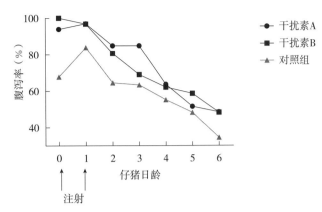

图5　注射干扰素后仔猪腹泻率随时间的变化情况

参考文献

［1］ Glass R I，Bresee J S，Parashar U，et al. Rotavirus vaccines at the threshold ［J］. Nature Medicine，
　　 1997，3（12）：1324-1325.

［2］ Gentsch J R，Laird A R，Bielfelt B，et al. Serotype diversity and reassortment between human and
　　 animal rotavirus strains：implications for rotavirus vaccine programs ［J］. The Journal of Infectious
　　 Diseases，2005，192（Suppl 1）：S146-S159.

［3］ 秦昭恒. 猪轮状病毒 VP7 蛋白单克隆抗体制备与其抗原表位的初步鉴定 ［D］. 哈尔滨：东北农业大
　　 学，2014.

［4］ 于恩庶，徐秉锟. 中国人兽共患病学 ［M］. 福州：福建科学技术出版社，1988.

［5］ Vlasova A N，Amimo J O，Saif L J. Porcine rotaviruses：epidemiology，immune responses and control
　　 strategies ［J］. Viruses，2017，9（3）：48.

［6］ Almeida P R，Lorenzetti E，Cruz R S，et al. Diarrhea caused by rotavirus A，B，and C in suckling
　　 piglets from southern Brazil：molecular detection and histologic and immunohistochemical
　　 characterization ［J］. J Vet Diagn Invest，2018，30（3）：370-376.

［7］ Shi H，Chen J，Li H，et al. Molecular characterization of a rare G9P ［23］ porcine rotavirus isolate from
　　 China ［J］. Arch Virol，2012，157（10）：1897-1903.

［8］ Saikruang W，Khamrin P，Chaimongkol N，et al. Genetic diversity and novel combinations of G4P ［19］
　　 and G9P ［19］ porcine rotavirus strains in Thailand ［J］. Veterinary Microbiology，2013，161（3-4）：
　　 255-262.

［9］ Amimo J O，Vlasova A N，Saif L J. Detection and genetic diversity of porcine group A rotaviruses in
　　 historic（2004）and recent（2011 and 2012）swine fecal samples in Ohio：predominance of the G9P ［13］
　　 genotype in nursing piglets ［J］. J Clin Microbiol，2013，51（4）：1142-1151.

［10］ Amimo J O，E1-Zowalaty M E，Githae D，et al. Metagenomic analysis demonstrates the diversity of
　　 thefecal virome in asymptomatic pigs in East Africa ［J］. Arch Virol，2016，161（4）：887-897.

［11］ Amimo J O，Otieno T F，Okoth E，et al. Risk factors for rotavirus infection in pigs in Busiaand Teso
　　 subcounties，Western Kenya ［J］. Trop Anim Health Prod，2017，49（1）：105-112.

［12］ Ward R L，Bernstein D I，Young E C，et al. Human rotavirus studies in volunteers：determination of infectious dose and serological response to infection［J］. J Infect Dis，1986，154（5）：871-880.

［13］ Goede D，Morrison R B. Production impact & time to stability in sow herds infected with porcine epidemic diarrhea virus（PEDV）［J］. Prev Vet Med，2016，123：202-207.

［14］ Lin J D，Lin C F，Chung W B，et al. Impact of mated female nonproductive days in breeding herd after porcine epidemic diarrhea virus outbreak［J］. PLoS One，2016，11（1）.

［15］ Bohl E H，Saif L J，Theil K W，et al. Porcine pararotavirus：detection，differentiation from rotavirus，and pathogenesis in gnotobiotic pigs［J］. J Clin Microbiol，1982，15（2）：312-319.

［16］ Estes M K，Kang G，Zeng C Q，et al. Pathogenesis of rotavirus gastroenteritis.［J］. Novartis Foundation Symposium，2001，238：82-96.

［17］ Collins J E，Benfield D A，Duimstra J R. Comparative virulence of two porcine group-A rotavirus isolates in gnotobiotic pigs［J］. Am J Vet Res，1989，50（6）：827-835.

［18］ Carpio M，Bellamy J E，Babiuk L A. Comparative virulence of different bovine rotavirus isolates［J］. Can J Comp Med，1981，45（1）：38-42.

［19］ Kapikian A Z. Overview of viral gastroenteritis［J］. Arch Virol Suppl，1996，12：7-19.

［20］ lssacs A，Lindenmann J. Virus inter fernce. I［J］. The Interferon，1957：258-267.

［21］ 解希帝，周雨霞. 干扰素研究进展［J］. 畜牧与饲料科学，2008，3：70-72.

［22］ MowenK A，Tang J，Zhu，et al. Arginine methylation of STATl mod γ Lates IFNα/β-induced transcription［J］. Cell，2001，104（5）：731-741.

［23］ Dmga B，Hyy A，Wei Z C，et al. Inhibitory effects of recombinant porcine interferon-α on porcine transmissible gastroenteritis virus infections in TGEV-seronegative piglets［J］. Veterinary Microbiology，2021，252：108930.

评估不同材质过滤棉对 ASFV 气溶胶的过滤效果

胡志强　樊铭玉　卞陆杰　李孝文

摘要：本试验旨在评估初效过滤棉与亚高效过滤棉对非洲猪瘟病毒（ASFV）气溶胶的过滤效果。通过模拟一线猪场的过滤棉防护装置，将初效和亚高效过滤棉进行不同的组合，利用空气采样器收集过滤后的空气样品，进行 ASFV Ct 值的检测。结果表明，在该试验条件下，初效过滤棉对 ASFV 气溶胶几乎没有过滤效果；亚高效过滤棉对 ASFV 气溶胶有一定的过滤作用，但是，并不能完全消除病毒；"初效＋高效"过滤棉模式比"高效＋初效"过滤棉模式效果更优。

关键词：ASFV；气溶胶；过滤棉

进入冬季后，猪舍内环境进入冬季模式，此时舍内通风量减少，换气周期变长，氨气浓度升高，粉尘气溶胶浓度加大，舍内空气质量变差，与之相对应的，北方猪场发生 ASFV 阳性的概率和数量大大提升。空气过滤措施是当前防止空气中携带病毒的有效手段。目前，过滤棉主要分为三个等级，初效过滤棉、亚高效（中效）过滤棉和高效过滤棉[1]。对于初效空气过滤棉来说，一般过滤小于 5 μm 的尘土颗粒，过滤率比较低，只作为初步的过滤，对空气要求不高的场所大多选择初效过滤棉，所以将其安装于与外部空气接触的位置，用于通风设备以及空气控制系统的吸入口；对于亚高效空气过滤棉来说，一般过滤小于 1 μm 的尘土颗粒，过滤率适中，通常作为初步过滤之后的第二级过滤或者在洁净要求不高的情况下作为之后一步的过滤；对于高效空气过滤棉来说，一般过滤小于 0.5 μm 的尘土颗粒，过滤率较高，一般情况下在过滤系统是用于最后一步过滤。气溶胶是指悬浮在气体介质中的固态或液态颗粒所组成的气态分散系统，这些固态或液态颗粒的密度与气体介质的密度可以相差微小，也可以很悬殊，气溶胶颗粒大小通常在 0.01～10 μm[2]。气溶胶已经被报道能够作为一些猪相关病毒的传播媒介，如蓝耳病病毒[3]、猪流行性腹泻病毒[4]、猪流感病毒[5]等。只有少数报道提出 ASFV 具有通过空气气溶胶传播的风险[6,7]。因此，如何从气溶胶的角度抵抗 ASFV 由外环境传入猪舍内环境是值得探究的课题。结合目前的猪场多采用初效过滤棉或亚高效过滤棉的情况，本试验通过设计气溶胶过滤模型探究不同材质的过滤棉对空气中 ASFV 的过滤效果，比较两种过滤棉和不同的过滤棉组合对病毒的过滤效果，旨在为过滤棉于实际生产中的使用提供参考依据。

1　试验材料与方法

1.1　试验材料

100 L塑料桶、共度雾化造雾器、铝箔通风管道、塑料量杯和艾美特单向抽风机；初效过滤棉、亚高效（中效）过滤棉、MD8空气采样器和空气采样凝胶膜；非洲猪瘟野毒病毒液保存于实验室。

1.2　试验设计

实验装置如图1所示，该装置由两个半密闭桶组成，分别是病毒释放箱和病毒收集箱，中间用铝箔通风管道连接，在通风管道中放置一台单向抽风机，方向从病毒释放箱至病毒收集箱。分别在病毒释放箱与通风管道接口处设置过滤棉①，在通风管道与抽风机接口处设置过滤棉②。通过在位置①和位置②设置不同类型的过滤棉，将试验分为8组，分别是阴性对照、阳性对照、单层亚高效过滤棉、双层亚高效过滤棉、亚高效＋初效过滤棉、单层初效过滤棉、双层初效过滤棉、初效＋亚高效过滤棉，过滤棉类别设置如表1所示。

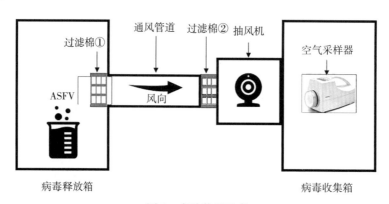

图1　实验装置示意

表1　过滤棉类别设置

组别	病毒液	过滤棉位置①	过滤棉位置②
阴性对照	清水	无	无
阳性对照	ASFV	无	无
单层亚高效过滤棉	ASFV	无	亚高效过滤棉
双层亚高效过滤棉	ASFV	亚高效过滤棉	亚高效过滤棉
亚高效＋初效过滤棉	ASFV	亚高效过滤棉	初效过滤棉
单层初效过滤棉	ASFV	无	初效过滤棉
双层初效过滤棉	ASFV	初效过滤棉	初效过滤棉
初效＋亚高效过滤棉	ASFV	初效过滤棉	亚高效过滤棉

1.3　试验方法

通过图 1 的装置，在病毒释放箱内，提前准备含有 ASFV 的病毒液，利用造雾雾化装置，将病毒液进行雾化，形成气溶胶；打开超声波雾化装置 10 min 后，待病毒液雾化稳定后，开启抽风机，风速为 8 m/s；然后，在病毒收集箱内，用空气采样器进行样品采集，每次收集空气 1 000 L，重复 3 次。

1.4　检测指标

收集到气溶胶样品后，每个样品加入 3 mL 生理盐水进行溶解，然后送到实验室进行 ASFV Ct 值的检测和统计。

2　试验结果

2.1　现场实验装置

现场实验装置如图 2 所示，总共有 3 组，一组进行单层亚高效过滤棉、双层亚高效过滤棉、亚高效＋初效过滤棉试验；二组进行单层初效过滤棉、双层初效过滤棉、初效＋亚高效过滤棉试验；三组进行阴性对照和阳性对照组试验。

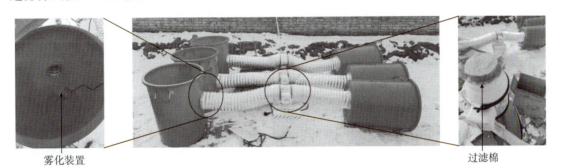

雾化装置　　　　　　　　　　　　　　　　　　　　　　　　　过滤棉

图 2　现场实验装置

2.2　病毒收集箱内空气样品 Ct 值检测

Ct 值检测结果如图 3 所示，Ct 值从高到低顺序，依次为单层亚高效过滤棉＞初效＋亚高效过滤棉＞双层亚高效过滤棉＞亚高效＋初效过滤棉＞双层初效过滤棉＞单层初效过滤棉。其中，虽然单层亚高效过滤棉的平均 Ct 值高于阳性对照组，但是无显著性差异（$P>0.05$）；单层亚高效过滤棉装置的 Ct 值显著高于单层初效过滤棉和亚高效＋初效过滤棉装置（$P<0.05$）；单层亚高效过滤棉和双层亚高效过滤棉装置之间无显著性差异（$P>0.05$）。有趣的是，初效＋亚高效过滤棉的组合效果显著优于亚高效＋初效过滤棉的组合（$P<0.05$）。另外，阴性对照组结果为阴性。

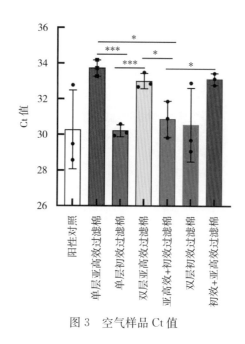

图 3 空气样品 Ct 值

3 试验讨论

首先，亚高效过滤棉虽然对病毒有一定的过滤作用，但是，并不能完全去除病毒。其次，初效＋亚高效过滤棉的组合效果优于亚高效＋初效过滤棉，可能是由于初效过滤棉能够对亚高效过滤棉起到一定的保护作用，甚至可能延长亚高效过滤棉的使用寿命；另外，双层亚高效过滤棉的效果并不比单层过滤棉的好，可能的原因是位于病毒释放箱的出风口的过滤棉被湿润的气溶胶破坏，没有了过滤效果，这也可能是亚高效＋初效过滤棉的组合对病毒气溶胶的过滤效果极差的原因之一。

4 结论

（1）初效过滤棉对 ASFV 气溶胶没有过滤效果。

（2）亚高效过滤棉对 ASFV 气溶胶有一定的过滤作用，但是并不能完全消除病毒。

（3）初效＋高效过滤棉模式比高效＋初效过滤棉模式效果更优。

参考文献

［1］ 王富强，郭晶，张卫红，等．GB/T 14295—2019《空气过滤器》内容解读与比较分析［J］．标准科学，2021（9）：61-65.

［2］ 王翔朴．卫生学大辞典［M］．北京：华夏出版社，1999．

［3］ Otake S，Dee S，Corzo C，et al. Long-distance airborne transport of infectious PRRSV and Mycoplasma hyopneumoniae from a swine population infected with multiple viral variants ［J］. Veterinary Microbiology，2010，145（3-4）：198-208.

［4］ Alonso C，Goede D P，Morrison R B，et al. Evidence of infectivity of airborne porcine epidemic diarrheavirus and detection of airborne viral RNA at long distances from infected herds ［J］. Veterinary Research，2014（45）：73.

［5］ Alonso C，Raynor P C，Davies P R，et al. Concentration，Size Distribution，and Infectivity of Airborne Particles Carrying Swine Viruses ［J］. PLoS One，2015，10（8）：e0135675.

［6］ De C F H C，Weesendorp E，Quak S，et al. Quantification of airborne African swine fever virus after experimental infection ［J］. Veterinary Microbiology，2013，165（3-4）：243-251.

［7］ Olesen A S，Lohse L，Boklund A，et al. Transmission of African swine fever virus from infected pigs by direct contact and aerosol routes ［J］. Veterinary Microbiology，2017（211）：92-102.

猪腹股沟淋巴结微创采样方法用于
非洲猪瘟早期诊断的技术规范

李孝文　樊铭玉　高文超　樊士冉　李敬涛　袁　朋　吴伟胜

1　范围

本技术规范规定了猪只非洲猪瘟（ASF）疾病检测中活体猪腹股沟淋巴结微创采集与实验室非洲猪瘟病原检测方法。

本技术规范适用于在猪群非洲猪瘟早期监测中对活体猪腹股沟淋巴结组织样品的采集及猪只非洲猪瘟病原检测。

2　规范性引用文件

《非洲猪瘟诊断技术》（GB/T 18648—2020）、《非洲猪瘟病毒实时荧光 PCR 检测方法》（T/CVMA 5—2018）、《非洲猪瘟检疫技术规范》（SN/T 1559—2010）以及《猪饲养标准》（NY/T 65—2004）对于本文件的应用是必不可少的，且仅注日期的版本适用于本文件。

3　活体猪腹股沟淋巴结组织采集

3.1　准备工作

3.1.1　物资工具　采样操作前准备人员穿戴一次防护服、一次性医用乳胶手套、一次性鞋套，用于隔离防护，避免人猪交叉污染；金属套猪器，用于保定猪只；3％次氯酸钠溶液或84 消毒液（有效氯 1％），消毒过的采样工具；记号笔，用于在离心管上标记样品信息；1.5 mL离心管。

3.1.2　腹股沟淋巴结微创采样器　腹股沟淋巴结微创采样器用于活体猪腹股沟淋巴结组织的微创采样，由采样针头、固定帽、握柄、推柄、撞针 5 个构件组成。采样针头由内孔径1.7 mm 不锈钢针头加工制作，针头前端设置有倒刺采样孔；推柄与直径 1.5 mm 撞针相连，撞针内套于采样针头内部（图 1）。

3.2　猪只保定

3.2.1　小体重（体重小于 20 kg）猪只保定　参照《猪饲养标准》（NY/T 65—2004）中的

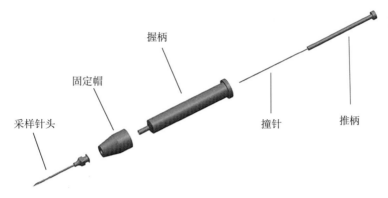

图 1　腹股沟淋巴结采样器结构

猪只日龄和体重的关系，针对小体重猪只，双手握住猪只后肢，猪只腹侧向外，人员双腿夹住猪头部或使头部自然下垂，暴露腹股沟位置（图 2）。

图 2　小体重猪只保定示意

3.2.2　中大猪（体重 20 kg 以上）**保定**　使用套猪器套住猪鼻固定，在后方将猪只侧卧放倒，并使用套猪器或绳索将悬空的猪后肢固定，暴露腹股沟（图 3）。

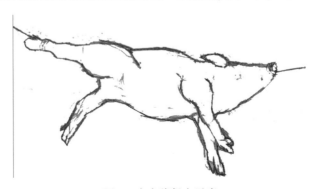

图 3　中大猪保定示意

3.3 淋巴结组织的采集操作步骤

（1）猪只保定后，手指固定腹股沟淋巴结部位。

（2）将采样针头穿透猪只腹部皮层，纵向刺入腹股沟淋巴结内。

（3）握住握柄，外抽，样品孔处钩出腹股沟淋巴结组织。

（4）按压采样器推柄，联动内部撞针，将钩取的腹股沟淋巴结组织从采样针头内推出，放入装有 0.5 mL 生理盐水的 1.5 mL 规格离心管中。

（5）使用记号笔，标记猪只耳号和采样日期等信息，冷链送往实验室检测或冷冻保存。

4 实验室淋巴结样品处理

取出保存或送检的含有淋巴结组织的 1.5 mL 离心管，用组织破碎仪破碎，反复冻融 3 次，然后 1 000 r/min 离心 3 min，取上清液进行核酸提取。

5 DNA 提取方法

（1）取 4 中离心的上清液 200 μL 于 1.5 mL 指形管中，依次加入 300 μL 超纯水 92 μL、10％ SDS 溶液和 8 μL 蛋白酶 K（20 mg/mL），涡旋混匀后置 55℃ 金属浴 30 min。

（2）加入体积比为 1∶1 的苯酚∶氯仿溶液 600 μL，剧烈震荡混匀，12 000 r/min 离心 15 min。

（3）吸取上清 400 μL，加入−20℃ 预冷的无水乙醇 800 μL，反复颠倒数次，−20℃ 静置 10 min 以上，12 000 r/min 离心 9 min。

（4）弃上清液，加入 70％ 乙醇 800 μL 洗涤，12 000 r/min 离心 5 min，弃上清液，瞬时离心，吸取残液并晾干，加入 30 μL RTE，37℃ 作用 30 min，立刻使用或−20℃ 保存备用。

6 荧光定量 PCR 检测猪只病毒病原

荧光定量 PCR 具体操作步骤参照《非洲猪瘟诊断技术》（GB/T 18648—2002）、《非洲猪瘟病毒实时荧光 PCR 检测方法》（T/CVMA 5—2018）及《非洲猪瘟检疫技术规范》（SNT 1559—2010）的规定说明操作，并进行结果判定。

6.1 荧光定量 PCR 具体操作步骤

（1）取出试剂盒，从中取出试验所需试剂，融化混匀并瞬时离心以去除管壁附着的液体。

（2）取需要数量的 PCR 管（需设置阳性对照及阴性对照），在各 PCR 管中分别加入 20 μL

的荧光 PCR 反应液，转移至样本准备区。

（3）分别加入样品处理后获得的模板各 2 μL，加样顺序为阴性对照、样品、阳性对照，混匀，瞬时离心转移至扩增区。

配液体系	
荧光 PCR 反应液	20 μL
模板	2 μL
总体积	22 μL

（4）将各 PCR 管放置在仪器样品槽相应位置，并记录放置顺序。

（5）设置仪器的相关参数，进行 PCR 扩增。

体系	22 μL 反应体系		
信号通道	FAM		
PCR 反应条件	阶段	条件	循环数
	预变性	95℃；30 s	1
	PCR	95℃：10 s	40
		60℃：30 s（此阶段采集荧光信号）	

6.2　结果判定

6.2.1　试验成立条件　阳性对照 Ct 值≤30，有明显的指数增长，呈典型的 S 形曲线。阴性对照无 Ct 值，无明显指数增长期和平台期。

6.2.2　判定标准　样品检测结果 Ct 值≤35，有明显指数增长，判定为阳性，表明样品中检测出非洲猪瘟病毒。样品检测结果 35＜Ct 值≤40，判定为可疑。应对样品进行复测，若复测结果 Ct 值仍在 35～40，有明显的指数增长，则判定为阳性，否则为阴性。

样品检测结果无 Ct 值或 Ct 值＞40，判定为阴性，表明样品中未检测出非洲猪瘟病毒。

猪群亚健康的评估与干预的研究进展

关　然　韩秋平

摘要：猪群亚健康是目前规模化猪场普遍存在的现象，其本质是机体处于一种结构退化、生理功能减退和体质失衡状态，特征是免疫力低下、代谢速度缓慢、抗病力降低、对环境的适应性差等，最终导致生产指标下降，如断奶后 7d 内发情率、受胎率均偏低、窝均活仔数偏少、仔猪断奶前死亡率偏高等。本研究通过收集和分析体系内核心猪场母猪群和断奶后猪群的临床指标，评估整体健康状态。对定义为亚健康的猪群制定综合干预措施，包括使用植物类饲料添加剂、发酵中药、免疫调节剂等产品，改善饲养条件，优化饲喂方法等。经过系统地扩大验证，制定猪群亚健康改善综合方案，以期降低疾病发生风险，同时创造经济价值。

关键词：猪群亚健康；生产指标下降；评估；干预措施

在猪生产中我们经常遇到一些令人困惑的问题，如注射免疫疫苗后不能产生很好的抗体水平，注射完猪繁殖与呼吸综合征（蓝耳病）疫苗后出现流产，猪苗从断奶到养殖场后死淘率很高，抗生素的使用效果越来越差、用量越来越大，突然降温时猪群发病等。这可能是猪群健康出现了问题。猪群的亚健康状态是一种介于健康与疾病之间的状态，猪没有明显的病理临床症状，但也没有健康状态的活力。

1　猪群亚健康的成因

猪群的生产成绩、母猪的繁殖性能两大指标需要至少一个生产周期才能了解，影响这两大指标的原因之一就是猪群的亚健康，而出现猪群亚健康的首要原因则是机体出现了慢性中毒[1]。出现慢性中毒的原因：一是体内的革兰氏阴性杆菌，如大肠杆菌、沙门杆菌、副猪嗜血杆菌等死亡后释放的细菌细胞内毒素；二是某些抗生素的使用时间过长、使用量过多而蓄积在机体中导致机体中毒；三是猪正常代谢产生的代谢次生产物即自由基致机体蓄积中毒；四是饲料中的霉菌毒素中毒[2]；五是饮水中的污染物而致的蓄积中毒。另外，还有一些导致猪群亚健康的原因，如饲料营养不均衡；温度、湿度、饲养密度等方面造成的环境应激；免疫和带猪消毒不规范等饲养管理问题。需要特别注意的是，凡是能引起免疫抑制的病原体都可导致猪群亚健康[3]，包括蓝耳病、猪瘟、猪圆环病、猪伪狂犬病、猪支原体病、流感等疾病对应的病原体。

2 猪群亚健康的临床表现

2.1 母猪

观察母猪断奶 7 d 的发情率、配种分娩率是否分别低于 85%；观察后备母猪初情期是否推迟，膘情是否正常，背膘厚断奶时和参配前不低于 16 mm，分娩前在 20～22 mm；观察母猪皮毛的光泽、颜色，母猪毛孔是否有出血点[4]，母猪的难产比例；观察其分娩的仔猪初生重是否正常（不低于 1.2 kg），体表是否发黄，窝产死胎、木乃伊是否平均超过 0.8 头等；观察母猪群粪便是否正常，评估颜色、硬度；观察哺乳期母猪乳房膨胀度、泌乳量、柔软度、是否有肿块；观察母猪阴道分泌物是否正常[5]。

2.2 生长猪

喂料过程中，观察猪群反应是否迟缓，采食量和饮水量与日龄是否相符；观察猪只眼睛是否灵活，是否有分泌物，是否有泪斑；观察猪只的呼吸频率和节奏，包括腹侧部肌肉的相应运动情况；观察猪只膘情、皮毛、粪便是否正常[6]；观察猪只姿势和运动是否正常，是否有犬坐弓背姿势、胸部起伏较大、呼吸速度较快且长时间伴随气喘等情况[7]；观察猪只料重比、日增重是否与日龄相符；观察猪只尾巴的状态，上翘、摇摆的为健康猪只；观察猪只是否出现生长速度缓慢、个体发育不匀、皮肤病发病率上升、反复腹泻或咳喘等[8]。

3 亚健康对猪群造成的危害

3.1 生产性能下降

仔猪和育肥猪采食量下降，导致其营养缺乏，生长速度缓慢，饲料报酬降低；母猪出现繁殖障碍，如发情迟缓、受胎率低、所产仔猪重量小、弱仔率、死胎率增多，预产期推迟或提前[9]；泌乳母猪食欲时好时坏，会造成泌乳量减少，哺乳仔猪腹泻增加，断奶时体重轻；母猪不爱运动，易造成便秘；猪群免疫系统低，抗病能力严重下降，极易感染疾病；同时，对特异性免疫接种的应答能力相对降低。

3.2 成为潜在疾病发生的导火索

如果机体的防御能力强于外界因素的刺激，则猪群不会发病，从而发展到亚健康状态或维持亚健康状态；若外界因素的刺激能力占优势，则猪群会迅速发展到疾病状态。因此，如何调整机体的亚健康状态，即成为整个猪群是否发病的关键。

3.3 导致机体免疫力下降，易继发其他疾病

当机体免疫力下降时，体内代谢紊乱，条件性致病菌就会引起猪只发病，慢性传染性疾病也会呈现急性发作。在应激因素长期刺激下，接种某些活疫苗也引起疫病的非典型性症

状，导致确诊困难，造成巨大的经济损失[10,11]。

4　改善猪群亚健康的措施

改善猪群亚健康的措施包括"三提高"和"三降低"，"三提高"是指提高饲料营养水平和均衡度，提高环境舒适度，提高猪的福利。通俗来讲，就是让猪吃好、喝好、吸入的空气好、住好和玩好。"三降低"是指降低内外环境病原载量，降低体内各种毒素，降低各种不良应激刺激[12]。

4.1　加强猪舍的环境控制

只有在适宜的环境条件下，猪群才能保持健康的体魄，其生产潜力才能得以充分发挥。现代化猪场更加重视基础设施建设，尤其在解决通风和保温技术环节。猪舍环境出现任何干扰，都会对猪群产生应激，诱发疾病。因此，猪舍要保持清洁卫生、防热降温、防寒采暖、防潮排水、通风换气，为猪提供适宜的温度与湿度，降低氨气、二氧化碳、二氧化硫等有害气体的浓度。规模化养猪场要彻底实现养猪生产各阶段的全进全出，至少做到产房和保育两个阶段的全进全出。饲养密度要适中，不同年龄的猪应分群饲养，从而减少和降低猪群之间蓝耳病和猪圆环病毒Ⅱ型等病原的接触感染机会，尽可能地降低猪群的感染率。

4.2　减少免疫抑制性疾病对猪群的危害

最大限度地保持蓝耳病和猪圆环病毒Ⅱ型阴性猪群，保持猪群中无蓝耳病和猪圆环病毒Ⅱ型感染，这对于一般规模猪场是不容易做到的。对于检疫蓝耳病和猪圆环病毒Ⅱ型阴性场或新建场，首先要求把好引种关，由核心种猪场或信誉好的种猪场提供所需的后备种猪，切实做好产地检疫，确认健康后方可引入，而且须有一整套完善的、科学的生物安全防疫体系做保障。

4.3　重视猪群疫病监测工作，严防外来病原入侵

对于猪瘟、蓝耳病、猪伪狂犬病、猪链球菌病、猪圆环病毒病的感染等，要加强对免疫抗体进行监测，或对强毒抗体或抗原进行检测[13]。规模化猪场一般每季度监测 1 次或 1 年监测 2 次。如果每次监测抗体阴性率没有明显变化，免疫抗体效价达标，证明该病在该场的控制是稳定的，说明饲养管理、卫生防疫工作是有效的；相反，则说明该猪场在某些方面还存在问题，应分析原因，积极采取措施，消除隐患。此外，应选择高效消毒液加强猪舍及环境的消毒，降低猪场环境病原微生物的数量。

4.4　通过饲料添加剂增强猪群的抵抗力，改善亚健康

Li 等[14]开展了有关菊粉添加剂改善新生仔猪的健康状况的研究，目的是探究母体摄入膳食纤维菊粉后与仔猪健康之间的关系。该研究有 40 头母猪参与试验，共饲喂 30 d。结论：

日粮中同时添加 1.6％菊粉可以调节妊娠母猪的肠道微生物，提高繁殖性能，从而改善仔猪的生产性能及健康状况。Shang 等[15] 在母猪妊娠后期往饲料中添加 15％麦麸（WB）和 10％甜菜浆（SBP）的混合物，哺乳期时添加 7.5％WB 和 5％ SBP 混合物，研究对仔猪生长性能和肠道功能的影响。试验期间共 30 头母猪，从妊娠期第 85 天到断奶期，饲喂时间约 51 d。结论：在母猪饲料中添加 WB 和 SBP 混合物可以改善仔猪的生长性能、肠道形态、屏障功能以及仔猪体内的微生物菌群等。

Parraguez 等[16] 研究将具有维生素 C 和维生素 E 抗氧化活性的多酚类草药产品（C-Power 和 Herbal-E50）作为饲料添加剂（每千克饲料中添加每种补充剂 290 mg）对母猪繁殖特性和仔猪性能的影响。试验共 1 027 头妊娠母猪，对照组 523 头，试验组 504 头，从配种第 1 天开始到妊娠 112 d 结束，共饲喂 112 d。试验结果显示：补充剂增加了活产仔猪的数量和总窝重，降低了低体重仔猪的发生率；补充剂降低了母猪及其仔猪的血浆丙二醛（MDA）水平，并且仔猪有表现出较高断奶体重的趋势。

Xin 等[17] 研究天然和发酵形式的草药对肥育猪的生长性能和营养消化率等的影响，研究表明：补充天然草药可以提高生长肥育猪的生长性能和养分消化率。补充发酵草药还可增强血清抗氧化状态，并积极改变脂肪酸谱。Cheng 等[18] 研究在母猪和仔猪日粮中添加甜菜碱对仔猪生长性能、血浆激素和脂质代谢的影响。结果显示：在仔猪日粮中添加甜菜碱可以提高其生长性能。此外，在母猪和仔猪日粮中添加甜菜碱可以通过改变血浆激素水平和脂肪酸组成以及调节与脂肪酸相关的基因表达来提高脂质代谢。

Zhang 等[19] 研究在妊娠后期和哺乳期添加酵母培养物（YC）和有机硒（Se）对经产母猪的繁殖性能、仔猪断奶前性能、抗氧化能力和免疫球蛋白分泌的影响。试验分为四组高营养（HN）组、低营养（LN）组、LN ＋ YC 组、LN ＋ YC ＋ Se 组，从妊娠第 85 天开始饲喂到泌乳第 35 天结束，共 160 头妊娠猪。试验结果显示：同时饲喂 YC Se 的母猪 MDA 的含量较低，谷胱甘肽过氧化物酶（GSH-Px）的活性较高，YC 和 Se 可提高母猪的抗氧化能力和奶水中的脂肪含量，从而提高母猪的繁殖性能和断奶体重。

近几年，该领域的研究者主要通过添加饲料添加剂以提高母猪的繁殖性能及仔猪的生产性能来改善猪群亚健康，并且取得了很好的效果。

5　中草药的应用与展望

近年来，由于受到免疫抑制、病毒变异及病菌耐药等多方面因素的影响，猪病病原愈发多元化、病症更加复杂化，极大地增加了猪病诊断和治疗难度。单纯采用抗生素等药物治疗猪病，易出现耐药性，将中草药应用于猪病防治中，具备显著的抗细菌、抗病毒以及抗应激的效果，可以提高猪群的生产性能，为生猪养殖业稳定发展提供了强有力的保障[20,21]。

5.1　控制传染性疾病

猪传染性疾病在一些地区广泛存在，出现这类疾病的根源是猪群免疫系统遭到破坏。免疫抑制的产生主要指动物体内存在多个导致疾病发生的因素，这类疾病的感染性通常极强。猪群感染疾病后会发生腹泻，中草药制剂中的郁金散、白头翁汤对猪群传染性疾病有十分重要的预防和治疗作用，止痢汤的使用效果优于郁金散、白头翁汤。使用这类中草药制剂时需严格按照医生要求抓取药物，具体药物包含石膏、栀子、黄芩、桔梗等，辅助配合西药治疗。仔猪的抵抗力、免疫力较差，容易患黄白痢，抗菌药物能抑制和杀灭致病性大肠杆菌，保护仔猪肠道，进而达到理想的治疗效果。

5.2　治疗母猪产后乳汁分泌不足

母猪分娩后通常会出现乳汁分泌不足等问题，常见于年龄较大的母猪[22]。母猪乳汁分泌不足轻则无法保证仔猪的健康生长，严重时还会造成仔猪成为僵猪，严重影响母猪和仔猪的生产性能。将中草药制剂应用到母猪产后治疗能够有效预防和控制母猪乳汁分泌不足问题。中医认为，乳汁是由母猪身体中的气血精微转化产生的，只要母猪气血充足，就能够保证乳汁充沛。因此，母猪分泌后可以食用中草药制剂来平衡母猪气血，从而有效防止母猪因气血不足而发生缺乳现象。

5.3　保育仔猪，抵抗腹泻

保育仔猪需要做好防腹泻工作，受生长环境和相关应激因素的影响，仔猪在生长过程中容易出现免疫力降低的问题，威胁健康生长。在保育期使用中草药制剂能够有效增强仔猪的免疫力，促进其健康生长。黄藤素具有抑菌消炎、抗炎解毒、增强免疫力的作用，可以降低腹泻发生率[23]；增强仔猪免疫力的中草药还包括黄芪、甘草、黄白、大青叶等，同时辅助使用微生物和多糖还能够促进血液循环，达到理想的抗菌和保健作用[24]。

5.4　中草药在猪咳喘病防治中的应用

仔猪断奶后，易于发生咳喘病，尤其是在春、秋、冬季，该病发生率更高，死亡率在30％左右[25]。患病猪会出现高热、咳喘、呼吸困难、消瘦、皮毛粗乱，重则呼吸衰竭而死。西药治疗猪咳喘病，治标不治本，副作用大。使用中草药防治猪咳喘，能够达到标本兼治的作用。取陈皮、甘草、苏子、瓜蒌各 6 g，连翘、桔梗、金银花各 10 g，将上述药物研磨成粉末状后拌入饲料中喂食病猪，早晚各 1 次，连服 3～5 d 即可，并取得了良好的防治效果。

5.5　改善猪群肠道健康

在猪病综合防治工作中科学合理地使用中草药不仅能够提高猪群的免疫功能，而且中草药中的活性成分，如生物碱、挥发油及有机酸等物质还会发挥免疫调节作用。如枸杞中的糖分能够有效增强猪群免疫系统中 B 淋巴细胞、T 淋巴细胞的免疫功能；挥发性油能够增强猪群的抵抗力，提高淋巴细胞的转化率，增强中枢淋巴器官和周围淋巴器官的细胞繁殖力，最终显著提高淋巴细胞的免疫功能[26]。

综上所述，在猪病综合防治工作中合理使用中草药能够达到理想的防治效果，因此，可以利用中草药制成植物类饲料添加剂或发酵中药的形式进行干预猪群的亚健康，调节猪群的生理机能，降低毒素含量，增强抵抗力，改善猪群亚健康，提高生产性能，增加经济收益。

参考文献

[1] 张晓爱，黄世才，吴晓林，等．中草药预混合制剂——律肝保防治母猪亚健康效果的临床试验报告[J]．中兽医学杂志，2017，6：7-9.

[2] 迁斌，颜运秋，徐少勇．母猪群亚健康的临床表现须关注[J]．北方牧业，2021，4：26-27.

[3] 陈星，李琼，刘进辉，等．清除猪体内"三毒"，缓解机体"亚健康"[J]．农业开发与装备，2019，7：46-52.

[4] 王世玉．母猪带毒的危害及解决措施[J]．农村养殖技术，2013，8：37-38.

[5] 张金辉．猪群亚健康知多少[J]．猪业科学，2014，31（11）：32-34.

[6] 薛佳俐．生长育肥猪福利养殖评价系统的建立[D]．北京：中国农业科学院，2020.

[7] 王志东，张振新，陈冬灵，等．微生态制剂和糖蜜在杜花二元杂育肥猪上的应用效果[J]．饲料研究，2020，43（7）：27-31.

[8] 东向森．猪群亚健康与免疫抑制病的认识[J]．北方牧业，2017，10：13.

[9] 赵浩．猪群亚健康的成因与对策[J]．北方牧业，2021，8：24-26.

[10] 王艳丰，张丁华．猪群亚健康状态产生原因及解决措施[J]．猪业科学，2014，31（11）：42-44.

[11] 张秀卫．猪群的亚健康状态及防控对策[J]．猪业观察，2018，4：46-47.

[12] 赵浩．猪群亚健康的成因与对策[J]．北方牧业，2021，8：24-26.

[13] 高宗良．规模化自繁自养猪场主要疫病监测及综合防控[D]．泰安：山东农业大学，2018.

[14] Li Hao，Ma Longteng，Zhang Longlin，et al. Dietary Inulin Regulated Gut Microbiota and Improved Neonatal Health in a Pregnant Sow Model[J]．Frontiers in Nutrition，2021，8.

[15] Shang Qinghui，Liu Sujie，Liu，et al. Maternal supplementation with a combination of wheat bran and sugar beet pulp during late gestation and lactation improves growth and intestinal functions in piglets[J]．Food & function，2021，12（16）：7329-7342.

[16] Parraguez Victor H，Sales Francisco，Peralta Oscar A，et al. Maternal Supplementation with Herbal Antioxidants during Pregnancy in Swine[J]．Antioxidants（Basel，Switzerland），2021，10（5）：658.

[17] Xin Jian Lei，Hyeok Min Yun，In Ho Kim. Effects of dietary supplementation of natural and fermentedherbs on growth performance，nutrient digestibility，blood parameters，meat quality and fatty acid composition in growing-finishing pigs[J]．Italian Journal of Animal Science，2018，17（4）：983-993.

[18] Cheng Yating，Song Mingtong，Zhu Qian，et al. Abul Kalam，Gao Qiankun，Kong Xiangfeng. Impacts of Betaine Addition in Sow and Piglet's Diets on Growth Performance，Plasma Hormone，and Lipid Metabolism of Bama Mini-Pigs[J]．Frontiers in Nutrition，2021，8.

[19] Zhang Shihai，Wu Zhihui，Heng Jinghui，et al. Combined yeast culture and organic selenium supplementation during late gestation and lactation improve preweaning piglet performance by enhancing the antioxidant capacity and milk content in nutrient-restricted sows[J]．Animal Nutrition，2020，6（2）：160-167.

［20］叶来飞.中草药在猪病防治中的使用效果［J］.中国动物保健，2021，23（10）：6-7.

［21］徐松琪.黄芪多糖和葛根提取物对哺乳母猪生产性能及血清生化免疫指标的影响［D］.东北农业大学，2018.

［22］庄建明.中草药在猪疾病防治中的应用探讨［J］.中国畜牧兽医文摘，2018，1：252.

［23］Wang Tongxin，Yao Weilei，Li Juan，et al. Dietary garcinol supplementation improves diarrhea and intestinal barrier function associated with its modulation of gut microbiota in weaned piglets［J］. Journal of Animal Science and Biotechnology，2020，11（3）：853-865.

［24］柴丽莉.中草药制剂在猪病防治中的应用［J］.国外畜牧学（猪与禽），2021，41（5）：69-70.

［25］孙晓荣.中草药在猪疾病防治中的应用探讨［J］.农业工程技术，2018，20：67-68.

［26］马云蕾.中草药在猪疾病预防中的应用［J］.畜牧兽医科学，2018，16：35-36.

Chapter

2

营养饲料篇

断奶母猪葡萄糖优饲对其繁殖性能的影响

辛海瑞

摘要：本试验旨在研究母猪断奶后葡萄糖优饲对其繁殖性能的影响。试验选取 8 个批次待断奶母猪 646 头（胎次＝1.94；体况评分：13.4±2.3），每个批次母猪均按胎次和背膘厚随机分为 2 个处理组：对照组（$n=322$）和葡萄糖组（$n=324$）。对照组不做处理，葡萄糖组在断奶 7 d 内饲喂 300 g/d 葡萄糖，分两次随母猪采食饲喂（2 次/d，150 g/次），若母猪发情，即停止饲喂葡萄糖。准确记录每头母猪断奶 7 d 内的采食量、发情信息、配种信息、受胎信息以及分娩信息。结果表明：①断奶母猪进行葡萄糖优饲显著提高 7 d 断配率 6.87%（$P<0.01$），且有缩短母猪断奶-发情间隔的趋势（$P=0.09$）；②断奶后葡萄糖优饲有提高母猪总产仔数（+0.44；$P=0.09$）和产活仔数（+0.46；$P=0.08$）的趋势。综上所述，在本试验条件下，断奶后饲喂 300 g/d 葡萄糖可提高母猪 7 d 断配率，且对总产仔数和产活仔数有一定积极影响，可应用于生产以提高母猪生产效率。

关键词：断奶母猪；葡萄糖；繁殖性能；分娩性能

7 d 断配率是母猪生产考核中的重要指标。断奶母猪群发情率低及发情间隔长严重影响年生产胎次、非生产天数和年提供的断奶仔猪头数（PSY）。当前造成母猪发情迟缓或不发情的因素很多，包括机体疾病、生殖激素分泌不平衡、环境温湿度高及遗传因素等。

为了提高断奶母猪的发情率以及减短发情间隔，欧洲国家尝试使用葡萄糖，饲喂葡萄糖可通过影响血糖水平来影响胰岛素分泌，而胰岛素在体内具有重要生理作用，其中包括通过促进促性腺激素分泌直接影响断奶母猪发情，还可以通过影响其他激素分泌间接影响发情[1]。有文献表明，葡萄糖经过糖酵解转变为乳糖并产生 ATP，通过磷酸戊糖途径产生烟酰胺腺嘌呤二核苷酸磷酸（NADP）和磷酸核糖，这两条途径在卵母细胞成熟和发育过程中起着重要作用[2]；体外试验发现，添加葡萄糖能够增加发育至第二次减数分裂中期的卵母细胞数量和提高囊胚率[3]。白红杰等[4]研究饲喂不同水平葡萄糖（0、200 g/d、600 g/d 及 1 000 g/d）对断奶母猪发情率和发情间隔的影响，结果发现与对照组相比，饲喂 600 g/d 和 1 000 g/d 葡萄糖均可显著提高母猪发情率（分别提高 3.57% 和 3.63%）及缩短断奶-发情间隔（分别减少 0.83 d 和 0.81 d），但二者之间无差异。因此，在提高母猪繁殖性能的前提下，饲喂 600 g/d 葡萄糖可节约饲养成本。廖昌韬等[5]研究日粮中添加 200 g/d 葡萄糖对断奶母猪发情配种的影响，发现葡萄糖组母猪 7 d 发情配种率提高 1.9%，断奶-发情间隔减少 0.26 d，淘汰率显著降低 4.04%。

本试验的目的是研究断奶 7 d 内饲喂葡萄糖对母猪繁殖性能的影响，其中包括 7 d 断配率、发情间隔、受胎率、分娩率及产仔情况等。

1 试验材料与方法

1.1 试验设计

本试验采用单因素试验设计，选取 8 个批次待断奶母猪 646 头（胎次＝1.94；体况评分：13.4±2.3），每个批次母猪按胎次和体况信息随机分为 2 个处理组：对照组（n＝322）和葡萄糖组（n＝324）。对照组不做处理；葡萄糖优饲组在断奶 7 d 内饲喂 300 g/d 葡萄糖，分 2 次随母猪采食饲喂（2 次/d，150 g/次）。从断奶第 2 天开始使用公猪进行诱情，若母猪发情，即停止饲喂葡萄糖。追踪每批母猪从断奶至分娩，准确记录每头母猪断奶 7 d 内的采食量、发情信息、配种信息、发情间隔、受胎信息、分娩信息、产仔信息以及淘汰情况。选取 2 批次共 105 头母猪（对照组 49 头；葡萄糖组 56 头），称量所产仔猪的初生重和出生窝重。

1.2 饲养管理

试验在湖北古岭的猪场开展，自 2021 年 3 月 17 日开始至 2021 年 8 月 25 日结束。日常管理按猪场常规管理规程进行饲养。母猪断奶 7 d 内，每天分 2 次饲喂：上午 2.4 kg 和下午 0.8 kg。当母猪配种后，按规定饲喂量饲喂相同的妊娠料。分娩前 3 d 进入分娩单元，饲喂相同的哺乳料。试验期间记录母猪异常情况。

1.3 7d 断配率和发情间隔

记录每头母猪断奶 7 d 内的发情和配种情况，计算母猪断奶-发情的间隔及 7 d 断配率。7 d 断配率＝断奶 7 d 内配种母猪数/总母猪数×100%。

1.4 35 d 受胎率

对参与试验的母猪在其妊娠 28～35 d 进行受胎情况检测，并追踪母猪在 35 d 前妊娠损失情况（空怀、返情等），计算 35 d 受胎率。受胎率＝妊娠 35 d 受胎母猪数/总配种母猪数×100%。

1.5 受胎-分娩率

追踪每头配种母猪的分娩情况，记录妊娠 35 d 至分娩期间母猪妊娠损失、死亡及淘汰情况，计算分娩率。分娩率＝分娩母猪数/总配种母猪数×100%。

1.6 母猪产仔信息

对参与试验的所有母猪在分娩当天记录其产仔信息，包括总产仔数、产活仔数、死胎数及木乃伊数。选取 105 头试验母猪，在其分娩后 6 h 内对每窝仔猪进行逐头称重，记录初生重和出生窝重，并计算仔猪窝内体重变异系数。

1.7 统计分析

利用 Excel 对原始数据进行记录，再利用 SAS 统计软件 Glimmix 程序对数据进行统计分析。日粮为固定效应，批次为随机效应。数据以平均数±标准误表示，$P<0.05$ 表示差异显著，$0.05 \leqslant P<0.10$ 表示趋于差异显著。

2 试验结果

2.1 葡萄糖优饲对母猪断奶-发情间隔的影响

葡萄糖优饲对母猪断奶-发情间隔的影响如图 1 所示。由图 1 可知，饲喂葡萄糖有缩短母猪断奶-发情间隔的趋势（$P=0.09$）。

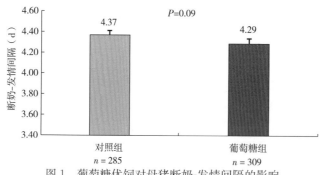

图 1 葡萄糖优饲对母猪断奶-发情间隔的影响

2.2 葡萄糖优饲对母猪 7 d 断配率、35 d 受胎率和分娩率的影响

由图 2 可知，断奶母猪通过葡萄糖优饲可以显著提高其 7 d 断配率（6.8%，$P<0.01$）；虽然 35 d 受胎率差异并不显著，但在数值上提高 2.3%。针对不同胎次分析发现，葡萄糖优饲组 1 胎母猪的 7 d 断配率提高 29%，2～3 胎母猪则提高 4.0%（图 3）。

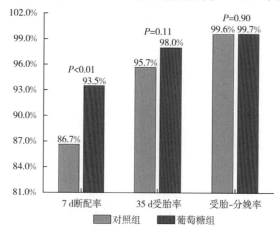

图 2 葡萄糖优饲对母猪 7 d 断配率、35 d 受胎率、受胎-分娩率的影响

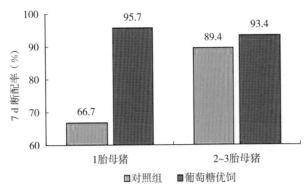

图 3　葡萄糖优饲对不同胎次母猪 7 d 断配率的影响

2.3　葡萄糖优饲对母猪产仔数的影响

由表 1 可知，饲喂葡萄糖有提高母猪总产仔数（＋0.44；$P=0.09$）和产活仔数（＋0.46；$P=0.08$）的趋势，但是对死胎数和木乃伊数无影响。

表 1　葡萄糖优饲对母猪产仔数的影响（头）

项目	对照组	葡萄糖组	SEM	P 值
母猪数	263	294		
总产仔数	12.96	13.40	0.22	0.09
产活仔数	12.21	12.67	0.21	0.08
死胎数	0.63	0.55	0.07	0.38
木乃伊数	0.13	0.18	0.04	0.22

2.4　断奶母猪葡萄糖优饲对其仔猪性能的影响

如表 2 所示，断奶母猪饲喂葡萄糖对仔猪初生重、出生窝重和窝内体重变异系数均无显著影响。

表 2　断奶母猪葡萄糖优饲对其仔猪性能的影响

项目	对照组	葡萄糖组	SEM	P 值
窝数	49	56		
初生重（kg）	1.58	1.54	0.04	0.34
出生窝重（kg）	19.26	18.54	0.55	0.35
窝内体重变异系数（%）	18.00	18.05	0.75	0.97

3　讨论

断奶母猪发情率低，其影响因素较多，包括营养、膘情、胎次、环境控制、霉菌毒素、光照、诱情、生殖系统感染、断奶日龄、卵巢囊肿以及一些疾病因素[6]。从营养角度对母猪情期进行调控主要有两种代谢途径：①营养改变激素分泌（如瘦素、胰岛素、胰岛素样生长因子），调控情期启动；②营养改变血液代谢底物浓度，参与情期启动调控[7]。泌乳母猪采食量低，泌乳量大，机体处于分解代谢，血液中胰岛素样生长因子-1（IGF-1）降低，情期启动和卵泡发育受阻[8]。而葡萄糖优饲可以影响血糖水平来促进胰岛素分泌，而胰岛素在体内通过促进促性腺激素分泌直接影响断奶母猪发情，还可以通过影响其他激素分泌间接影响发情[1]。因此，本试验通过对断奶母猪进行葡萄糖优饲，结果发现葡萄糖优饲可以显著提高其 7 d 断配率 6.8％；35 d 受胎率差异虽不显著，但在数值上也提高了 2.3％。

葡萄糖作为最重要的单糖，其代谢对卵母细胞成熟和发育起着重要作用。有研究表明，在后备母猪（70～80 kg）基础日粮中添加 200 g/d 葡萄糖，促进其血清中胰岛素、孕激素和雌激素的含量，且上调与子宫繁殖相关基因的表达[9]。Wientjes 等[10]报道，在母猪断奶-发情期间，饲喂 150 g 葡萄糖和 150 g 乳糖，可以提高胰岛素和 IGF-1 水平，促进促黄体生成素的分泌和卵泡发育。Van den Brand 等[11]则发现，断奶后 150 g 葡萄糖优饲，可以提高仔猪均匀度。本试验也通过持续追踪参与试验母猪产房数据，探究 300 g/d 葡萄糖优饲在提高总仔数、活仔数以及均匀度方面是否存在优势，结果发现葡萄糖优饲有提高总产仔数、产活仔数的趋势，这可能与葡萄糖促进卵母细胞成熟和发育有关，但未对初生仔猪的均匀度产生显著性影响。Plush 等[12]在不同季节对断奶母猪的日粮中添加 5％葡萄糖，也得出相同的结论，发现葡萄糖优饲可显著提高母猪分娩总产仔数（1.0 头）和产活仔数（1.4 头）。

4 结论

综上所述，在本试验条件下，断奶后饲喂 300 g/d 葡萄糖可以显著提高母猪 7 d 断配率，且有提高总产仔数和产活仔数的趋势。

参考文献

[1] 于森瑛，周虚. 胰岛素在能量影响猪卵泡发育中的作用［J］. 黑龙江动物繁殖，2005，13（2）：4.

[2] Bao Y，Shuang L，Jeong-Woo K，et al. The role of glucose metabolism on porcine oocyte cytoplasmic maturation and its possible mechanisms［J］. PloS One，2016，11（12）：e0168329.

[3] Castillo-Martín M，Yeste M，R Morató，et al. Cryotolerance of in vitro-produced porcine blastocysts is improved when using glucose instead of pyruvate and lactate during the first 2 days of embryo culture［J］. Reproduction Fertility and Development，2013，25（5）：737-745.

[4] 白红杰，范磊，王丽英，等. 3 种葡萄糖投饲量对断奶母猪发情率及断奶发情间隔的影响［J］. 东北农业科学，2016，41（6）：93-96.

[5] 廖昌韬，王燕桃，罗旭芳，等. 葡萄糖对诱导断奶母猪发情配种的效果评估［C］. 中国畜牧兽医学会动物微生态学分会第十一次全国学术研讨会暨第五届会员代表大会论文集. 2014：148.

[6] 迟兰，薛忠，朱广琴. 影响断奶母猪发情率的原因及应对措施［J］. 中国猪业，2021，16（3）：40-43.

［7］ 吴德，卓勇，吕刚，等. 母猪情期启动营养调控分子机制的探讨 ［J］. 动物营养学报，2014，26
（10）：3020-3032.

［8］ Xu J，Kirigiti M A，Grove K L，et al. Regulation of food intake and gonadotropin-releasing hormone/
luteinizing hormone during lactation：role of insulin and leptin ［J］. Endocrinology，2009，150（9）：
4231-4240.

［9］ 王璟，张家庆，白献晓，等. 葡萄糖对后备母猪繁殖相关基因表达的影响 ［J］. 河南农业科学，
2017，46（1）：140-143.

［10］ Wientjes J G M，Soede NM，Van der Peet-Schwering C M C，et al. Piglet uniformity and mortality in
large organic litters：Effects of parity and pre-mating diet composition ［J］. Livestock Science，2012，
144（3）：218-229.

［11］ Van den Brand H，Soede N M，Kemp B. Supplementation of dextrose to the diet during the weaning to
estrus interval affects subsequent variation in within-litter piglet birth weight ［J］. Animal Reproduction
Science，2006，91（3-4）：353-358.

［12］ Plush K，Glencorse D，Alexopoulos J，et al. Effect of dextrose supplementation in the pre-ovulatory
sow diet to reduce seasonal influences on litter birth weight variation ［J］. Animals，2019，9
（12）：1009.

教槽料对哺乳-保育仔猪生长性能的影响

尚庆辉　张博儒

摘要：本研究包括 2 个试验，分别研究教槽料对仔猪哺乳期（试验 1）和保育期（试验 2）生长性能的影响。试验 1：选取 5 个批次共 218 头母猪（长白×大白；胎次：1.3±0.2），每批按胎次、窝仔数和分娩日期分为 2 个处理组。对照组：不饲喂教槽料；教槽料组：在哺乳 14～21 d，每窝饲喂颗粒教槽料。仔猪在 21 d 断奶。试验 2：选取 164 头同一断奶批次的母猪（长白×大白；胎次：2.92±0.94），按胎次、窝仔数和分娩日期分为 2 个处理组（同试验 1）。断奶后选取 1 560 头体重为（6.29±1.20）kg 仔猪（对照组和教槽料组各 780 头），按体重分为 2 个处理组（同哺乳期），每个处理组 26 个重复，每个重复 30 头。结果表明：①教槽料组每天窝采食量为 69 g/d；②教槽料对母猪体况和 7 d 断配率无显著影响；③教槽料对仔猪断奶重及哺乳期日增重无显著影响，对小体重仔猪（14 d 体重<3.5 kg）日增重及断奶时弱仔率（断奶体重<4 kg 仔猪比例）也无显著影响；④教槽料对仔猪保育期生长性能、腹泻率和死淘率无显著影响。综上所述，在本试验条件下，教槽料不影响仔猪哺乳阶段和保育阶段的生长性能，不建议产房阶段饲喂教槽料。

关键词：教槽料；仔猪；生长性能；死淘率

仔猪在断奶阶段常遭受来自营养、环境及心理等多方面的应激（如母仔分离、液体母乳到固体饲料的改变、分娩舍到保育舍的改变等），易发生断奶后采食量下降及腹泻等，严重影响仔猪生长性能[1]。在实际生产中，在哺乳阶段饲喂教槽料已经成为一项常规操作，其主要目的是使仔猪提前熟悉饲料，适应母乳到饲料的转变，同时补充额外的营养，从而提高仔猪断奶重及断奶后采食量，最终缓解断奶应激[2]。然而，目前关于教槽料饲喂效果的研究结果并不一致。Lee 等[3]报道，饲喂教槽料可显著提高 24 d 断奶仔猪的哺乳期日增重及断奶重。而又有很多研究发现，饲喂教槽料对仔猪断奶前后的生长性能并无显著影响[4,5]。教槽料饲喂效果不一致的原因受断奶日龄、窝仔数及教槽料类型等多种因素的影响。Shea 等[6]报道，对于 21 d 断奶仔猪，饲喂教槽料组仔猪中仅有 4% 是真正采食教槽料的。在我国，猪场普遍实行 21 d 断奶，窝仔数大多为 10～12 头。因此，本试验的目的是研究在 21 d 断奶条件下，教槽料对哺乳-保育仔猪生长性能的影响。

1　试验材料与方法

1.1　试验设计及日粮

试验一：在湖北罗汉和河北张家口选取 5 批次健康长白×大白母猪（平均胎次 1.3），每个批次母猪按胎次和分娩日期随机分为 2 个处理组，每个处理组 20～30 个重复，每个重复 1 头。试验自仔猪哺乳 14 d 开始至 21 d 断奶结束。对照组不饲喂教槽料；教槽料组在仔猪哺乳 14～21 d，每天对每窝仔猪饲喂 2 次颗粒教槽料；饲喂标准为第 14 天、65 g/窝，第 15 天、80 g/窝，第 16 天、100 g/窝，第 17 天、120 g/窝，第 18 天、140 g/窝，第 19 天、160 g/窝，第 20 天、185 g/窝，第 21 天、210 g/窝。

试验二：选取同一断奶批次的 3 个产房，每个产房母猪按照胎次和分娩日期随机分为 2 个处理组（处理组同试验一，即分为对照组和教槽料组）。每个产房中每个处理组约 27 个重复，每个重复 1 头仔猪。两个处理组的仔猪通过不同颜色耳号牌进行区分。仔猪断奶后转至保育场，从中选取 1 560 头初始体重为（6.29±1.20）kg 的杜×长×大断奶仔猪（对照组和教槽料组各 780 头），按体重随机分为 2 个处理组：对照组和教槽料组（仔猪在哺乳阶段和保育阶段保持同一处理组），每个处理组 26 个重复，每个重复 30 头仔猪。试验分为两个阶段：前期（1～13 d）和后期（14～33 d）。两个阶段饲喂相同的日粮。

1.2　饲养管理

试验一：试验在湖北罗汉和河北张家口的猪场完成。实验母猪于产房单元进行，母猪分娩前 2 d 上产床，仔猪在 21 d 断奶。产房温度维持在 24℃以上，全程自由采食和饮水，常规保健。

试验二：哺乳阶段的试验在德州禹城的猪场进行。饲养管理同试验一。

保育阶段的试验在德州临邑的猪场完成。保育舍为封闭、半漏缝水泥地板式猪舍，配有地暖，地面温度在 28～32℃。仔猪饲养于 8.25 m×3.97 m 栏内，每栏配有一个 6 料位不锈钢料槽（1.8 m×0.9 m）以及一个圆形补料槽（直径 0.6 m）。试验期间，仔猪自由采食和饮水，按照正常程序进行免疫，每天观察仔猪的健康状况。

1.3　母猪体况测定

试验一：采用背膘仪和体况卡尺分别测定母猪哺乳 14 d 和 21 d 的背膘厚和体况评分，计算母猪的背膘损失和体况评分损失。

1.4　母猪 7 d 断配率测定

试验一：记录母猪断奶后 7 d 内的配种情况，计算 7 d 断配率。

1.5　仔猪生长性能测定

试验一：在哺乳 14 d 和 21 d，对每窝仔猪逐头称重，计算窝增重及仔猪日增重。每天记录每窝教槽料的添加量及剩余量，计算每窝仔猪日采食量。

试验二：于保育 1 d、13 d 和 33 d，对每栏仔猪逐头称重，计算仔猪平均日增重、平均日采食量和料重比。

1.6　死淘率测定

试验一：准确记录试验期间每窝仔猪的死淘情况，计算死淘率。

试验二：在保育阶段，准确记录每栏仔猪的死淘情况，计算死淘率。

1.7　断奶仔猪腹泻率测定

在保育阶段，每天早晨准确记录每栏中腹泻仔猪的头数，按如下公式计算腹泻率：

腹泻率＝腹泻头数×腹泻天数／（仔猪头数×试验天数）×100％

1.8　统计分析

采用 Excel 对原始数据进行整理，利用 SAS 9.2 统计软件对仔猪生长性能数据进行独立样本 T 检验分析。数据以平均数±标准误表示，$P<0.05$ 表示差异显著。

2　试验结果

2.1　教槽料窝均采食量（试验一）

哺乳阶段教槽料窝均采食量见图 1。由图 1 可知，窝均采食量随仔猪日龄的增加而提高；7 d 饲喂期内的窝均采食量为（69±4）g/d。

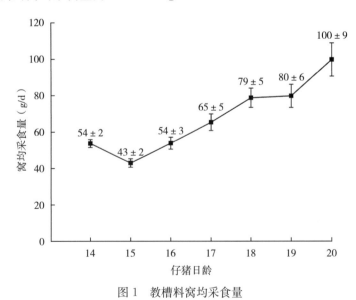

图 1　教槽料窝均采食量

2.2 母猪体况和 7 d 断配率（试验一）

　　教槽料对母猪体况和 7 d 断配率的影响见表 1。由表 1 可知，教槽料对母猪体况损失和 7 d 断配率均无显著影响。

表 1　教槽料对母猪体况和 7 d 断配率的影响

项目	对照组	教槽料组	P 值
背膘损失（mm）	0.68±0.20	0.81±0.22	0.67
体况损失	1.20±0.12	0.92±0.15	0.14
7 d 断配率（%）	72.30±6.36	73.50±5.84	0.86

2.3 哺乳仔猪生长性能（试验一）

　　教槽料对哺乳仔猪生长性能的影响见表 2。由表 2 可知，教槽料对仔猪断奶重（21 d 体重）和日增重均无显著影响。此外，两个处理组间小体重仔猪日增重及 21 d 弱仔比例也无显著差异。

表 2　教槽料对哺乳仔猪生长性能的影响

项目	对照组	教槽料组	P 值
体重（kg）			
哺乳 14 d	3.87±0.28	3.89±0.29	0.66
哺乳 21 d	5.64±0.43	5.72±0.40	0.21
14～21 d 平均日增重（kg）			
所有仔猪	0.26±0.01	0.27±0.01	0.33
小体重仔猪（体重<3.5 kg）	0.24±0.13	0.25±0.09	0.86
21 d 弱仔比例（体重<4.0 kg）（%）	10.4	10.0	

2.4 保育猪生长性能（试验二）

　　表 3 为教槽料对保育猪生长性能的影响。由表 3 可知，在 1～13 d、14～33 d 及 1～33 d，教槽料对保育猪平均日增重、平均日采食量和料重比均无显著影响。此外，两个处理组间死淘率和腹泻率也无显著差异。

表 3　教槽料对断奶仔猪生长性能的影响

项目	对照组	教槽料组	P 值
前期（1～13 d）			
平均日增重（g）	126±3.50	129±4.20	0.67
平均日采食量（g）	226±6.07	230±6.23	0.62
料重比	1.81±0.05	1.81±0.05	0.93
腹泻率（%）	10.63	10.65	

（续）

项目	对照组	教槽料组	P 值
死淘率（%）	0.77	0.77	
后期（14~33 d）			
平均日增重（g）	265±9.35	259±9.60	0.71
平均日采食量（g）	386±12.32	379±13.25	0.73
料重比	1.46±0.02	1.47±0.02	0.91
腹泻率（%）	17.76	17.60	
死淘率（%）	5.82	5.69	
全期（1~33 d）			
平均日增重（g）	209±6.44	207±6.96	0.82
平均日采食量（g）	322±9.50	320±10.21	0.89
料重比	1.54±0.01	1.55±0.01	0.74
腹泻率（%）	14.89	14.81	
死淘率（%）	6.54	6.41	

3　讨论

仔猪断奶后营养获取由易消化的液体母乳转换到难消化的固体饲料，加之环境等方面的应激，极易发生采食量下降和腹泻，严重影响其生长性能[7]。一直以来，饲喂教槽料被作为一种可通过提高仔猪断奶重及断奶后采食量来缓解其断奶应激的手段[8]。然而，本试验发现，断奶前一周饲喂教槽料不仅对 21 d 断奶仔猪的哺乳期日增重和断奶重无显著影响，且对断奶后的生长性能也无显著影响。许多研究也报道，饲喂教槽料对仔猪断奶前后生长性能无显著影响[4,9,10]。然而，也有部分研究发现饲喂教槽料可显著提高仔猪断奶后的采食量和日增重[11,12]。结果不一致的原因有以下几点：①仔猪教槽料的采食量变异大。多数报道教槽料具有显著效果的文献是将教槽料组仔猪分为采食和未采食两部分进行分析，发现真正采食教槽料的仔猪生长性能显著提高。而本试验是以教槽料组所有仔猪为研究对象，这可能是未观察到教槽料效果的原因之一。②断奶日龄的差异。Shea 等[6]报道，对于 21 d 断奶仔猪，采食教槽料的仔猪所占比例仅为 4%，当断奶日龄延长至 28 d 时，该比例上升至 34%。Bandara 等[12]也发现对于 28 d 断奶仔猪，教槽料组有 37% 的仔猪真正采食教槽料。本试验中仔猪是在 21 d 断奶，因此采食教槽料仔猪所占比例可能较低，因而未观察到教槽料的有益效果。③教槽料的形式。Christensen 等[13]研究发现液体教槽料可显著提高仔猪的采食量以及断奶重。而本试验中使用的是颗粒教槽料，这也可能是研究结果不一致的原因之一。

4　结论

综上所述，在本试验条件下，教槽料对仔猪哺乳及保育阶段生长性能均无显著影响，且对弱仔（14 d 体重小于 3.5 kg）断奶重及日增重也无影响，因此不建议在产房添加教槽料。

参考文献

［1］ Campbell J M，Crenshaw J D，Polo J. The biological stress of early weaned piglets ［J］. Journal of Animal Science Biotechnology，2013，4（1）：19.

［2］ Lee S A，Febery E，Wilcock P，et al. Application of creep feed and phytase super-dosing as tools to support digestive adaption and feed efficiency in piglets at weaning ［J］. Animals，2021，11（7）：2080.

［3］ Lee S I，Kim I H. Creep feeding improves growth performance of suckling piglets ［J］. Revista Brasileira de Zootecnia，2018，47：e20170081.

［4］ Middelkoop A，Choudhury R，Gerrits W J J，et al. Effects of creep feed provision on behavior and performance of piglets around weaning ［J］. Frontiers in Veterinary Science，2020，7：520035.

［5］ Park B，Ha D，Park M，et al. Effects of milk replacer and starter diet provided as creep feed for suckling pigs on pre- and post-weaning growth ［J］. Animal Science Journal，2014，85（9）：872-878.

［6］ Shea J，Beaulieu D. Creep feeding in the farrowing room：do the outcomes depend on weaning age ［C］. In Proceedings of the 33rd Annual Centralia Swine Research Update，Kirkton，ON，Canada，29 January 2014；Centralia Swine Research Update：Exeter，ON，Canada，2014.

［7］ Heo J M，Opapeju F O，Pluske J R，et al. Gastrointestinal health and function in weaned pigs：a review of feeding strategies to control post-weaning diarrhoea without using in-feed antimicrobial compounds ［J］. Journal of Animal Physiology and Animal Nutrition，2013，97（2）：207-237.

［8］ Wensley M R，Tokach M D，Wooorth J C，et al. Maintaining continuity of nutrient intake after weaning. I. Review of pre-weaning strategies ［J］. Translational Animal Science，2021，5（1）：b21.

［9］ Martins S M M K，Ferrin M O，Poor A P，et al. Gruel creep feed provided from 3 days of age did not affect the market weight and the sow's catabolic state ［J］. Livestock Science，2020，231：103883.

［10］ Sulabo R C，Jacela J Y，Tokach M D，et al. Effects of lactation feed intake and creep feeding on sow and piglet performance ［J］. Journal of Animal Science，2010，88（9）：3145-3153.

［11］ Bruininx E M，Binnendijk G P，van der Peet-Schwering C M，et al. Effect of creep feed consumption on individual feed intake characteristics and performance of group-housed weanling pigs ［J］. Journal of Animal Science，2002，80（6）：1413-1418.

［12］ Bandara N，Shea J，Grillis D，et al. Creep feed provision in the farrowing room provides benefits to piglets showing evidence of intake ［J］. Annual research report prairie swine centre，2011：38.

［13］ Christensen B，Huber L. The effect of creep feed composition and form on pre- and post-weaning growth performance of pigs and the utilization of low-complexity nursery diets ［J］. Translational Animal Science，2021，5（4）：1-14.

妊娠后期母猪不同饲喂水平对其繁殖性能及仔猪生长性能的影响

辛海瑞

摘要： 本试验旨在探究妊娠后期不同饲喂水平对母猪繁殖性能及仔猪生长性能的影响。本研究采用单因素试验设计，选取 PIC 父母代后备母猪 476 头；二元经产母猪 612 头（胎次 1.36）；海波尔母猪 579 头（胎次 2.73），根据母猪体况、胎次及分娩日期随机分为两个处理组：对照组和妊娠后期补料组。对照组：妊娠 85 d 后不增加饲喂量，即后备母猪每天饲喂 2.0 kg，经产母猪每天饲喂 2.2 kg；妊娠后期补料组：妊娠 85 d 后开始提高饲喂量，即后备母猪每天饲喂 2.6～2.8 kg，经产母猪每天饲喂 2.8 kg。试验母猪进入产房单元后，持续记录母猪体况信息、分娩信息、采食量、断奶后发情率及仔猪生长性能数据。结果表明：①妊娠后期补料对二元经产母猪和海波尔母猪无显著性影响，但显著降低 PIC 后备母猪的仔猪初生重（$P=0.05$）；②两处理间弱仔比例无显著性差异，但妊娠后期补料有增加海波尔母猪死胎数的趋势（$P=0.08$）；③与妊娠后期补料组相比，对照组 PIC 后备母猪在泌乳 16～22 d 平均采食量显著升高（$P=0.02$）；④妊娠后期补料对仔猪断奶重和窝增重均未产生显著性影响。综上所述，在本试验条件下，经产母猪妊娠后期增加饲喂水平，造成一定的饲料浪费，并未产生预期收益。

关键词： 饲喂水平；妊娠后期；繁殖性能；仔猪生长性能

妊娠母猪体况管理是猪场管理的核心指标之一。体况过瘦会导致母猪乏情，返情率上升，繁殖性能下降，使用年限降低。体况过肥不仅浪费饲料，而且会降低泌乳期采食量，影响死胎率和断奶窝重[1]。妊娠母猪体况管理主要基于饲喂程序的调整，当前生产中存在妊娠后期加料攻胎的饲喂模式，即在妊娠后期通过增加饲喂量，以期提高仔猪的初生重。而又有文献表明，妊娠后期增加饲喂水平导致母猪泌乳期采食量下降，对仔猪的断奶重也不会有积极影响[2]。

随着母猪遗传选育的推进，母猪窝产仔猪数增加，胎儿个体生长和发育的空间减少，进而导致仔猪的初生重降低。Cromwell 等[3]旨在通过提高母猪妊娠后期饲喂量来提高后代仔猪的初生重，结果发现，妊娠后期饲喂量提高 1.4 kg，其后代仔猪初生重提高 40 g。Shelton 等[4]和 Soto 等[5]发现，提高初产母猪妊娠后期饲喂量可以提高仔猪初生重，但在经产母猪却未观察到相同效果。越来越多的文献发现，提高妊娠后期饲喂量未能改善仔猪初生重，但却导致母猪体重和背膘厚显著增加[6-10]。

综述文献发现，妊娠后期提高母猪饲喂水平，对初生重和母猪泌乳期采食量是否有积极作用，仍需进一步验证。因此，本试验拟探究经产和后备母猪妊娠后期不同饲喂水平对其体况、繁殖性能及仔猪生长性能的影响。

1 试验材料与方法

1.1 试验设计

本研究采用单因素试验设计，选取 PIC 父母代后备母猪 476 头（85 d 体况卡尺评分15.7，背膘厚 14.1 mm）；二元父母代经产妊娠母猪 612 头（体况卡尺评分 14.9，背膘厚14.0 mm，平均胎次 1.36）；海波尔母猪 579 头（体况卡尺评分 14.8，背膘厚 14.6 mm，平均胎次 2.73）。分为两个处理组：对照组和妊娠后期补料组。对照组：妊娠 85 d 后不增加饲喂量；妊娠后期补料组：妊娠 85 d 后开始提高饲喂量。具体饲喂量见表 1。

表 1　试验处理分组

母猪信息	试验处理	饲喂量（kg/d）
后备母猪	对照组	2.0
	妊娠后期补料组	2.6～2.8
经产母猪	对照组	2.2
	妊娠后期补料组	2.8

1.2 饲养管理

后备母猪在夏津试验基地进行饲养，时间为 2020 年 8 月 14 日至 11 月 2 日；经产母猪在湖北猪场进行饲养，时间为 2021 年 3 月 9 日至 7 月 10 日；海波尔母猪在四川猪场进行饲养，时间为 2021 年 7 月 23 日至 12 月 20 日。

日常管理按猪场常规管理规程进行。所有母猪妊娠 85 d 之前饲喂相同的妊娠日粮及规定的饲喂量。85～112 d 按试验分组，执行不同饲喂标准。分娩前 3 d 进入分娩单元，饲喂相同的泌乳料，仔猪平均哺乳日龄为 21 d，届时仔猪断奶，母猪下产床回到配怀舍。泌乳全程母猪自由采食和饮水，常规保健。记录试验期间母猪情况，包括免疫情况、发病原因及治疗信息。

1.3 母猪体况测定

准确测定每头母猪在妊娠 85 d 上产床时和断奶时的背膘厚和体况评分。利用背膘仪对每头母猪的背膘厚进行测定；利用体况卡尺测定每头母猪的体况评分。计算母猪在泌乳阶段的背膘损失和体况评分损失。

1.4 母猪采食量测定

在试验期间，分别选取一个单元的 PIC 后备母猪和二元经产母猪进行人工饲喂，记录每天加料量和废料量，分娩后每 5 d 结料一次，计算母猪分娩后 1～5 d、6～10 d、11～15 d、16～20 d 以及 1～20 d 内平均采食量。

1.5　母猪分娩数据记录

准确记录母猪分娩数据，包括分娩时长、总仔数、健仔数、弱仔数、死胎数、畸形数、木乃伊胎数。

1.6　仔猪生长性能测定

母猪分娩后 24～48 h 内调栏，所有处理的弱仔猪由奶妈猪饲养，不做哺乳期生长性能的统计。仔猪出生后逐头称重，计算仔猪初生重和初生窝重，21 d 后称量断奶重。

1.7　断奶后 7 d 发情率

追踪试验母猪断奶到发情的间隔天数，计算 7 d 断配率。

1.8　统计分析

利用 Excel 表格对原始数据进行记录，再利用 SAS 统计软件 Glimmix 程序对数据进行统计分析。饲喂水平为固定效应，试验开始时体况和背膘为协变量，其中对断奶窝重和断奶重分析时再增加调栏后仔猪数为协变量。对初生仔猪弱仔率分析时增加母猪分娩总仔数为协变量。数据以平均数±标准误（SEM）表示，$P<0.05$ 表示差异显著，$0.05 \leqslant P<0.10$ 表示趋于差异显著。

2　试验结果

2.1　妊娠后期不同饲喂量对仔猪初生重的影响

母猪妊娠后期补料不能对仔猪初生重产生积极影响（图 1）。妊娠后期补料对二元经产母猪和海波尔母猪无显著性影响，但显著降低后备母猪的仔猪初生重（$P=0.05$）。

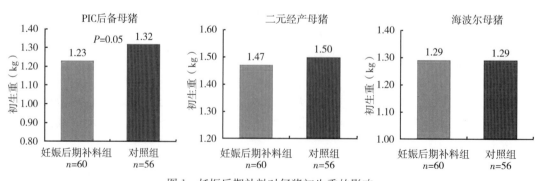

图 1　妊娠后期补料对仔猪初生重的影响

2.2　妊娠后期不同饲喂量对仔猪出生弱仔率的影响

　　表2显示的是妊娠后期补料对二元经产母猪的初生弱仔率的影响，而表3显示的是妊娠后期补料对海波尔母猪初生弱数和死胎数的影响。结果发现，两组间弱仔比例无显著性差异，但妊娠后期补料有增加海波尔母猪死胎数的趋势（$P=0.08$）。

表2　妊娠后期不同饲喂量对二元经产母猪的初生弱仔率的影响

试验处理	不同体重仔猪占比			
	<600 g	$600\sim800$ g	$800\sim1\,000$g	$<1\,000$ g
对照组（$n=60$，691）	0.81	3.54	4.24	8.60
妊娠后期补料组（$n=60$，697）	1.53	2.61	5.66	9.80
SEM	0.32	0.48	0.65	1.01
P 值	0.35	0.30	0.27	0.64

表3　妊娠后期不同饲喂量对海波尔母猪初生弱仔数和死胎数的影响

项目	对照组	妊娠后期补料组	SEM	P 值
母猪数（头）	288	291		
弱仔数（头）	0.77	0.92	0.05	0.19
死胎数（头）	0.70	0.86	0.05	0.08
弱仔率（%）	5.69	6.76	0.38	0.18
死胎率（%）	4.90	5.88	0.31	0.10

2.3　妊娠后期不同饲喂量对母猪采食量的影响

　　图2为妊娠后期补料对PIC后备母猪哺乳后期采食量的影响。由图2可知，与妊娠后期补料组相比，对照组PIC后备母猪在泌乳16～22 d的平均采食量显著升高（$P=0.02$）。但二元经产母猪两组间母猪泌乳阶段的采食量未有显著性差异（表4）。

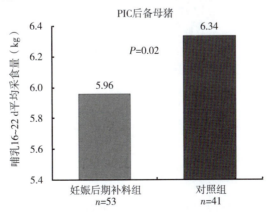

图2　妊娠后期补料对PIC后备母猪泌乳后期采食量的影响

表 4　妊娠后期不同饲喂量对二元经产母猪泌乳期采食量的影响

泌乳期	平均日采食量（kg）			
	对照组（n＝28）	妊娠后期补料组（n＝29）	SEM	P 值
1～5 d	5.21	5.04	0.12	0.42
6～10 d	6.33	6.08	0.14	0.23
11～15 d	7.04	6.74	0.16	0.18
16～20 d	7.19	7.03	0.16	0.36
1～20 d	6.44	6.22	0.12	0.20

2.4　妊娠后期不同饲喂量对仔猪断奶重的影响

如图 3 所示，妊娠后期补料对仔猪断奶重和窝增重（PIC 后备母猪和二元经产母猪）均无显著影响。

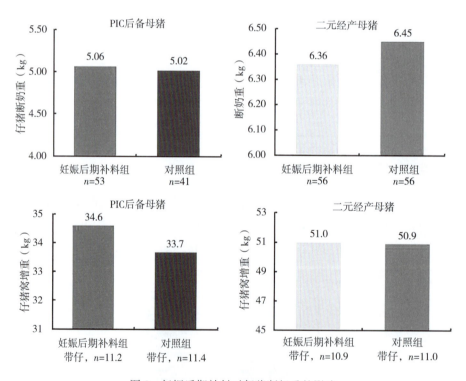

图 3　妊娠后期补料对仔猪断奶重的影响

2.5　妊娠后期不同饲喂量对母猪 7 d 断配率的影响

图 4 显示的是妊娠后期补料对母猪断奶后 7 d 断配率（PIC 后备母猪和二元经产母猪）的影响。对母猪下一胎次繁殖性能而言，两组间的 7 d 断配率未有显著性差异。

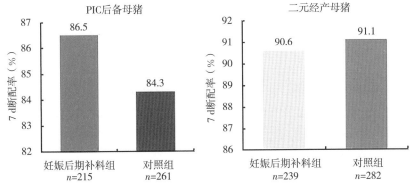

图 4　妊娠后期补料对母猪 7 d 断配率的影响

3　讨论

生产上，通过提高母猪妊娠后期饲喂量，以期获得更高的仔猪初生重。但对于这一结论，诸多研究结果并不一致，这些研究可能与氨基酸摄入量的增加相混淆，而这可能要比能量消耗的增加更重要[11]。PIC 公司汇总近十年相关文献，对母猪增重和仔猪初生增重进行加权平均，发现妊娠后期增加采食量，后备母猪增重 7.3 kg，仔猪初生重增加仅 12.6 g；经产母猪增重 8.4 kg，仔猪初生重反而降低 1.3 g。

本试验以当前猪产业妊娠料配方结构和营养浓度为前提，就妊娠后期是否需要补料（攻胎）开展试验，结果发现，妊娠后期补料组与对照组相比未体现出任何优势，却增加了 24 kg 饲料成本，这与 Shelton 等[4]的结果一致。Miller 等[6]研究同样发现，妊娠后期提高采食量对仔猪初生重、生长速度和死亡率均无显著影响，主要好处是降低母猪哺乳期背膘损失。Mallmann 等[12]研究发现，后备母猪妊娠 90 d 至分娩期间分别按照 1.8 kg/d、2.3 kg/d、2.8 kg/d 和 3.3 kg/d 进行饲喂，其增重和死胎率会随着饲喂量的增加而提高。Gonçalves 等[9]发现，与低能量日粮相比，摄入高能量日粮母猪的死胎率增加，但对初生窝重和初生个体重无显著影响。

Goodband 等[11]发现，在妊娠后期提供额外的能量会增加母猪的增重，并导致哺乳期采食量降低。另有文献报道，妊娠后期提高母猪的采食量会对泌乳期采食量产生不利影响，从而导致母猪泌乳期体重损失和背膘损失增加[2]。在本试验中，妊娠后期补料未对经产母猪泌乳期采食量有显著性影响，但却显著降低后备母猪泌乳后期（16～20 d）的采食量。仔猪生长性能，特别是窝增重和断奶重，与母猪采食量密切相关。但 Choi 等[13]发现，在母猪妊娠最后两周补饲未对断奶窝重产生影响。Liu 等[14]同样发现，妊娠后期较大的采食量（超过维持所需的能量）对仔猪初生重和断奶重没有影响。在本试验中，妊娠后期补料对仔猪断奶重和窝增重（PIC 后备母猪和二元经产母猪）均未产生显著性影响。

有研究发现，妊娠后期提高后备母猪采食量，仅增加其下一胎次的受孕率，并没有发现其他好处[4]。本试验中，妊娠后期补料组后备母猪的下一胎次 7 d 断配率，在数值上提高了

2.2%，但差异不显著。

4　结论

综上所述，在本试验条件下，按照当前妊娠料营养水平，母猪妊娠后期补料不会对仔猪的初生重产生积极影响；正常饲喂不补料也不会导致初生弱仔水平升高，且有利于维持母猪泌乳阶段采食量，减少母猪泌乳阶段的体况损失。

参考文献

[1] Zhou Y，Xu T，Cai A，et al. Excessive backfat of sows at 109 d of gestation induceslipotoxic placental environment and is associated with declining reproductive performance [J]. Journal of Animal Science，2018，96（1）：250-257.

[2] Mallmann A L，Betiolo F B，Camilloti E，et al. Two different feeding levels during late gestation in gilts and sows under commercial conditions：impact on piglet birth weight and female reproductive performance [J]. Journal of Animal Science，2018，96（10）：4209-4219.

[3] Cromwell G L，Hall D D，Clawson A J，et al. Effects of additional feed during late gestation on reproductive performance of sows：a cooperative study [J]. Journal of Animal Science，1989，67（1）：3-14.

[4] Shelton N W，Neill C R，Derouchey J M，et al. Effects of increasing feeding level during late gestation on sow and litter performance [J]. Kansas State University Swine Day Report of Progress，2009，10：38-50.

[5] Soto J，Greiner L，Connor J，et al. Effects increasing feeding levels in sows during late gestation on piglet birth weights [J]. Journal of Animal Science，2011，89（Suppl 2）：239.

[6] MillerH M，Foxcroft G R，Aherne F X. Increasing food intake in late gestation improved sow condition throughout lactation but did not affect piglet viability or growth rate [J]. Animal Science，2000，71（1）：141-148.

[7] Lawlor P G，Lynch P B，O connell M K，et al. The influence of over feeding sows during gestation on reproductive performance and pig growth to slaughter [J]. Archiv Fur Tierzucht，2007，50（Special 1）：82-91.

[8] Yang Y X，Heo S，Zheng J，et al. Effects of dietary energy and lysine intake during late gestation and lactation on blood metabolites, hormones, milk composition and reproductive performance in multiparous sows [J]. Archives of Animal Nutrition，2008，62：10-21.

[9] Gonçalves M A，Gourley K M，Dritz S S，et al. Effects of amino acids and energy intake during late gestation of high-performing gilts and sows on litter and reproductive performance under commercial conditions [J]. Journal of Animal Science，2016，94（5）：1993-2003.

[10] Rooney H B，O'driscoll K，O'doherty J V，et al. Effect of increasing dietary energy density during late gestation and lactation on sow performance, piglet vitality, and lifetime growth of offspring [J]. Journal of Animal Science，2020，98（1）：skz379.

[11] Goodband R D，Tokach M D，Goncalves M，et al. Nutritional enhancement during pregnancy and its effects on reproduction in swine [J] . Animal Frontiers，2013，3（4）：68-75.

[12] Mallmann A L，Camilotti E，Fagundes D P，et al. Impact of feed intake during late gestation on piglet birth weight and reproductive performance：a dose-response study performed in gilts [J] . Journal of Animal Science，2019，97（3）：1262-1272.

[13] Choi Y H，Hosseindoust A，Kim M J，et al. Additional feeding during late gestation improves initial litter weight of lactating sows exposed to high ambient temperature [J]. Revista Brasileira de Zootecnia，2019，48：e20180028.

[14] Liu Z H，Zhang X M，Zhou Y F，et al. Effect of increasing feed intake during late gestation on piglet performance at parturition in commercial production enterprises [J]. Animal Reproduction Science，2020，218：106477.

Chapter

3

生产技术篇

PG600 与 PMSG 对断奶母猪促发情效果的对比试验

郑智伟

摘要： 本试验旨在研究外源性激素对断奶母猪发情率的影响。试验选取 3 个批次即将断奶母猪 548 头，每个批次母猪均按胎次、背膘厚、哺乳天数随机分为 3 个处理组：PG600 组（$n=165$）、PMSG 组（$n=189$）和对照组（$n=194$）。对照组不做额外处理，其余两组断奶 24 h 后颈部肌内注射两种不同的外源性激素，使用剂量参考产品说明书。准确记录每头母猪断奶 7 d 内的发情信息和配种信息。结果表明：①断奶母猪外源性激素 PG600 可提高断奶母猪 7 d 断配率 11%（$P<0.05$），但是断奶至发情间隔无显著差异；②PG600 试验处理能够提升 1～3 胎母猪的发情率，分别提升了 12.79%、16.99% 和 1.58%。综上所述，在本试验条件下，断奶后使用外源性激素可提升低胎次断奶母猪 7 d 内发情率，尤其是 3 胎以内，可应用于实际生产以提高母猪生产效率。

关键词： 断奶母猪；激素；PG600；PMSG；断奶发情率

规模化和集约化养猪在用地、环保、养殖效率、疫病防控等方面有明显优势，其能够提高生产效率的关键技术是满负荷均衡的批次化生产。规模化猪场广泛采用 7 d 批的生产节律，每个生产节律内的配种母猪由断奶发情母猪、后备母猪及以往批次的返情、流产、长期不发情母猪构成。一般情况下断奶母猪可在断奶后 4～5 d 发情配种，但由于寄养不当、营养不良等因素使母猪不发情或发情不正常，这会延迟配种并增加非生产天数。此外，流产率增加、分娩率降低和产仔数减少均与较长的断奶至发情间隔有关[1,2]，因此提高断奶发情猪的发情率可降低非生产天数并提高生产成绩。母猪发情与卵泡发育息息相关，值得注意的是，母猪在断奶时卵巢卵泡的大小存在显著差异[3]，断奶时卵泡较小的母猪往往有较长的断奶至发情间隔[4]，母猪断奶时卵泡大小的差异在分娩 7 d 就已经存在，这些差异在整个泌乳期间保持不变[5]。为减少母猪断奶时卵巢小卵泡的发生率，可在泌乳期开始采取措施，或者断奶后采取措施，使小卵泡尽快发育为大卵泡。

母猪的卵泡发育和排卵是一个复杂的生理变化过程，受到很多的内源激素调控，如下丘脑分泌的促性腺激素释放激素（gonadotropin-releasing hormone，GnRH）、垂体分泌的促卵泡素（follicle stimulating hormone，FSH）和促黄体素（luteinizing hormone，LH）以及卵巢分泌的雌二醇（estrodio，E2）和孕酮（progesterone，P4）等。调节母猪发情和卵泡发育的外源生殖激素有孕马血清促性腺激素（pregnant mare serum gonadotropin，PMSG）和人绒毛膜促性腺激素（human chorionic gonadotropin，hCG）或猪促黄体素

（porcine luteotropic hormone，pLH）等[6,7]。

PMSG（也称为 eCG）来源于马属动物胎盘的杯状结构，主要存在与血清中，属于糖蛋白类激素，由 α 和 β 两个亚基组成，α 亚基与 LH 和 FSH 类似，而 β 亚基兼有 LH 和 FSH 两种作用，因此 PMSG 具有 LH 和 FSH 的活性。对雌性动物而言，PMSG 的主要生理功能是作用于卵巢，促进卵泡发育成熟并排卵及黄体生成[8]。hCG 是由胎盘合体滋养细胞分泌的一种具有促进性腺发育的糖蛋白激素，结构包括 α、β 两个亚基。因 α 亚基与 LHFSH 结构类似，hCG 主要作用是促进卵泡排卵生产黄体，但促卵泡成熟作用甚微[9]。

PG600 是一种 400 IU PMSG 和 200 IU hCG 混合的激素制剂，兼具促卵泡发育与促进排卵的双重功效。据报道，两种激素协同作用可诱导卵巢处于静止状态的乏情期后备母猪、正常断奶母猪恢复正常发情周期，能够正常配种产仔，而且能够促进排卵，具有一定的超数排卵效果，能显著地提高母猪的窝产仔数[10]。

为了比较 PMSG 与 PG600 两种激素产品对不同胎次母猪使用的效果差异，本试验以发情率低的 1～3 胎二元断奶母猪为试验对象，研究两种激素产品在规模化猪场的应用效果，旨在提高规模化猪场经产母猪繁殖效率，并针对不同胎次乏情断奶母猪提供使用 PG600 和 PMSG 两种激素产品推荐的数据依据。

1　试验材料与方法

1.1　试验设计

本试验于 2021 年 9—10 月在五河猪场开展。该试验在 1 条 3 000 头标准的父母代场线开展，选取 3 批次 1～3 胎二元断奶母猪共计 548 头，在断奶前一天记录分娩胎次、哺乳天数、产仔数、断奶背膘厚后将其分为对照组、PG600 处理组、PMSG 处理组，组间母猪的上述指标无差异，次日将断奶母猪放到相同配种单元，消除栋舍差异。激素处理及饲喂方法如表 1 所示。

表 1　试验分组及饲喂方法

组别	处理	饲喂方法
对照组	断奶当天记为 0 d，断奶第 2 天开始使用公猪诱情，上、下午各一次。公猪赶至母猪前鼻对鼻接触，员工采用查情五步法刺激母猪不少于 20 s	断奶当天不饲喂，保证母猪的充足饮水。断奶第 1 天开始饲喂母猪妊娠料，饲喂前保证料槽水深不超过 2 cm，8:00 饲喂 2 kg 妊娠料，17:00 饲喂 2 kg 妊娠料，如果母猪料槽舔食干净则再饲喂 0.25 kg 饲料，少加勤添直至仅剩余少许剩料。饲喂后给予母猪至少 40 min 的进食时间，之后将剩料均摊给进食量大的母猪。料槽放水，水深不超过 4 cm。断奶 4 d 后母猪发情导致采食量下降，酌情降低饲喂量以避免浪费饲料
PG600 处理组	查情方式同对照组，母猪断奶后 24 h 颈部肌内注射 1 头份 PG600	
PMSG 处理组	查情方式同对照组，母猪断奶后 24 h 颈部肌内注射 1 头份 PMSG	

注：剔除瘸腿、哺乳天数少于 14 d、背膘厚小于 10 mm、带仔数小于 6 头、发情后排脓的母猪。

试验共进行 3 批次，每批次参与试验的母猪均匀分为 3 组，剔除了胎次、产仔数、断奶背膘厚、哺乳天数对激素处理措施的影响。PG600 组处理 1～3 胎母猪数分别为 45 头、86 头、34 头；PMSG 处理组 1～3 胎母猪数分别为 42 头、93 头、54 头；自然发情组 1～3 胎母猪数分别为 41 头、101 头、52 头。其他分组细节见表 2。

表 2　各批次母猪分组细节

批次	组别	胎次	活仔数（头）	背膘厚（mm）	哺乳天数	头数
	PG600	1.95±0.7	12.45±3.72	12.93±2.07	23.71±3.53	56
1	自然发情	2.09±0.7	12.11±3.15	13.66±2.0	23.26±2.47	65
	PMSG	2.05±0.71	13.39±2.65	12.69±2.12	24.08±2.74	62
	PG600	1.93±0.7	13.07±3.11	12.44±1.81	23.91±3.18	54
2	自然发情	2.05±0.69	11.77±2.8	13.75±2.06	22.88±1.94	65
	PMSG	2.08±0.72	13.41±3.08	12.59±2.09	23.89±2.86	64
	PG600	1.93±0.69	12.8±2.0	12.81±1.8	23.85±3.34	55
3	自然发情	2.03±0.69	12.36±2.79	13.36±1.92	23.08±2.35	64
	PMSG	2.06±0.72	12.68±2.97	12.84±2.24	24.11±2.68	63

1.2　仔猪日常饲养管理

（1）温度　环境控制器设定，猪舍纵向温度为 18℃，横向温度为 21℃，猪舍无额外的加热装置。

（2）光照　猪舍光照 16 h，日光灯管照明，试验期间每天 7：00 开灯，23：00 关灯，光照度为 50～100 lx。

（3）粪污清理　大棚式猪场采用水泡粪工艺，每隔 15 d 排污 1 次。

1.3　试验指标

（1）采食量　每天跟踪断奶母猪采食情况，记录剩料多于 2 kg 的母猪。

（2）发情记录　记录母猪发情日期、发情表现、断配间隔信息。

（3）配种记录　记录母猪配种时间、配种次数、配种异常情况。

1.4　统计方法

所有数据用 Excel 进行初步整理后，采用 SPSS 软件 22.0 版本分析。使用二元 Logistics 回归对数据分析，因变量为是否发情，发情编码为 1，分类变量为试验处理，连续变量为哺乳天数、产仔数、胎次。统计结果以平均值表示，$P<0.05$ 为差异性显著的判断标准。

2　试验结果

2.1　PG600 与 PMSG 对断奶母猪发情率和断配间隔的影响

3 批次试验结果如图 1 所示，每一批次试验中，PG600 处理组母猪发情率最高，比

PMSG 处理组高 2%～9%，比自然发情组高 5%～24%。除第一批次试验中自然发情组高于 PMSG 处理组 4% 外，其余批次中 PMSG 处理组高于自然发情组 11%～15%。PG600 处理组 7 d 发情率为（74.52±3.85）%，PMSG 处理组为（69.83±1.86）%，自然发情组为（63.37±9.27）%。使用二元 Logistics 回归分析激素处理对发情的影响，发现激素处理项系数的 Wald 检验差异显著（$P=0.038$），即激素处理项系数不为 0，不同激素处理的发情率有显著差异，PG600 处理组相对于 PMSG 处理组的发情优势比为 1.36，PG600 处理组相对于自然发情组的发情优势比为 1.82。PG600 处理组的断配间隔为（4.94±0.30）d，PMSG 处理组的断配间隔为（5.29±0.15）d，自然发情组的断配间隔为（4.98±0.08）d，方差检验发现不同处理间断配间隔不存在显著性差异。以上分析结果发现 PG600 与 PMSG 处理可提升断奶母猪发情率，但不会改变断配间隔时间，且 PG600 的促发情效果优于 PMSG。

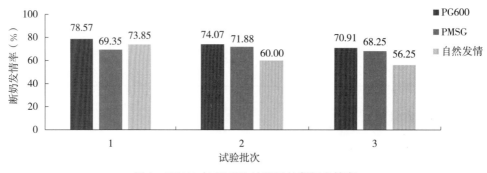

图 1　PG600 与 PMSG 处理后的断奶发情率

2.2　PG600 与 PMSG 对不同胎次断奶母猪发情率的影响

本试验统计分析了两种激素产品处理对不同胎次母猪的促发情效果，如图 2 所示，PG600 能够提升 1～3 胎母猪的发情率，分别提升了 12.79%、16.99%、1.58%，PMSG 可提升 2 胎母猪 14.61% 的发情率，但是处理 1 胎母猪和 3 胎母猪后，发情率并没有提升。此外分析发现，两种激素产品并没有提升 3 胎母猪的断奶发情率。母猪发情率也受到胎次的影

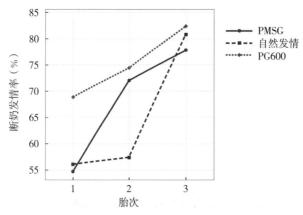

图 2　PG600 与 PMSG 对不同胎次母猪断奶发情率的影响

响，随着胎次的提高，自然发情组母猪发情率提升了 1.3%～23%。PG600 对 1～3 胎母猪的促发情效果均优于 PMSG。

2.3　PG600 与 PMSG 对断奶母猪受胎率的影响

PG600 处理组、PMSG 处理组和自然发情组的受胎率分别为 79.34%、83.97% 和 72.73%，如图 3 所示。

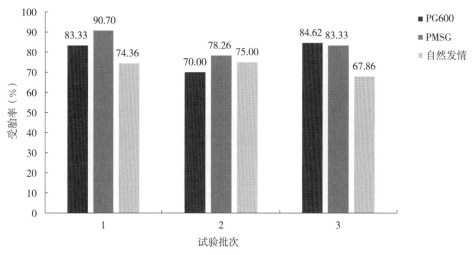

图 3　PG600 与 PMSG 处理后的受胎率

3　经济测算

根据试验结论，断奶发情率提升 12%，断奶母猪超期未发情 28 d 后淘汰；为了满足配种目标，每个批次需要额外补充 11% 后备母猪入群；由于断奶后超期未发情母猪增加额外的非生产天数，全年预计增加 20 840 d，预计需要耗费饲料 26.8 万元；由于额外补充后备母猪，需要额外投入后备母猪驯化成本共计 38.3 万元；1 个 3 000 头标准的父母代场线预计每年节省 65 万元；目前 60% 的运营猪场的断奶发情率指标有很大的提升空间，每年可节省成本 1.3 亿元。

3 000 头场线年投入疫苗成本 20 万元，因此使用激素时要慎重考虑，应以提升管理技术为主，针对低胎次发情效果差的母猪场可作为应急使用。

参考文献

[1] Lopes T P，Sanchez-Osorio J，Bolarin A，et al. Relevance of ovarian follicular development to the seasonal impairment of fertility in weaned sows [J]. Veterinary Journal，2014，199：382-386.

[2] Knox R. 124Factors influencing follicle development in gilts and sows and management strategies used to regulate growth for control of estrus and ovulation [J]. Journal of Animal Science，2018，96：343-343.

[3] Britt J H，Armstrong J D，Cox N M，et al. Control of follicular development during and after lactation in

sows ［J］. J Reprod Fertil Suppl，1985，33：37-54.

［4］Lucy M C，Liu J，Boyd C K，et al. Ovarian follicular growth in sows ［J］. Reprod Suppl，2001，58：31-45.

［5］Willis H J，Zak L J，Foxcroft G R. Duration of lactation，endocrine and metabolic state，and fertility of primiparous sows ［J］. Journal of Animal Science，2003，81：2088.

［6］Guthrie H D，Bolt D J，Cooper B S. Effects of gonadotropin treatment on ovarian follicle growth and granulosal cell aromatase activity in prepuberal gilts ［J］. Journal of Animal Science，1990，11：3719-3726.

［7］Manjarin R，Cassar G，Sprecher D J，et al. Effect of eCG or eCG Plus hCG on oestrus expression and ovulation in prepubertal gilts ［J］. Reproduction in Domestic Animals，2010，44：411-413.

［8］张春礼，刘恩柱. PMSG 的生物学特性及在生命科学研究领域中的应用 ［J］. 试验动物科学，2009，26（04）：47-50.

［9］王秀萍，焦琳，郭红燕. 人绒毛膜促性腺激素的临床应用进展 ［J］. 药学服务与研究，2010，10：5.

［10］张有强，姜增固. PG600 诱情配种在养猪生产中的应用 ［J］. 广东畜牧兽医科技，2003，28：2.

日粮中添加葡萄糖对断奶母猪发情率的影响

乜 豪

摘要：本试验旨在研究母猪断奶后添加不同剂量葡萄糖对其发情率的影响。选取 4 个批次共计 426 头母猪进行不同剂量葡萄糖饲喂，每个批次母猪均按胎次和背膘厚随机分为 3 组，即对照组（$n=75$）、150 g 葡萄糖组（$n=173$）、300 g 葡萄糖组（$n=178$），对照组不做处理，葡萄糖组在断奶 7 d 内按要求剂量饲喂葡萄糖，随母猪采食饲喂，若母猪发情，即停止饲喂葡萄糖。准确记录每头母猪发情信息和配种信息。结果表明：①断奶母猪进行葡萄糖优饲可以显著提高母猪 7d 断配率 8.1％（$P<0.01$）；②分胎次分析，对 1 胎母猪效果显著提升 11％以上（$P<0.01$）。综上所述，在本试验条件下，断奶后饲喂 300 g/d 葡萄糖可提高母猪 7 d 断配率，可应用于生产以提高母猪生产效率。

关键词：断奶母猪；葡萄糖；断奶 7 d 内发情率

生猪养殖历史悠久，猪肉更是我国人民必不可少的一种肉类[1]。当前形势下，猪肉价格相较前两年下降较多，故降低成本、提高产量成为当前规模化养猪的一个亟待解决的问题，想要最大化利用母猪，就要从降低非生产天数入手，而提高断配率成为降低成本的一项有效措施[2]。影响断奶母猪发情率的因素有很多，例如：营养与膘情、胎龄结构、环境因素、霉菌毒素、断奶日龄、光照、诱情等。陈裕明[3]研究发现，断奶后发情率与哺乳期和断奶后采食量呈正相关。中等体况的母猪断奶后的发情率最高，母猪哺乳期的采食量和哺乳期长短对母猪断奶后的体况和发情率有直接影响。郭金彪[4]指出，母猪断奶体况偏瘦或偏肥均导致断奶后发情延迟。断奶母猪群发情率低、发情间隔长会严重影响年生产胎次和年提供的断奶仔猪头数。为了提高断奶母猪繁殖性能，欧洲养猪发达国家在断奶母猪中尝试使用葡萄糖。葡萄糖通过血糖浓度影响胰岛素浓度水平，而胰岛素在动物体内具有重要生理作用，其可以通过促进促性腺激素分泌而直接影响断奶母猪发情，还可通过影响其他激素而间接影响母猪发情。白红杰[5]提出，断奶母猪最佳葡萄糖日饲喂量为 600 g。廖昌韬等[6]研究日粮中每天添加 200 g 葡萄糖对断奶母猪发情配种的影响，发现葡萄糖组母猪 7 d 发情配种率提高 1.9％，断奶-发情间隔减少 0.26 d，淘汰率显著降低 4.04％。实际生产中，每天饲喂 600 g 葡萄糖的量比较大，故本试验研究断奶母猪日粮中添加葡萄糖和多维以及不同添加量的葡萄糖对发情率的影响，旨在探索实际生产中葡萄糖的精准用量。

1　试验材料与方法

1.1　试验设计

本试验于 2021 年 11—12 月在河北张家口的猪场进行。选取断奶母猪 4 批，共计 426 头，进行 4 次为期 28 d 的重复试验（其中有 2 批母猪的试验无对照组）。本试验断奶母猪按照产房单元、胎次、品系、体况和带仔数等因素随机分为 3 个处理组，每个处理组 35 头左右，分为对照组、试验组一和试验组二；后 2 批试验无对照组，只有试验组一和试验组二，每个处理组有 50 头左右母猪。试验分组及饲喂方法如表 1 所示。

表 1　试验分组及饲喂方法

组别	处理	饲喂方法
对照组	不添加葡萄糖	
试验组一	添加 150 g 葡萄糖	每天饲喂 2～3 次，葡萄糖添加 1 次
试验组二	添加 300 g 葡萄糖	

1.2　日常饲喂管理

（1）饲喂　饲喂新好 026 料号饲料，每天饲喂 2～3 次（11 月下旬改为每天饲喂 3 次）。

（2）诱情管理　每天公猪诱情 2.5～3 h，每头公猪刺激约 30 头母猪。

（3）温度、饮水　温度由环境控制器控制，饮水充足，光照 16 h 以上。

1.3　试验指标

（1）背膘厚　由背膘仪和背膘卡尺测定，于母猪断奶下产床后进行测定。

（2）带仔数　断奶前 1～2 d 去产房单元统计。

（3）发情率　查情时根据查情人员对断奶母猪发情标记进行记录，试验时间为断奶后 7 d 内。

（4）胎次　慧养猪 2.0 管理系统导出。

（5）哺乳天数　慧养猪 2.0 管理系统导出。

1.4　统计方法

所有数据经 Excel 进行初步整理后，采用 SPSS 软件 21.0 版本对数据进行二元 Logistic 回归分析。统计结果以平均值表示，$P < 0.05$ 为差异显著性判断标准。

2　试验结果

2.1　不同添加量葡萄糖对断奶母猪发情率的影响

试验过程中对发情母猪进行标记，每一组试验结束后进行统计分析。前 2 批试验的平均结果显示，添加 300 g 葡萄糖对提升母猪发情率相对于对照组效果显著；添加 150 g 葡萄糖对发情率也有提升（表 2）。结合 4 批试验结果得出，添加 150 g 和 300 g 葡萄糖对提升母猪发情率相对于对照组效果显著。

2.2 不同添加量葡萄糖对 1～3 胎断奶母猪发情率的影响

本试验结束后，对胎次进行统计后分析，由表 3 看出，添加 150 g 和 300 g 葡萄糖对 1 胎母猪发情率的影响显著高于对照组；2 胎和 3 胎断奶母猪的发情率与对照组相比未发现显著差异。

表 2　不同添加量葡萄糖对断奶母猪发情率的影响（头）

处理		对照组	试验组一	试验组二
第 1 批	总数	37	37	37
	发情数量	30	33	36
	占比	81.08%	89.19%	97.30%
第 2 批	总数	38	37	37
	发情数量	32	32	34
	占比	84.21%	86.49%	91.89%
	平均	82.65%[a]	87.84%[ab]	94.59%[b]
第 3 批	总数	0	53	52
	发情数量	0	50	47
	占比	0	94.34%	90.38%
第 4 批	总数	0	46	48
	发情数量	0	42	45
	占比	0	91.30%	93.75%
总体平均		82.67%[a]	90.75%[b]	93.10%[b]

注：同列数据肩标字母不同表示差异显著 $P<0.05$，反之无差异，下同。

2.3 不同添加量葡萄糖对不同哺乳天数断奶母猪发情率的影响

本试验结束后，对哺乳天数进行统计后分析，葡萄糖添加组中，哺乳天数在 21～24 d，发情率较高，但结果不显著；哺乳天数在 21～24 d，添加 150 g 和 300 g 葡萄糖相对于对照组，发情率均有提升，但效果不显著；在同一哺乳天数阶段，添加 300 g 葡萄糖组的发情率都较高，效果较好（表 3）。

表3   不同添加量葡萄糖对1~3胎和不同哺乳天数断奶母猪发情率的影响（头）

处理		胎次			哺乳天数		
		1	2	3	18~20	21~24	25 及以上
对照组	总数	26	15	34	7	46	22
	发情数量	17	13	32	7	39	16
	占比	65.38%[a]	86.67%	94.12%	100.00%	84.78%	72.73%
试验组一	总数	86	33	54	38	84	54
	发情数量	74	29	54	33	77	47
	占比	86.05%[b]	87.88%	100.00%	86.84%	91.67%	87.04%
试验组二	总数	84	37	53	36	89	50
	发情数量	77	33	52	33	83	44
	占比	91.67%[c]	89.19%	98.11%	91.67%	93.26%	88.00%
总体平均		81.03%	87.91%	97.41%	92.84%	89.90%	82.59%

3 讨论

本试验在断奶母猪日粮中添加葡萄糖和多维，结果表明添加300 g葡萄糖能够提升断奶母猪3%左右的发情率。在同一试验组中，3胎和2胎断奶母猪的发情率显著高于1胎母猪，符合实际生产情况。哺乳天数与断奶母猪发情率呈现负相关的趋势，而且哺乳天数在18~20 d的断奶母猪发情率最高，平均在90%以上，怀疑是因为这个数据是从慧养猪2.0管理系统中导出，是以场线内技术员的数据录入为准的，而实际母猪分娩时间要早1~2 d，此处有偏差可能会导致出现数据统计的问题。白红杰[5]研究发现，添加葡萄糖短时间内血糖浓度提升，刺激胰岛素分泌，胰岛素可增强母猪GnRH的释放频率，进而影响发情；试验得出葡萄糖最佳添加量为600 g。而本试验结果比较符合预期，但是未达到理想结果，场线刚刚达到满负荷运转，猪只整体状态较差，11月中旬以后，场线实际断配率较之前提升较大，基本可达到85%以上，故后续又进行了不同添加量葡萄糖对断奶母猪发情率影响的试验，结果良好。前2批试验结果显示，添加300 g葡萄糖对断奶母猪的发情率相较于对照组提升显著，而综合4批试验结果来看，仍然有这种效果，且后2批试验总体断配率达到90%以上，与之前的断配率相比较提升较大。添加葡萄糖后对1胎母猪发情率的影响显著，提升非常明显，添加300 g葡萄糖组提升效果最显著；对2胎和3胎断奶母猪发情率没有影响。添加葡萄对于不同哺乳天数断奶母猪发情率的影响不显著，试验组21~24 d阶段发情率更好一些，对照组18~20 d阶段样本量太少，参考价值不大。

另外，日粮中增加葡萄糖摄入量可诱导体内胰岛素分泌量增加，胰岛素可影响GnRH的活性和作用，GnRH又是启动生殖的重要激素。近年来的研究表明，胰岛素通过以下三种途径影响GnRH分泌：①早期生长应答蛋白-1（early growth response protein-1，Egr-1）途径，在胰岛素存在下，Egr-1可与GnRH启动子结合，提高启动子的活性，促进GnRH

基因表达[7]；②胞外信号调节激酶（ERK）途径，GnRH 在垂体促进促黄体素（luteinizing hormone，LH）和促卵泡素（follicle-stimulating hormone，FSH）分泌时主要依靠 ERK 作用，而胰岛素可影响 ERK 的活性，从而间接促进 GnRH 基因表达[8,9]；③神经肽 Y（NPY）途径，NPY 可以刺激或抑制 GnRH 脉冲发生器，而胰岛素可以调节 NPY 的活性[10,11]。另外，胰岛素分泌还可以促进卵巢发育和提高优势卵泡的数量[12]。

综合以上结果，断奶母猪日粮中添加 300 g 葡萄萄可以有效节省饲料成本，提升断奶母猪发情率。

4　经济测算

根据本试验结论，断奶发情率提升 12%，断奶母猪超期未发情 28 d 后淘汰；为了满足配种目标，每个批次需要额外补充 11% 后备母猪入群；由于断奶后超期未发情母猪增加额外的非生产天数，全年预计增加 20 840 d，预计需要耗费饲料 26.8 万元；由于额外补充后备母猪，需要增加后备母猪驯化成本共计 38.3 万元；1 个 3 000 头标准的父母代场线预计每年省 65 万元；目前运营的猪场有 60% 的提升空间，每年可节省成本 1.3 亿元；单头母猪添加葡萄糖的成本为 7.08 元，3 000 头场线每年投入成本 4 万元左右。

参考文献

［1］迟兰，薛忠，朱广琴．影响断奶母猪发情率的原因及应对措施［J］．中国猪业，2021，16（3）：40-43.

［2］林杰，康乐，居瑞海．影响母猪发情的因素分析和防治措施［J］．河南畜牧兽医：综合版，2015，11：3.

［3］陈裕明，黄文焕，王祖昆．影响母猪发情因素的探讨［J］．广东畜牧兽医科技，2000，25（3）：3.

［4］郭金彪，陈景仁．母猪哺乳期长短和体况对繁殖效率的影响［J］．养猪，1998，4：19-20.

［5］白红杰，范磊，王丽英，等．3 种葡萄糖投饲量对断奶母猪发情率及断奶发情间隔的影响［J］．东北农业科学，2016，41（6）：4.

［6］廖昌韬，王燕桃，罗旭芳，等．葡萄糖对诱导断奶母猪发情配种的效果评估［C］//中国畜牧兽医学会动物微生态学分会会员代表大会，2014.

［7］Sara A，Sally R，Andrew W，et al. Egr-1 binds the GnRH promoter to mediate the increase in gene expression by insulin［J］. Molecular and Cellular Endocrinology，2007，270（1/2）：64 -72.

［8］Liu F，Usul I，Evans L G，et al. Involvement of both Gq/11 and Gs proteins in gonadotropin-releasing hormone receptor-mediated signaling in LβT2 cells［J］. Journal of Biological Chemistry，2002，277（35）：32099-32108.

［9］Navratil A M，Song H，Hernandez J B，et al. Insulin augments gonadotropin-releasing hormone induction of translation in LβT2 cells［J］. Molecular and Cellular Endocrinology，2009，311（1/2）：47-54.

［10］Hill J W，Elmquist J K，Elias C F. Hypothalamic pathway linking energy balance and reproduction［J］. American Journal of Physio Endocrinology and M metabolism，2008，294：E827 -E832.

[11] Xu J，Kirigiti M A，Grove K L，et al. Regulation of food intake and gonadotropin-releasing hormone/ luteinizing hormone during lactation：role of insulinand leptin ［J］. Endocrinology，2009，150：4231-4240.

[12] Zhai H L，Wu H，Xu H，et al. Trace glucose and lipid metabolism in high androgen and high-fat diet induced polycystic ovary syndrome rats ［J］. Reproductive Biology and Endocrinology，2012，10（5）：2-9.

延长公猪诱情时间对断奶母猪发情率的影响

郑智伟

摘要：本试验旨在研究母猪断奶后延长公猪诱情时间对其繁殖性能的影响。试验选取 5 个批次即将断奶的母猪 309 头，每个批次母猪均按胎次、背膘厚、哺乳天数随机分为 2 个处理组：对照组（$n=155$）和加强诱情组（$n=155$）。对照组不做额外处理，加强诱情组（试验组）在断奶第 2 天开始，每天查情结束后，每 15 头断奶母猪前放置 1 头公猪 1 h。准确记录每头母猪断奶 7 d 内的发情信息、配种信息。结果表明：①断奶母猪第 2 天延长公猪诱情时间可以提高断奶母猪 7 d 断配率 6.5%（$P<0.05$），但是断奶至发情间隔无显著差异；②试验处理能够提升 1~3 胎母猪的发情率，分别提升了 7.98%、4.04% 和 12.41%。综上所述，在本试验条件下，断奶后延长公猪诱情时间可提高断奶母猪 7 d 内发情率，且对 3 胎以内断奶母猪效果较好，可应用于生产以提高母猪生产效率。

关键词：断奶母猪；公猪；诱情时间；断奶发情率

满负荷均衡生产是规模化猪场高效运行的必然选择。标准猪场广泛采用 7 d 批的生产节律，7 d 内配种数目达标是最基本的要求，然而猪场遭受断奶猪长期不发情的困扰。2021 年 1—11 月的生产数据显示，断奶母猪的平均发情率为 51.44%，远低于 90% 的标准。断奶母猪发情率低会对生产带来种种不良影响，如增加母猪的非生产天数，相应地增加饲料成本，降低母猪 PSY。此外，断奶空怀母猪的增加占用了大量的栏位，增加了猪群排布工作的困难以及查情和诱情的工作难度。为提升生产效益，有必要探讨和研究母猪断奶发情率提升，以达到高效生产的目标。

断奶后 7 d 内的公猪诱情是刺激母猪断奶后早期排卵与检测母猪发情的基础且重要工作，Walton 及 Langendijk[1,2] 肯定了公猪刺激的重要性。公猪接触可以增加断奶后排卵和发情的母猪数量，与没有接触公猪的母猪相比，接触公猪不会抑制发情行为。促黄体素分泌减少可能导致母猪断奶后不发情、不排卵[3]；公猪刺激在参与调节卵巢活动的神经内分泌通路方面很重要，已证明引入公猪可引起母猪释放促黄体素[2]。不同的刺激方式对促进母猪发情有不同的影响，Langendijk 与 Gerritsen[4,5] 比较了人工压背刺激、公猪与母猪单独接触、公猪与母猪接触后的人工压背刺激、假公猪模型的视觉刺激、假公猪模型的视觉刺激与公猪叫声、假公猪模型的视觉刺激与公猪叫声及公猪气味和真实公猪刺激，发现使用真实公猪并结合人工压背刺激能获得 90% 以上的发情率，该研究也肯定了含雄甾烯类固醇的白色泡沫唾液的重要性，支持了 Perry[6] 的没有颌下唾液腺的公猪无法在发情后备母猪中引起接受行为的研究结果。对断奶母猪的刺激强度也会影响母猪发情，Langendijk[4]研究比较了4种刺激强度在2种圈养模式下的促发情效果，发现随着与公猪接触越密切、刺激强度越大，母猪发情效果越好。然而Hemsworth[7]研究发现，每天3次查情后，公猪存放位置靠近或不靠近母猪，对断奶母猪发情率没有影响。

综上所述，母猪断奶后的诱情是提升发情率的必要工作，然而公猪与母猪的接触时间延长对发情率提升是否有效还存在争议。因此，本试验将探讨延长公猪与母猪的接触时间能否提升不同胎次断奶母猪的发情率。

1   试验材料与方法

1.1   试验设计

本试验于2021年12月至2022年1月在江西猪场开展。该试验在2条3 000头标准的父母代场线同时开展，选取5批次断奶母猪共计331头，在断奶前1 d记录分娩胎次、哺乳天数、产仔数、断奶背膘厚后将其均匀分为两组，保证两组母猪的上述指标无差异，两组中同胎次母猪比例一致，次日将母猪断奶并分配到2个配种单元，分别作为试验组与对照组。试验分组及饲喂方法如表1所示。

表1   延长公猪诱情时间试验分组及饲喂方法

组别	处理	饲喂方法
试验组	断奶当天记为0 d，断奶第2天开始使用公猪诱情，上、下午各1次，公猪赶至母猪前鼻对鼻接触，员工采用查情五步法刺激母猪不少于20 s。查情结束将发情母猪转群后，将公猪锁在母猪面前的喂料通道前，公猪可与母猪鼻对鼻自由接触。1 h后将公猪赶回公猪栏，该处理上、下午各1次，重复至断奶第7天	断奶当天不饲喂，保证母猪的充足饮水。断奶第1天开始饲喂母猪妊娠料，饲喂前保证料槽水深不超过2 cm，8：00饲喂2 kg妊娠料，17：00饲喂2 kg妊娠料，如果母猪料槽舔食干净则再饲喂0.25 kg饲料，少加勤添直至仅剩余少许剩料。饲喂后给予母猪至少40 min的进食时间，之后将剩料均摊给进食量大的母猪，料槽放水，水深不超过4 cm。断奶4 d后母猪发情导致采食量下降，酌情降低饲喂量避免浪费饲料
对照组	断奶当天记为0 d，断奶第2天开始使用公猪诱情，上、下午各1次，公猪赶至母猪前鼻对鼻接触，员工采用查情五步法刺激母猪不少于20 s	

注：剔除瘸腿、哺乳天数少于14 d、背膘厚小于10 mm、带仔数小于6头、发情时排脓的母猪。

试验共进行5批次，每批次参与试验的母猪均匀分为2组，剔除胎次、产仔数、断奶背膘厚、哺乳天数对公猪诱情处理措施的影响。试验组1～4胎母猪数分别为85头、82头、111头和37头，对照组1～4胎分别为80头、91头、113头和31头。其他分组细节见表2。

表2   各批次母猪分组细节

批次	组别	胎次	活仔数（头）	背膘厚（mm）	哺乳天数	头数
1	试验组	1.33±0.48	10.79±2.96	14.44±2.28	22.21±0.59	24
	对照组	1.29±0.46	10.38±2.75	15.5±2.04	22.08±0.58	24
2	试验组	3.2±0.54	10.82±2.1	17.1±2.11	20.0±2.1	49
	对照组	3.23±0.52	10.27±2.47	16.96±2.02	19.88±2.05	48

（续）

批次	组别	胎次	活仔数（头）	背膘厚（mm）	哺乳天数	头数
3	试验组	1.32±0.48	9.54±2.36	15.18±2.57	22.07±2.18	28
	对照组	1.29±0.46	9.18±1.98	15.64±2.31	22.36±1.68	28
4	试验组	1.74±0.45	10.63±4.08	17.0±1.96	23.41±1.39	27
	对照组	1.89±0.32	11.41±2.45	18.04±2.12	22.81±1.69	27
5	试验组	3.3±0.54	11.0±1.36	17.41±2.24	21.52±1.28	27
	对照组	3.19±0.62	10.96±1.6	15.67±2.87	21.22±1.65	27

1.2　日常饲养管理

（1）温度　环境控制器设定，猪舍纵向温度为18℃，横向温度为21℃，该猪舍无额外的加热装置。

（2）光照　猪舍光照16 h，日光灯管照明，试验期间每天7：00开灯，23：00关灯，光照度为50～100 lx。

（3）粪污清理　大棚式猪场采用水泡粪工艺，每隔15 d排污1次。

1.3　试验指标

（1）采食量　每天跟踪断奶母猪采食情况，记录剩料多于2 kg的母猪。

（2）发情记录　记录母猪发情日期、发情表现、断配间隔信息。

（3）配种记录　记录母猪配种时间、配种次数、配种异常情况。

1.4　统计方法

所有数据经Excel初步整理后，采用SPSS软件22.0版本分析。使用二元Logistics回归对数据进行分析，因变量为是否发情，发情编码为1，分类变量为试验处理，连续变量为哺乳天数、背膘厚、产仔数、胎次。统计结果以平均值表示，$P<0.05$为差异显著性判断标准。

2　试验结果

2.1　延长公猪诱情时间对断奶母猪发情率和断配间隔的影响

本试验进行了5批次，每一批次试验组的发情率均高于对照组4.17%～7.41%，试验组7 d发情率为（91.21±4.36）%，对照组7 d发情率为（84.71±4.87）%，二元Logistics回归分析发现试验处理项系数差异显著（$P=0.041$），试验组相对于对照组的发情优势比为2.16。试验组的断配间隔为（4.2±0.24）d，对照组的断配间隔为（4.25±0.25）d，卡方独立性检验发现不存在显著差异（表3）。以上分析结果发现，延长公猪诱情时间可提升断奶母猪发情率，但不会改变断配间隔时间。

表 3   各批次母猪发情信息

批次	组别	不同断配间隔发情头数					7 d 发情头数	未发情头数	发情率（%）
		3 d	4 d	5 d	6 d	7 d			
1	试验组		13	8	1		22	2	91.67
	对照组		5	12	3	1	21	3	87.50
2	试验组	2	21	19	4		46	3	93.88
	对照组	2	24	15	1		42	6	87.50
3	试验组		16	11			27	1	96.43
	对照组		9	11	4	1	25	3	89.29
4	试验组		15	9			24	3	88.89
	对照组	1	11	10			22	5	81.48
5	试验组		7	12	4		23	4	85.19
	对照组		11	7	3		21	6	77.78

2.2   延长公猪诱情时间对不同胎次断奶母猪发情率的影响

本研究统计分析了试验处理对不同胎次母猪促发情的效果，如图 1 所示，试验处理能够提升 1~3 胎母猪的发情率，分别提升 7.98%、4.04% 和 12.41%，4 胎母猪延长公猪诱情后受胎率下降。此外，随着胎次的提高，试验组的发情率呈下降趋势，但仍维持在 90% 以上。对照组的发情率随胎次增加呈先增后减再增高的趋势，仅四胎母猪发情率超过 90%。数据分析发现，延长公猪诱情时间不仅可提高低胎次断奶母猪发情率，还可使各胎次维持平稳的、较高的断奶发情率。

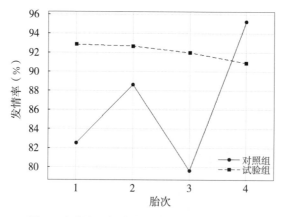

图 1   试验处理与胎次对母猪发情的交互作用

3   讨论

在断奶母猪中，发情检测期间延长公猪与母猪接触时间似乎对于诱导母猪的发情至关重要。在本试验中，延长公猪与母猪接触时间试验组增加了断奶母猪 7 d 发情率（$P < 0.05$）。然而，延长公猪与母猪接触时间并没有缩短母猪断奶到发情的时间间隔，该结果与 Knox[8] 的研究结果一致。公猪接触对发情母猪数量的影响可能是由垂体活动的增加介导的，如公猪唾液中的雄烯酮、雄烯醇会刺激母猪中枢神经系统释放参与调节促黄体素的激素和神经肽，卵泡期后期的卵泡生长取决于促黄体素的浓度[9]。

在本试验中，部分母猪可能对公猪刺激不敏感，发情表现不明显，可能是代谢影响了卵泡发育。众所周知，母猪断奶时身体状况较差，导致发情延迟或无发情期的发生率相对较高[10]。本试验统计了母猪断奶时的平均背膘厚，背膘厚 12～13 mm 的母猪发情率为 77.78%，14～15 mm 的发情率为 80.36%，16～17 mm 的发情率为 88.10%，随后随着背膘厚增加，母猪发情率开始下降，背膘厚 21～22 mm 的母猪发情率下降至 79.03%，再次证实了母猪断奶时体况的重要性，推测可能是母猪的身体状况更好，使它们对公猪刺激更敏感，这些刺激可能会增加促黄体素分泌，导致这些母猪排卵发情。

本试验将公猪放在母猪限位栏前的通道内，允许公猪自由移动并能与母猪鼻对鼻接触，公猪与母猪数目比例为 1∶（20～50），当母猪数量太多时公猪可能无法与每头母猪充分接触，可能在一定程度上降低了诱情的作用。在过去的一项研究中，对比了没有公猪、有 1 头公猪和同时有 4 头公猪的情况下进行的发情检测，结果与 4 头公猪接触并没有提升母猪发情率。因此，使用公猪诱情是必要的，其间使用过多的公猪不一定会提高母猪发情率，而诱情期间没有公猪接触会降低母猪发情率。找到延长公猪与母猪接触时间时公猪与母猪合适的比例是未来的一项工作。

4　经济测算

根据本试验结论，断奶发情率提升 12%，断奶母猪超期未发情 28 d 后淘汰；为了满足配种目标，每个批次需要额外补充 11% 后备母猪入群；由于断奶后超期未发情母猪增加额外的非生产天数，全年预计增加非生产天数 20 840 d，预计需要额外耗费饲料 26.8 万元；由于额外补充后备母猪，需要额外投入后备母猪驯化成本共计 38.3 万元；一个 3 000 头标准父母代场线预计每年节省 65 万元；目前运营的猪场有 60% 的提升空间，每年可节省成本 1.3 亿元。

无其他额外投资成本，若猪场公猪数量不足，需购买查情公猪进行操作，1 头公猪每年的成本为 5 000 元（包含饲料、生物资产折旧、药品、疫苗等）。

参考文献

[1] Walton J S. Effect of Boar Presence Before and After Weaning on Estrus and Ovulation in Sows [J]. Journal of Animal Science，1986，62：9-15.

[2] Langendijk P，Brand H，Soede N M，et al. Effect of boar contact on follicular development and on estrus expression after weaning in primiparous sows [J]．Journal of Animal Science，2000c，54：0-1303.

［3］ Van dW，Booman P. Post-weaning anoestrus in primiparous sows：LH patterns and effects of gonadotropin injection and boar exposure［J］. Vet Q，1993，15：162-166.

［4］ Langendijk P，Soede N M，Kemp B. Effects of boar contact and housing conditions on estrus expression in weaned sows.［J］. Journal of Animal Science，2000，78：871-878.

［5］ Gerritsen R，Langendijk P，Soede N，et al. Effects of artificial boar stimuli on the expression of oestrus in sows［J］. Applied Animal Behaviour Science，2005，92：37-43.

［6］ Perry G C，Patterson R，Macfie H，et al. Pig courtship behaviour：pheromonal property of androstene steroids in male submaxillary secretion［J］. Animal Production，1980，31：191-199.

［7］ Hemsworth P H，Hansen C. The effects of continuous boar contact on the oestrus detection rate of weaned sows［J］. Applied Animal Behaviour Science，1990，28：281-285.

［8］ Knox R V，Miller G M，Willenburg K L，et al. Effect of frequency of boar exposure and adjusted mating times on measures of reproductive performance in weaned sows.［J］. Journal of Animal Science，2002，80：892-899.

［9］ Driancourt M，Locatelli A，Prunier A. Effects of gonadotrophin deprivation on follicular growth in gilts［J］. Reproduction Nutrition Development，1995，35：663.

［10］ Vesseur P C，Kemp B，den Hartog LA. Factors affecting the weaning-to-estrus interval in the sow［J］. Journal of Animal Physiology and Animal Nutrition，1994，72：225-233.

Chapter

4

环保工程与资源化利用篇

规模化养猪过程中臭气的减排措施研究进展*

周思邈　柴小龙　梁晓飞　李　同　闫之春

摘要： 目前，养殖场周边存在着日益严重的养殖场臭气污染问题，特别是大型养猪场周边，这给养殖场的畜禽、养殖人员和周围居民造成较为严重的健康影响。如何有效处理养殖场臭气，已成为大型养殖场面临的棘手问题。本文综述了养殖场臭气的主要成分、影响因素、减排措施及工程应用案例，以期为规模化养猪场减少臭气排放提供参考。

关键词： 猪舍；臭气；氨气；减排措施

近几年，我国生猪养殖业快速发展，特别是大型集约化、规模化养殖场的大量建设，给养殖场周围的空气质量带来严重影响。随着国家对环境保护越来越重视，以及养殖场周围居民对臭气投诉增多，如何有效解决养殖场臭气成为大型养殖场面临的棘手问题。养殖过程中臭气减排的主要方式可以概括为 3 个方面：源头减排、过程控制和末端处理。源头减排即通过提高饲料利用率，减少粪尿中氮素排放，这是实现减少臭气排放的关键；过程控制是在养殖过程中注重猪只粪尿管理，减少臭气释放；末端处理是将臭气进行收集处理，减少臭气向周围环境扩散。本文总结了国内外关于养殖场臭气的相关研究，从臭气主要成分、各组分浓度、臭气产生的主要影响因素以及减排措施等方面进行综述，以期对养殖场臭气的减排提供思路。

1 养殖场臭气的主要成分及影响因素

一切刺激嗅觉器官，引起人们不愉快感觉及损害生活环境的异味气体都可以称为臭气。畜禽粪便中含有大量微生物，其对粪尿中有机物的分解是养殖臭气的主要来源；此外动物表皮、毛发、饲料、垫料等富含蛋白质的废弃物的厌氧分解也会产生臭气[1]。

1.1 养殖场臭气的主要成分

养殖场臭气成分复杂，会受到养殖畜禽种类、生产管理方式、粪尿处理措施等因素的影响，大体上可以分为 4 类：NH_3 和挥发性胺类、H_2S 和含硫化合物、芳香族化合物（吲哚类和酚类）、挥发性脂肪酸类[1]。Schiffman 等[2]通过气相色谱和质谱（GC/MS）分析了猪场臭气，发现猪场臭气中的挥发性有机物（VOC）和固定气体多达 331 种，包括酸类、醇

* 文章发表于《中国畜牧杂志》2022 年第 58 卷第 1 期。

类、胺类、酰胺类、酯类、醚类、酚类、芳族化合物、其他含氮化合物、含硫化合物、类固醇和其他化合物等。在臭气组分中，浓度较高、方便现场实时监测、对人畜健康影响较大的有害气体主要是 NH_3 和 H_2S，这也是我国国家标准《恶臭污染物排放标准》（GB 14554—1993）[3]规定的主要监控指标。《规模猪场环境参数及环境管理》（GB/T 17824.3—2008）[4]中规定猪舍内 NH_3 浓度须低于 25 mg/m³，H_2S 浓度须低于 10 mg/m³。当 NH_3、H_2S 超过一定浓度后，会对猪只的生长和健康产生负面影响，如降低猪只的生产性能、抵抗力，影响疫苗的免疫效果，甚至会诱发疾病等[5]。

1.2　养殖场臭气的主要影响因素

畜禽粪便腐败分解产生恶臭物质的成分和浓度会因尿液的混入量、水分含量、环境温度、清粪方式、通气量、pH 以及贮存时间等因素的不同而存在很大差异[6]。国内外许多研究学者对猪舍内各臭气浓度（主要是 NH_3）的检测结果见表 1。

表 1　猪舍 NH_3 浓度统计

生长阶段	清粪方式	通风方式	时间段	NH_3（mg/m³）	参考文献
育肥猪		机械通风	夏季	3.31±0.31	[7]
育肥猪		半机械半自然通风	秋季	4.91±0.56	[7]
育肥猪	人工干清粪	自然通风	1 月	10.09 ± 4.60	[13]
育肥猪	人工干清粪	自然通风	7 月	3.44 ± 2.34	[13]
育肥猪	水泡粪	机械通风	夏季	3.15±1.02	[8]
育肥猪	水泡粪	机械通风	春季	3.60±1.67	[8]
育肥猪	水泡粪	机械通风	秋季	3.88±0.38	[8]
育肥猪	水泡粪	机械通风	冬季	8.41±0.98	[8]
育肥猪	深坑式	自然通风	3—6 月	1.60~7.76	[14]
育肥猪	固液分离式	自然通风	3—6 月	2.36~7.22	[14]
育肥猪	垫料式	自然通风	3—6 月	0.61~3.88	[14]
育肥猪	垫料式	自然通风	春天	5.9±2.7	[15]
育肥猪	垫料式	自然通风	夏天	6.8±3.4	[15]
育肥猪	垫料式	自然通风	夏季	6.8±3.4	[16]
育肥猪	干清粪	自然通风	夏季	16.7±3.9	[16]
育肥猪	垫料式	自然通风	春季	5.9±2.7	[16]
育肥猪	干清粪	自然通风	春季	14.5±2.3	[16]
保育猪	人工干清粪	自然通风	8 月至次年 1 月	1.21~4.88 2.94±1.48	[10]
育肥猪	人工干清粪	自然通风	8 月至次年 1 月	1.55~5.24 3.26±1.49	[10]
妊娠猪	人工干清粪	机械通风	8 月至次年 1 月	1.22~6.35 3.48±2.20	[10]
妊娠猪	深坑式	机械通风	5—10 月	8.48±4.13	[17]
妊娠猪	塞流式	机械通风	5—10 月	4.08±1.71	[17]

通风量是影响猪舍 NH_3 浓度的主要因素，自然通风模式下，通风量较低，NH_3 积聚在猪舍内难以扩散，致使浓度水平较高；机械通风模式下，较高的通风量可将大量臭气带出猪舍，从而降低舍内 NH_3 浓度[7,8]。在通风模式不变的情况下，猪只的生长阶段、环境温度和猪只活动强度也是影响 NH_3 浓度水平的重要因素[7,9]。不同生长阶段的猪只因饲料配方、饲养密度的不同会造成粪便中含氮量的差异，饲养密度会造成粪便产量的不同，两种因素叠加会造成舍内 NH_3 浓度的明显差异。舍内环境温度会影响粪尿中微生物的代谢活性，进而影响舍内 NH_3 含量。

纪英杰等[7]监测了夏秋两季舍内 NH_3 浓度的变化，结果表明猪舍夏季机械通风模式下 NH_3 浓度为 3.31 mg/m³，秋季采用半机械通风半自然通风模式时，通风量约为夏季的 1/3，舍内 NH_3 浓度达到 4.91 mg/m³。刘杨等[8]测得规模猪场机械通风育肥舍春季、夏季、秋季和冬季的平均 NH_3 浓度分别为 3.60 mg/m³、3.15 mg/m³、3.88 mg/m³ 和 8.41 mg/m³，对应的夏季通风量分别是春季、秋季和冬季通风量的 2.08、2.34 和 3.04 倍。王文林等[9]监测夏季猪只不同生长阶段（保育、育肥Ⅰ、育肥Ⅱ）以及妊娠母猪与分娩母猪舍内的 NH_3 浓度分别为 0.97 mg/m³、3.37 mg/m³、5.45 mg/m³、2.19 mg/m³ 和 1.44 mg/m³，表明各生长阶段氨排放存在差异。此外，研究还发现猪舍内氨气排放具有明显的日变化规律，夏季主要受温度和猪只活动强度的影响，秋季主要受机械通风运行模式的影响[7]。许稳等[10]连续检测了 8 月至次年 1 月育肥舍、妊娠舍、哺乳舍和保育舍的 NH_3 浓度，发现舍内 NH_3 平均浓度呈现冬季＞秋季＞夏季的季节变化趋势，这与不同季节的通风量有关。总的来看，规模化猪场在夏季和冬季面对的臭气问题不同，夏季猪场面临的主要是臭气减排和臭气处理问题，主要因为夏季为降低舍内和猪只体表温度，猪舍会进行大量的机械通风，在这个过程中会将大量的臭气污染物排放到舍外；冬季猪舍因保温需要减少换气量，主要面对舍内氨气浓度较高问题。

猪舍清粪方式会影响粪尿的存在形式，如粪尿是否分离、粪便含水量、粪便在舍内贮存时间，进而造成舍内 NH_3 含量差异。汪开英等[11]在动物人工气候室分析了 3 种不同栏舍地板结构对 NH_3 排放的影响，发现地面结构类型对猪舍 NH_3 排放影响显著，常规养殖方式下半缝隙地面和实心地面的猪舍 NH_3 排放量相对生物发酵床猪舍较高。与干清粪相比，水冲粪和水泡粪都存在耗水量大、污水产生量大及其污染物浓度高、舍内有害气体含量高等问题，从清洁生产的角度考虑，干清粪工艺是规模化猪场的必然选择[12]。

2 养殖场臭气的减排措施

2.1 平衡氨基酸组分降低蛋白摄入

在满足猪只营养需求的前提下，通过减少饲料中粗蛋白含量，减少粪尿中氮元素的排放量，可以实现从源头上减少臭气产生。根据氨基酸平衡理论，必需氨基酸之间及其与非必需氨基酸之间达到最佳平衡时，才能够达到最佳利用效率。一般在以谷物和豆粕为主的植物性饲料中赖氨酸、蛋氨酸等含量较低，不能满足猪只需要，因此，采取额外补加赖氨酸、蛋氨酸以及谷氨酸，降低日粮的蛋白水平等方式，可以提高饲料蛋白利用率，大大节省蛋白质用量，降低饲养成本，提高猪只生产效率，同时还能够减少粪氮和尿氮的排放，有效改善猪舍小环境，减轻粪污处理压力[18,19]。

Kerr 等[20]进行了 33 组猪只的代谢试验，试验数据显示在平衡氨基酸的基础上，日粮蛋白水平每降低 10 g/kg，氮的排泄能降低 8.4％。而 Leek 等[19,21]研究发现日粮蛋白水平每降低 10 g/kg，氮的排泄能降低 8.7％和 6.7％。Liu 等[22]研究也证明，通过氨基酸来补充日粮中降低的粗蛋白能有效减轻 NH_3 排放，在标准日粮的基础上降低 2.1％～3.8％、4.4％～7.8％的粗蛋白后，NH_3 排放量较对照组分别降低了 33％和 57.2％，且粪便排泄量和粪便中的 NH_4^+-N 浓度也明显降低。刘作华等[23]试验发现，日粮蛋白质含量每降低 3％～4％，猪舍内 NH_3 浓度显著降低 26.55％～57.85％。不同研究者之间的数值差异可能与试验猪体重、对照组日粮的粗蛋白水平、日粮粗蛋白降低的幅度以及氨基酸的平衡模式有关，更为准确的减排效果有待进一步研究。

2.2　添加饲料添加剂

养殖场臭气主要由微生物在不完全厌氧条件下分解粪尿中的蛋白质和碳水化合物产生，因此通过改善动物肠道和粪尿中的微生物菌群组成，或者降低产生恶臭气体微生物的活性，也可以实现减少臭气排放的目的。益生菌、益生元、有机酸、丝兰提取物、纤维、中草药等饲料添加剂可以通过调节消化道微生物菌群，改善营养物质的消化和吸收，进而减少臭气排放。

廖新俤等[24]研究发现在猪只日粮中添加 0.05％的活菌制剂可以起到猪舍 NH_3 减排效果，两条生产线上连续 10 d 的监测发现舍内 NH_3 浓度分别降低了 40.28％和 56.46％。Murphy 等[25]研究表明，在育肥猪日粮中添加不同剂量的苯甲酸（0 g/kg、10 g/kg、20 g/kg 和 30 g/kg）可降低总氮和尿氮的排出率，并且可降低尿氮/粪氮比，从而降低猪粪尿中 NH_3 的排放。王保黎等[26]研究发现，在每吨饲料中添加 90 g 的丝兰属植物提取物，能够显著降低猪舍内 NH_3 产生量。人们对丝兰属植物提取物的除臭机理还未达成一致意见，有研究者认为主要是其含有脲酶抑制剂，能够抑制脲酶活性；也有人认为，其主要通过与 NH_3 和 H_2S 相结合进而抑制有害气体散逸[26,27]。然而，大多数研究对饲料添加剂提高蛋白利用率，降低 NH_3 排放的效果是充分肯定的。在饲料减抗的背景下，通过向饲料中添加有益微生物、丝兰属植物提取物、中草药成分等提高猪只抵抗力，同时减少 NH_3 排放，对于规模化养殖具有重要的应用价值。

2.3　控制舍内温度和湿度

猪舍内 NH_3 含量受饲养密度、饲养管理水平、地面结构、舍内温湿度、通风情况、粪污清除等多因素影响。舍内温度升高会促进 NH_3 释放，因为温度升高会使脲酶活性增高，同时使水相的 NH_4^+ 迅速转化为气相的 NH_3。暴雪艳等[28]在封闭式分娩舍内进行的气体监测表明，舍内气体中 NH_3 和 CO_2 浓度与圈舍温度显著正相关（$r=0.898$）。猪舍内的温湿度控制在适宜范围有利于猪只的生长增重，还可以有效降低猪舍 NH_3 浓度，提高舍内空气质量。

2.4　粪尿分离和控制清粪频率

规模化生猪养殖过程中产生的 NH_3 主要来源于尿液中尿素的水解和粪便中有机物氮的分解[1,22]。猪体内形成的尿素有 $75\%\sim80\%$ 随尿液排出体外，尿素分解为 NH_3 的过程很快，在常温下只需几小时，而猪粪中的有机物氮分解产生 NH_3 是一个相对缓慢的过程，通常需要数周的时间且产生的 NH_3 量也较少。因此，及时进行粪尿分离，并将尿液排出猪舍进行集中处理，可有效减少猪舍氨气排放量，也可降低粪污的后续处理难度。

Panetta 等[29]在实验室条件下研究发现，将尿液从粪便中分离可以使 NH_3 排放量降低 99%。Lachance 等[30]研究发现，将水泡粪清粪频率由每周清空 1 次增加到每周 3 次，能够使 NH_3 排放降低 46%；刮粪板清粪工艺下，每周刮粪次数由 3 次增加到 7 次反而使 NH_3 排放量增加 11%；对比不同清粪工艺对舍内 NH_3 浓度的影响发现，每周 3 次刮粪板清粪比水泡粪每周清空 1 次，NH_3 排放量低 49%。牛欢等[31]发现机械清粪过程会加速 NH_3 释放，NH_3 浓度会提高 $70\%\sim75\%$，在清粪 1 h 后 NH_3 浓度会降低到刮粪前浓度的 $80\%\sim85\%$。Misselbrook 等[32]和 Wood 等[33]认为，粪污表面结壳可以减少粪便内的 NH_3 向舍内挥发，清粪时伴随着粪便翻滚，会破坏粪便表面已形成的壳状物，增加粪便内 NH_3 挥发，造成舍内 NH_3 浓度增大。过高的清粪频率会让更多新鲜粪便暴露在空气中，从而促进 NH_3 的产生并挥发到舍内。李新建等[34]研究发现，使用局部漏缝地板可有效地降低氨气释放量，育肥舍内漏缝地板占总面积 37% 时，NH_3 释放量比采用全漏缝地板减少 40%。综上所述，在规模化养猪生产过程中，建议采用局部漏缝地板配合干清粪工艺进行粪尿分离，并合理控制清粪频率，以减少粪尿内 NH_3 挥发。

2.5　使用粪便除臭剂

通过降低粪沟粪尿 pH，在粪沟中添加吸附类物质和优势除臭菌种等措施也可以减少粪尿 NH_3 的产生。当温度保持不变时，pH 决定了 NH_4^+ 和 NH_3 之间的平衡关系，较低的 pH 会降低氨水比例，减少 NH_3 挥发。pH $7\sim10$ 的范围内会最大限度地增加 NH_3 释放量，pH 降低到 7 以下会抑制 NH_3 挥发，pH 在 4.5 左右时几乎没有可以测量的游离氨。直接将微生物除臭菌剂或者微生物发酵液喷洒到粪沟内，利用微生物之间的协同、繁殖、共生作用转化分解粪便中的有机物质，利用微生物的硝化和反硝化作用来转换氮元素，以达到降低 NH_3 排放的目的。

2.6　生物过滤

生物过滤法应用于机械通风猪舍一般需要将通风系统与过滤装置进行组合，猪舍排出的污浊气体经处理后再排入环境中，是一种重要的末端处理方式。气体进入生物过滤器后，经气、水界面传递到附着于填料表面的生物膜中，膜中微生物以有机气体污染物作为其生长繁殖所需的基质，将大分子结构的有机气体污染物经不同转化途径氧化分解为简单无害或少害

的 CO_2、NO_3^-、H_2O 等无机物，达到净化的目的。生物滤器内的结构、过滤腔内填充物以及运行参数都会对处理效果产生影响。

　　Taghipour 等[35]比较了生物过滤器内部分区（1 层和 3 层）对不同浓度 NH_3 降解性能的影响，发现分 3 层时生物滤池的最大清除能力、处理效率、压降损失等方面均优于 1 层结构。Dumont 等[36]对比了气体在过滤器中的停留时间、加载速率对 NH_3 去除效率的影响，发现夏季停留时间为 12 s 时 NH_3 去除效率在 90%～100%，停留时间减小到 6 s 时 NH_3 去除效率降为 30%～50%；外界环境（温度和湿度）也会对处理效率产生影响，秋季停留时间为 12 s 时 NH_3 去除率约为 80%，小于夏季相同运行条件下的处理效率。在相对干燥的秋季要特别注意生物滤池的加湿处理，为达到良好的 NH_3 去除效果，可以在填料中接种优势除臭除氨菌种，增加生物过滤器中微生物的丰富度和数量。Yasuda 等[37]用以混合岩棉为填料的生物滤池来降解畜禽养殖场中的 NH_3，结果表明经生物滤池处理后 NH_3 的出口浓度小于 3.8 mg/m^3。Melse 等[38,39]利用生物过滤池来减少畜禽养殖场中甲烷的释放量的试验发现，在进气的平均温度为 12℃左右、填料的 pH 为 6.8 时、NH_3 的进气速率为 83 mg/（m^3·h）、H_2S 的进气速率为 0.6～16 mg/（m^3·h）的条件下，NH_3 去除率为 90%～100%，H_2S 的去除率达 100%。

3　猪场除臭应用案例分析

　　一些研究者根据猪场臭气排放特点，依据水对 NH_3 等臭气的溶解特性和微生物对有机臭气的代谢特性，采用循环水喷淋、生物滤床等方式进行猪场除臭的中试试验，验证了其应用的可行性和除臭效率。如 Hartung 等[40]在生产中对 2 个生物滤池进行长期（6.5年）监测，2 个微生物滤池对 NH_3 的平均去除率分别为 15%、36%，将滤料的含水率从 20%升高到 40%可提高生物滤器的清洗效率。中国农业大学工学院等单位在北京某猪场进行了生物氧化法除臭试验（图 1），试验装置尺寸为 6 m×3 m×3 m，处理风量为 10 000 Nm^3/h，NH_3 处理效率为 90%[41]。华南农业大学在江西某种猪场进行了 2 层喷淋滤墙除臭试验（图 2），并给出了除臭效率和运行成本的分析，在喷淋液中使用硫酸时 NH_3 去除率达到 85%，每天耗电量 292kW·h；不加硫酸时 NH_3 去除率达到 50%，每天耗电量 290kW·h[42]。此外，现在已有一些猪场采用在猪舍排风口安装湿式洗涤器来净化猪舍排出废气中的粉尘、NH_3、H_2S 及其他臭气（图 3）。猪舍湿式洗涤器主要是由过滤器和喷淋系统组成，过滤器面积较大，置于风机后合适距离位置处，使其与猪舍废气充分接触；喷淋系统循环利用液体充分润湿过滤器。仅使用水的条件下，湿式洗涤器对猪舍废气中粉尘和 NH_3 的去除效果与通风量大小有关，冬季通风时可以降低 60%，春秋通风时降低 30%，夏季通风时降低 20%[43]。此外，为提高废气的去除效果，可以考虑在循环水中添加除臭菌剂或气味掩蔽剂。

　　美国北卡罗来纳州做了一种多孔阻风墙＋植物条带的简易除臭试验（图 4）。在风机外搭建了一个 2.8 m×5.4 m×2.9 m 的箱体，采用孔隙率为 60%的防蝇网覆盖；与风机相对的前下方留有 0.25 m 高的开口用以减少风阻，开口后端种植植物。试验发现，该方式对通风风压损失较小（<13 Pa），对总固形物颗粒降低的效率约为 46%（36%～58%），但对 NH_3 的去除率较低，约为 13%（5%～16%），对于 H_2S 几乎没有去除效果[43]。

传统生物过滤器应用到猪场除臭需要考虑猪舍大量通风需求和过滤器较大风阻之间的矛盾，解决猪舍通风和臭气在滤床中滞留时间的匹配问题。目前已有猪场采用喷淋墙式的除臭方式虽可以有效减少风阻，但因滤层较薄，臭气和喷淋水接触时间短，还存在除臭效率低、耗水量大的问题。如何在保证猪舍通风量的前提下，通过滤料结构设计增加气液接触，借助微生物、除臭剂进一步提高除臭效率，是下一阶段应该持续优化解决的问题。

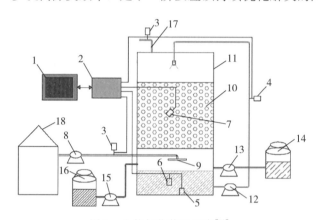

图 1   生物氧化装置示意[41]

1. 触摸屏 2. PLC 与模拟扩展模块 3. 压力传感器 4. 液位电磁流量计 5. 液位传感器
6. pH 传感器 7. 温湿度传感器 8. 风机 9. 气体分布器 10. 填料 11. 生物氧化反应器
12. 循环泵 13. 加碱泵 14. 碱液 15. 加水泵 16. 水箱 17. 出气口 18. 畜禽舍

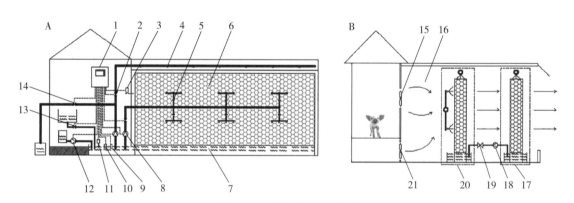

图 2   两层喷淋滤墙除臭[42]
A. 主视图 B. 侧视图

1. 控制电箱 2. 喷淋电磁阀 3. 压差传感器 4. 螺旋形喷嘴 5. 手控球阀 6. 塑料网格填料
7. 蓄液池 8. 酸洗泵 9. pH 传感器 10. 电导率传感器 11. 液位传感器 12. 硫酸计量泵 13. 供水电磁阀
14. 排废电磁阀 15. 末端负压风机组 16. 废气收集压力室
17. 水洗净化单元 18. 水洗泵 19. 回水电磁阀 20. 酸洗净化单位 21. 地沟风机组

图 3　现场安装湿式洗涤器

图 4　多孔阻风墙＋植物条带除臭装置[43]
1. 采样歧管　2. 开口　3. 风速计　4. 植物挡墙

4　讨论及建议

通过饲料营养管理、粪尿清理管理以及末端过滤处理，实现猪场夏季臭气减排，冬季舍内 NH₃ 浓度控制，是规模化猪场臭气减排过程中应着重关注的事项。在满足猪只不同饲养阶段营养需求的条件下，通过优化饲料配方，降低饲料粗蛋白含量，补充必需氨基酸优化氨基酸组成，添加有益微生物提高蛋白转化吸收水平，降低粪尿中含氮量，是减少臭气排放的根本方法。在粪便管理过程中推荐使用半漏缝地板，配合使用干清粪工艺，做到粪尿的干湿分离；对于采用水泡粪工艺的猪舍，要适当减少粪尿贮存天数，增加猪舍通风量，控制舍内氨气含量。末端处理工艺建议在猪舍排风口处采用湿式洗涤器，配合喷淋系统和除臭微生物，降解臭气。

参考文献

[1] 闫志英，许力山，李志东，等. 畜禽粪便恶臭控制研究及应用进展 [J]. 应用与环境生物学报，2014，20（2）：322-327.
[2] Schiffman S S, Bennett J L, Raymer J H. Quantification of odors and odorants from swine operations in North Carolina [J]. Agricultural and Forest Meteorology, 2001, 108（3）：213-240.

［3］国家环境保护局，国家技术监督局 . GB 14554—1993 恶臭污染物排放标准［S］. 北京：中国标准出版社，1993.

［4］中华人民共和国国家市场监督管理总局，中国国家标准化管理委员会 . GB/T 17824. 3—2008 规模猪场环境参数及环境管理［S］. 北京：中国标准出版社，2008.

［5］王悦，赵同科，邹国元，等 . 畜禽养殖舍氨气排放特性及减排技术研究进展［J］. 动物营养学报，2017，29（12）：4249-4259.

［6］徐廷生，雷雪芹，赵芙蓉，等 . 养殖场粪污的恶臭成分及其产生机制［J］. 中国动物保健，2001，29：36-37.

［7］纪英杰，沈根祥，徐昶，等 . 典型季节规模化猪场氨排放特征研究［J］. 农业环境科学学报，2019，38（11）：2573-2582.

［8］刘杨，尚斌，董红敏，等 . 规模猪场机械通风育肥舍氨气产生及排放研究［J］. 农业环境科学学报，2020，39（9）：2058-2065.

［9］王文林，刘筱，韩宇捷，等 . 规模化猪场机械通风水冲粪式栏舍夏季氨日排放特征［J］. 农业工程学报，2018，34（17）：214-221.

［10］许稳，刘学军，孟令敏，等 . 不同养殖阶段猪舍氨气和颗粒物污染特征及其动态［J］. 农业环境科学学报，2018，37（6）：1248-1254.

［11］汪开英，代小蓉，李震宇，等 . 不同地面结构的育肥猪舍 NH_3 排放系数［J］. 农业机械学报，2010，41（1）：163-166.

［12］刘安芳，阮蓉丹，李厅厅，等 . 猪舍内粪污废弃物和有害气体减量化工程技术研究［J］. 农业工程学报，2019，35（15）：200-210.

［13］朱志平，董红敏，尚斌，等 . 育肥猪舍氨气浓度测定与排放通量的估算［J］. 农业环境科学学报，2006（4）：1076-1080.

［14］Kima K Y，Kob H J，Kim H T，et al. Quantification of ammonia and hydrogen sulfide emitted from pig buildings in Korea［J］. Journal of Environmental Management，2008，88（2）：195-202.

［15］Dong H，Kang G，Zhu Z，et al. Ammonia，methane，and carbon dioxide concentrations and emissions of a hoop grower-finisher swine barn［J］. American Society of Agricultural and Biological Engineers，2009，52（5）：1741-1747.

［16］朱志平，康国虎，董红敏，等 . 垫料型猪舍春夏育肥季节的氨气和温室气体状况测试［J］. 中国农业气象，2011，32（3）：356-361.

［17］Rahman S，Newman D. Odor，ammonia，and hydrogen sulfide concentration and emissions from two farrowing-gestation swine operations in North Dakota［J］. Applied Engineering in Agriculture，2012，28（1）：107-115.

［18］孙国荣，任继明，薛惠琴，等 . 低蛋白氨基酸平衡日粮对断奶仔猪生产性能和氮排放的影响［J］. 上海农业学报，2011，27（3）：114-117.

［19］Leek A B G，Hayes E T，Curran T P，et al. The influence of manure composition on emissions of odour and ammonia from finishing pigs fed different concentrations of dietary crude protein［J］. Bioresource Technology，2007，98（18）：3431-3439.

［20］Kerr B J，McKeith F K，Easter R A. Effect on performance and carcass characteristics of nursery to finisher pigs fed reduced crude protein，aminoacid-supplemented diets［J］. Journal of Animal Science，

1995，73（2）：433-440.

[21] Leek A B G，Callan J J，Henry R W. The application of low crude protein wheat-soyabean diets to growing and finishing pigs：1. The effect on the growth performance and carcass characteristics of boars and gilts［J］. Irish Journal of Agricultural and Food Research，2005，44（2）：233-245.

[22] Liu S，Ni J Q，Radcliffe J S，et al. Mitigation of ammonia emissions from pig production using reduced dietary crude protein with amino acid supplementation［J］. Bioresource Technology，2017，233：200-208.

[23] 刘作华，黄健，邓红，等. 低蛋白和杂粕日粮对生长猪生产性能、养分消化、血液指标和猪舍氨气的影响［J］. 饲料工业，2015，36（21）：45-47.

[24] 廖新俤，吴楚泓，雷东锋. 活菌制剂改进猪氮转化和减少猪舍氨气的研究［J］. 家禽生态，2002，23（2）：20-22.

[25] Murphy D P，O'Doherty J V，Boland T M，et al. The effect of benzoic acid concentration on nitrogen metabolism，manure ammonia and odour emissions in finishing pigs［J］. Animal Feed Science and Technology，2011，163：194-199.

[26] 王保黎，樊信鹏，卢丽，等. 丝兰属植物提取物降低猪舍氨气浓度的试验［J］. 畜禽业，2010，257：52-55.

[27] 梁国旗，王旭平，王现盟，等. 樟科、丝兰属植物提取物对仔猪排泄物中氨和硫化氢散发的影响［J］. 中国畜牧杂志，2009，45（13）：22-26.

[28] 暴雪艳，张静，张楠，等. 冬季分娩猪舍内主要有害气体分布规律及其影响因素探究［J］. 山西农业科学，2018，46（06）：1019-1023.

[29] Panetta D M，Powers W，Lorimor J C. Management strategy impacts on ammonia volatilization from swine manure［J］. Journal of Environmental Quality，2005，34（3）：1119-1130.

[30] Lachance I，Jr，Godbout S，et al. Separation of pig manure under slats：to reduce releases in the environment［C］. Florida：Tampa Convention Center，2005.

[31] 牛欢，张政，颜培实. 冬季机械清粪牛舍与人工清粪牛舍空气环境分析［J］. 畜牧与兽医，2015，47（6）：26-31.

[32] Misselbrook T H，Brookman S K E，Smith K A，et al. Crusting of stored dairy slurry to abate ammonia emissions：pilot-scale studies［J］. Journal of Environmental Quality，2005，34（2）：411-419.

[33] Wood J D，Gordon R J，Wagner-Riddle C，et al. Relationships between dairy slurry total solids，gas emissions，and surface crusts［J］. Journal of Environmental Quality，2012，41（3）：694-704.

[34] 李新建，吕刚，任广志. 影响猪场氨气排放的因素及控制措施［J］. 家畜生态学报，2012，33（1）：86-93.

[35] Taghipour H，Shahmansoury M R，Bina B，et al. Comparison of the biological NH_3 removal characteristics of a three stage biofilter with a one stage biofilter［J］. International Journal of Environmental Science & Technology，2006，3（4）：417-424.

[36] Dumont E，Hamon L，Lagadec S，et al. NH_3 biofiltration of piggery air［J］. Journal of Environmental Management，2014，140：26-32.

[37] Yasuda T，Kuroda K，Fukumoto Y，et al. Evaluation of full-Scale biofilter with rockwool mixture

treatingammonia gas from livestock manure composting [J] . Bioresource Technology, 2009, 100 (4): 1568-1572.

[38] Melse R W, Van der Werf A W. Biofiltration for mitigation of methane emission from animalhusbandry [J] . Environmental Science & Technology, 2005, 39 (14): 5460-5468.

[39] Melse R W, Mol G. Odour and ammonia removal from pig house exhaust air using a biotrickling Filter [J] . Water Science and Technology, 2004, 50 (4): 275-282.

[40] Hartung E, Jungbluth T, Bascher W. Reduction of ammonia and odor emissions from a piggery with biofilters [J] . Transactions of the ASAE, 2001, 44 (1): 113-118.

[41] 郭建斌, 牛红林, 韩玉花, 等. 基于PLC的养殖场氨气生物氧化装置设计与试验 [J] . 农业机械学报, 2017, 48 (3): 310-316.

[42] 王昱, 吴鹏, 曾志雄, 等. 规模化猪舍废气复合净化系统设计与试验 [J] . 农业机械学报, 2020, 51 (4): 344-352.

[43] Ajami A, Shah S B, Wang-Li L, et al. Windbreak wall-vegetative strip system to reduce air emissions from mechanically ventilated livestock barns: Part 2-swine house evaluation [J] . Water, Air, & Soil Pollution, 2019, 230 (12): 289-317.

聚落化猪场不同猪舍及清粪工艺对臭气排放的影响[*]

梁晓飞　柴小龙　周思邈　李　同　闫之春

摘要：为了解聚落化猪场模式不同猪舍（育成舍、妊娠舍、分娩舍和育肥舍）及清粪工艺（干清粪、水泡粪）条件下氨气（NH_3）、硫化氢（H_2S）和颗粒物（$PM_{2.5}$和PM_{10}）的排放规律及差异，在山东省德州市选择了一个规模为 17 500 头的聚落化猪场进行了为期 15 d 的试验。试验过程中对不同猪舍的风机口处臭气和颗粒物浓度进行了连续检测（10：00—15：00），同时还检测了干清粪、水泡粪妊娠舍各个风机口处臭气和颗粒物浓度。结果表明，干清粪妊娠舍风机口氨气的浓度是 $0.71\sim1.06$ mg/m³，硫化氢的浓度是 $0.08\sim0.24$ mg/m³，$PM_{2.5}$ 浓度是 $3\sim10$ μg/m³，PM_{10} 浓度是 $11\sim50$ μg/m³；水泡粪妊娠舍风机口氨气浓度为 $1.63\sim5.34$ mg/m³，硫化氢的浓度为 $0.21\sim0.99$ mg/m³，$PM_{2.5}$ 浓度为 $10\sim24$ μg/m³，PM_{10} 浓度为 $25\sim53$ μg/m³。臭气浓度和颗粒物浓度在妊娠舍内分布不均匀，其中水泡粪妊娠舍内臭气和颗粒物浓度分布不均匀性更显著；干清粪方式可以显著降低妊娠舍排风口处 NH_3、H_2S 以及颗粒物浓度；排风口处 NH_3 和 H_2S 浓度与猪只饲养密度成正比，水泡粪工艺下育成舍、妊娠舍、分娩舍和育肥舍的臭气浓度依次增大。

关键词：清粪方式；NH_3；H_2S；$PM_{2.5}$；PM_{10}

生猪养殖在我国畜牧业中占有重要地位，聚落化的生猪养殖模式在维持高效率生产的同时，带来了诸多的环境问题，如大气污染、水体富营养化等。在养殖过程中，会产生大量的悬浮颗粒物、氨气（NH_3）和硫化氢（H_2S）等有害物质，有关资料显示，一个年出栏量 10.8 万头的猪场，每小时向大气排放菌体 15 亿个，氨气 159 kg，硫化氢 14.5 kg，饲料粉尘 25.9 kg[1]。这些有害物质不仅能够导致猪只呼吸系统疾病、过敏反应等，对饲养员及周边居民的健康也会产生一定的负面影响。恶臭污染是世界七大环境公害之一，《恶臭污染物排放标准》中规定了 8 种恶臭物质，氨气和硫化氢均在其中[2]。猪舍中的氨气、硫化氢和颗粒物排放速率取决于猪舍结构、猪只类型、猪的饲养方式和饲料中粗蛋白含量、粪便管理及舍内环境条件等[3-6]。

目前，对聚落化养殖模式下不同猪舍类型、不同清粪方式所产生有害物质的研究较少。因此，本文通过检测聚落化养殖猪场妊娠舍、分娩舍、育肥舍、育成舍以及妊娠舍不同清粪方式下的氨气、硫化氢、$PM_{2.5}$ 和 PM_{10} 的浓度特征，以了解中国典型聚落化猪场养殖模式下

---

* 文章发表于《中国畜牧杂志》2021 年第 57 卷第 9 期。

相关指标的排放情况，为猪舍臭气治理和技术开发提供相关的参考数据。

1 试验材料与方法

1.1 试验设施

本试验在山东省德州市夏津县苏留庄猪场（北纬37°05′26.13″，东经116°07′46.03″）进行，该场的母猪规模为17 500头，其中选择了妊娠舍、分娩舍、育成舍和育肥舍四种不同类型的猪舍进行测量。

妊娠舍简图如图1所示，猪舍长70 m，宽40 m，猪只样本量为1 200头左右，饲养阶段为成年母猪，体重为135～180 kg。猪舍东西走向，东墙为湿帘，西墙有15个风量为58 000 m³/h的轴流风机，南北两面墙各有2个无级变速风机。妊娠舍夏季最大需风量为320 000 m³/h。

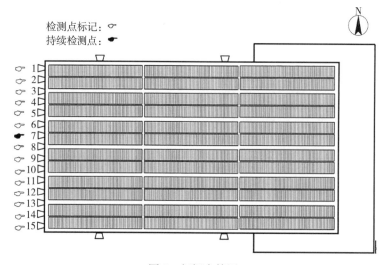

检测点标记：
持续检测点：

图1 妊娠舍简图

分娩舍简图如图2所示，猪舍长140 m，宽32 m，南北走向，共分为10个单元（图2中只展示了1个单元），每个单元有60头左右临产母猪，体重为150～190 kg。西面墙设计为进风湿帘，东面墙装有出风的风机口，每个单元共5个风机：4个风机的风量为58 000 m³/h，1个风机的风量为22 200 m³/h。分娩舍夏季最大需风量为120 000 m³/h。

育成舍/育肥舍简图如图3所示，每栋舍长69 m，宽37 m，分为2个单元，共132个大栏。育成舍每个单元的猪只样本量约为1 500头，体重6～30 kg；育肥舍每单元猪只样本量约为750头，体重30～150 kg。东面墙设计为进风湿帘，西面墙设计为出风风机口，西面墙共16个风机，其中14个风机的风量为58 000 m³/h，2个风机的风量为22 200 m³/h，南北两面墙各2个22 200 m³/h的风机。育成舍夏季最大需风量为160 000 m³/h，育肥舍夏季最大需风量为290 000 m³/h。

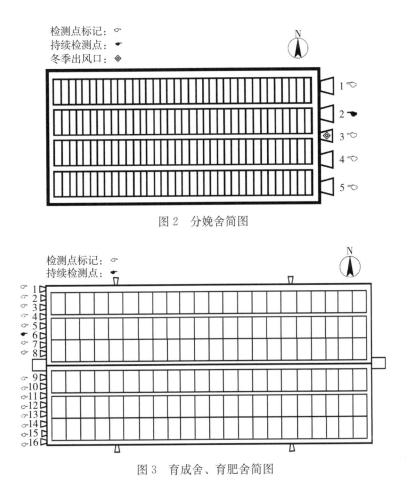

图2　分娩舍简图

图3　育成舍、育肥舍简图

1.2　猪的饲养方式

妊娠舍饲养阶段为配种前到临产，早（8：00）晚（16：00）各喂料1次。分娩舍饲养阶段为分娩前3 d到仔猪断奶出栏（日龄0～3周），饲喂方式为自由饮水、采食。育成舍、育肥舍饲养阶段为仔猪断奶到出栏（日龄1～6个月），育成舍饲喂方式为自由饮水、采食，育肥舍采用定时饮水、采食的饲喂方式。

1.3　猪舍环境管理

妊娠舍有干清粪和水泡粪两种清粪方式，分娩舍、育成舍和育肥舍均为水泡粪工艺。干清粪通过V形排粪沟下方的导尿管及时排出猪尿液，粪便通过刮粪板沿反方向刮出，每天早上（8：00左右）、中午（12：00左右）、下午（17：00左右）各清粪1次。水泡粪通过拔塞的方式进行清粪，清粪周期为25 d左右（本次试验开始时为清粪后的第17天），分娩舍的水泡粪粪池深度为0.8 m，妊娠舍、育成舍和育肥舍的水泡粪粪池深度为1.0 m。猪舍通风形式为机械通风，夏季采用纵向通风模式，室外空气通过湿帘进入猪舍内，猪舍气体从湿帘对面的风机口排出；冬季采用横向通风模式，空气从舍内顶部的新风系统进入，从南、北侧墙的 4 个小风机排出，分娩舍从东墙上小风机排出（图 2 中已标注）。通过猪舍内的温湿度传感器控制风机的开启和关闭时间，保持夏季舍内温度在 20～22℃。

1.4　样品采集与分析

试验在 2020 年 6 月 11—25 日进行，检测点为各猪舍外风机口处，距离风机水平距离 2.0 m，距离地面垂直高度 1.5 m。在 2 个时间点（10：00，16：00）分别测定不同猪舍外风机口处 NH_3、H_2S、$PM_{2.5}$ 和 PM_{10} 的浓度；然后，对不同类型猪舍分别选择一个数据差异性不大的风机口按照相同的测量方式对 NH_3、H_2S 浓度进行连续监测（10：00—15：00）。氨气和硫化氢检测设备为便携式多功能二合一气体检测仪［Smart pro 10-C4-D-PID，H_2S 检测范围为（0～50 mg/m³）/（0.01 mg/m³），电化学原理，NH_3 检测范围为（0～100 mg/m³）/（0.01 mg/m³），电化学原理］，$PM_{2.5}$ 和 PM_{10} 检测设备为激光 $PM_{2.5}$ 检测仪（JSA2-$PM_{2.5}$，$PM_{2.5}$ 检测范围为 0～2 999 μg/m³，$PM_{2.5}$ 分辨率为 0.1 μg/m³，PM_{10} 检测范围为 0～5 999μg/m³，PM_{10} 分辨率为 0.1 μg/m³，原理为激光检测法）。

1.5　统计分析

使用 Excel 进行数据处理，差异显著性用 LSD 法进行比较，显著水平设为 $P=0.05$。

2　试验结果与分析

2.1　清粪方式对臭气和颗粒物浓度的影响

采用干清粪和水泡粪工艺妊娠舍不同风机口处氨气和硫化氢浓度规律如图 4 所示，干清粪妊娠舍排风口处氨气和硫化氢的浓度均低于水泡粪妊娠舍，干清粪妊娠舍氨气浓度为 0.71～1.06 mg/m³，硫化氢的浓度为 0.08～0.24 mg/m³，水泡粪妊娠舍氨气浓度为1.63～5.34 mg/m³，硫化氢的浓度为 0.21～0.99 mg/m³。干清粪妊娠舍不同风机口处氨气和硫化氢

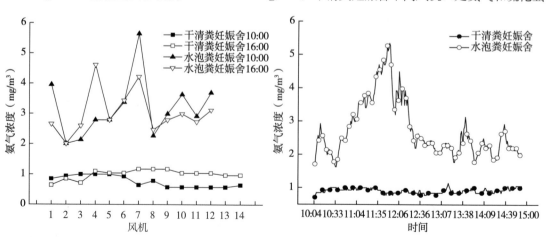

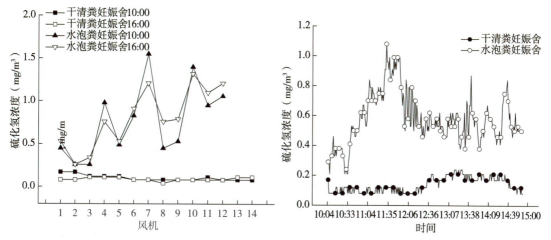

图 4　不同清粪工艺妊娠舍各风机口处氨气和硫化氢气体浓度

浓度相较于水泡粪妊娠舍更稳定。分别对两种清粪工艺妊娠舍的某一个风机口进行氨气和硫化氢的连续检测（10：00—15：00）后发现，水泡粪妊娠舍的氨气和硫化氢浓度在11：00—12：00时间段内显著升高，而干清粪妊娠舍的氨气和硫化氢浓度变化不显著（$P > 0.05$）。

　　猪舍中排放的颗粒物，特别是 $PM_{2.5}$ 和 PM_{10} 会造成大气污染。同时颗粒物还会成为臭气的吸附载体加剧臭气的扩散。图 5 为不同清粪方式的猪舍风机口处测得 $PM_{2.5}$ 和 PM_{10} 浓度，如图 5 所示，干清粪妊娠舍 $PM_{2.5}$ 和 PM_{10} 浓度均小于水泡粪妊娠舍。水泡粪妊娠舍 $PM_{2.5}$ 的浓度为 $10 \sim 24 \ \mu g/m^3$，PM_{10} 的浓度为 $25 \sim 53 \ \mu g/m^3$；干清粪妊娠舍 $PM_{2.5}$ 的浓度为 $3 \sim 10 \ \mu g/m^3$，PM_{10} 的浓度为 $11 \sim 50 \ \mu g/m^3$。

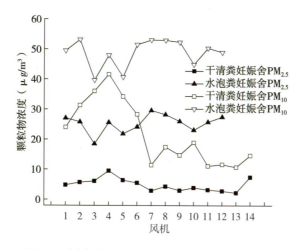

图 5　不同清粪工艺妊娠舍各风机口处颗粒物浓度

2.2　不同猪舍类型对臭气及颗粒物浓度的影响

　　为比较不同猪舍类型的氨气、硫化氢和颗粒物浓度，连续检测（10：00—15：00）五种类型猪舍（干清粪妊娠舍、水泡粪妊娠舍、水泡粪分娩舍、水泡粪育成舍和水泡粪育肥舍）风机口处的氨气、硫化氢和颗粒物的浓度，检测结果如图6所示。水泡粪清粪工艺下不同类型猪舍的氨气浓度差异性显著（$P>0.05$），其中育肥舍的氨气浓度最高（$1.21\sim10.04$ mg/m^3），育成舍氨气浓度最低（$0.78\sim3.49$ mg/m^3）；水泡粪清粪工艺下不同类型猪舍的硫化氢浓度差异性显著（$P>0.05$），分娩舍的硫化氢浓度最高（$1.08\sim4.52$ mg/m^3），妊娠舍硫化氢浓度最低（$0.21\sim0.99$ mg/m^3）。特别地，干清粪妊娠舍的氨气浓度（$0.71\sim1.06$ mg/m^3）和硫化氢浓度（$0.08\sim0.24$ mg/m^3）均低于水泡粪猪舍。

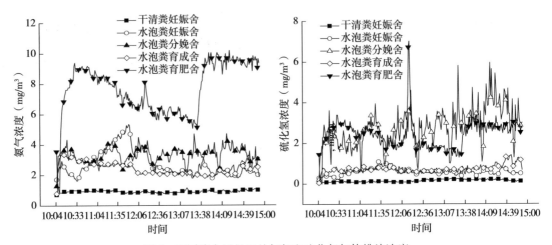

图6　不同猪舍风机口处氨气和硫化氢气体排放浓度

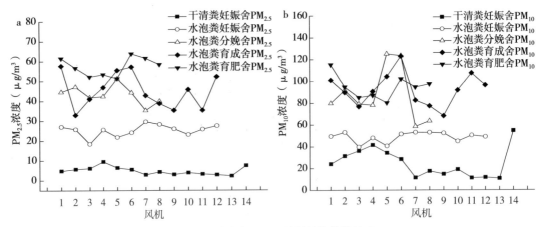

图7　不同猪舍风机口处颗粒物排放浓度

图7显示了各猪舍不同风机口处 PM$_{2.5}$ 和 PM$_{10}$ 的浓度，分娩舍、育成舍和育肥舍的 PM$_{2.5}$ 和 PM$_{10}$ 的浓度较高。其中水泡粪育肥舍风机口处 PM$_{2.5}$ 和 PM$_{10}$ 浓度最高，分别为 $44\sim60$ μg/m^3、$74\sim100$ μg/m^3；水泡粪妊娠舍 PM$_{2.5}$ 和 PM$_{10}$ 浓度最低，分别为 $10\sim24$ μg/m^3、$25\sim53$ μg/m^3。

3　讨论

3.1　清粪方式对风机口处臭气和颗粒物浓度的影响

清粪方式是影响猪舍内臭气排放的重要因素，氨气和硫化氢主要来源于粪尿中蛋白质、氨基酸、尿素等物质的微生物降解过程[7,8]；可吸入颗粒物（PM_{10}）与入肺颗粒物（$PM_{2.5}$）主要来源于饲料、粪便、动物皮肤、体毛和羽毛[7]，吸入的 $PM_{2.5}$ 会破坏猪的呼吸系统和心血管系统的结构和功能。粉尘大，氨气和硫化氢等臭气浓度过高都会增加肺部疾病、心血管疾病和神经系统疾病的发病率，使猪只的机体免疫和生长性能下降[9,10]。汪开英等[11]研究了猪舍地面类型、温度和猪体重与氨气排放的相关性，发现清粪方式对臭气排放有很大影响，生物发酵床氨气的排放量最低，对氨气的减排有一定的效果。刘安芳等[12]研究发现水泡粪工艺因为粪尿长时间留在舍内易形成厌氧发酵产生大量有害气体，使得舍内臭气浓度较高。干清粪通过 V 形排粪沟下方的导尿管能够及时排出猪的尿液，粪便则通过刮粪板沿反方向刮出，每天上、下午各清粪 1 次，清粪更及时，因此干清粪妊娠舍氨气和硫化氢浓度普遍都低于水泡粪妊娠舍。同时，氨气和硫化氢又是 $PM_{2.5}$ 和 PM_{10} 形成的前体物，故水泡粪的 $PM_{2.5}$ 和 PM_{10} 的浓度偏高[13-15]。

通过连续检测（10：00—15：00）水泡粪妊娠舍风机口的氨气和硫化氢浓度，发现 11：00—12：00 时间段内氨气和硫化氢的浓度明显升高，可能是由于中午舍内温度升高，猪只活动剧烈，粪尿排放集中，并且微生物活性增强而导致氨气浓度较高[4]。

3.2　不同猪舍风机口处臭气及颗粒物浓度比较

不同猪舍对应的猪只日龄不同，不同日龄猪只的饲喂管理方式、猪舍构造以及通风模式都不同。Huaitalla 等[16]利用红外光探测仪在线监测了不同猪舍内 PM_{10}、$PM_{2.5}$ 和 PM_1 在夏季和冬季的实时浓度，发现不同猪舍内检测指标差异显著。许稳等[17]研究表明妊娠舍内氨气浓度高于育肥舍，这可能跟猪舍通风方式不同有关，因许稳等采集气体的育肥舍为自然通风，而一般猪舍为机械通风。此次试验中发现，育肥舍的氨气浓度最高，分娩舍、妊娠舍和育成舍氨气浓度依次减小，这可能与猪只饲养密度和喂食饲料的不同相关：育肥舍为大栏圈养，饲养密度大，猪只自由饮水、采食，活动量较大，新陈代谢能力强，从而氨气排放浓度较高[7]。育肥舍和分娩舍风机口处硫化氢浓度也明显高于其他猪舍，分娩舍和育肥舍饲料中粗蛋白含量高于其他饲养阶段的饲料，使得粪便中的蛋白含量也较高，而硫化氢主要由粪便中蛋白分解产生[18,19]，因此硫化氢浓度较高。

分娩舍、育成舍和育肥舍的 $PM_{2.5}$ 和 PM_{10} 浓度显著高于妊娠舍（$P<0.05$），这也与猪只日龄和状态、猪舍构造以及猪只管理有关[20]。妊娠舍为限位栏，饲养密度小，猪只活动量小，$PM_{2.5}$ 和 PM_{10} 浓度小。育成舍和育肥舍为大栏圈养，自由采食，饲养密度大，猪只活动量大，因此 $PM_{2.5}$ 和 PM_{10} 浓度高。分娩舍虽然也是限位栏，但分娩母猪自由采食代谢快、排泄多，另外仔猪保温板上铺撒的干燥粉也会引起 $PM_{2.5}$ 和 PM_{10} 浓度的升高。除此之外，

妊娠舍饲喂方式特殊，饲料和水都在同一个槽内，一定程度上减少了饲料颗粒的扩散，所以分娩舍、育成舍和育肥舍 $PM_{2.5}$ 和 PM_{10} 浓度高于妊娠舍[21]。

4　结论

（1）干清粪与水泡粪相比，干清粪工艺能够有效降低风机口处 NH_3、H_2S、$PM_{2.5}$ 和 PM_{10} 浓度，且统计学水平差异性显著。

（2）不同类型猪舍的 NH_3 和 H_2S 浓度差异性显著，水泡粪育肥舍风机口处 NH_3 浓度最高，水泡粪育成舍风机口处 NH_3 浓度最低；水泡粪分娩舍风机口处 H_2S 浓度最高，水泡粪妊娠舍风机口处 H_2S 浓度最低；分娩舍、育成舍和育肥舍风机口处 $PM_{2.5}$ 和 PM_{10} 浓度显著高于妊娠舍（$P<0.05$）。

（3）水泡粪妊娠舍风机口处 NH_3 和 H_2S 浓度在 10：00—12：00 时间段内有明显的升高过程。

参考文献

[1] 杨柳，邱艳君. 畜禽养殖场恶臭气体的生物控制研究 [J]. 中国沼气，2013，31（2）：30-33.

[2] Lu Wenjing，Duan Zhenhan，Li Dong，et al. Characterization of odor emission on the working face of landfill and establishing of odorous compounds index [J]. Waste Management，2015，42（8）：74-81.

[3] Liu S，Ni J Q，Radcliffe J S，et al. Mitigation of ammonia emissions from pig production using reduced dietary crude protein with amino acid supplementation [J]. Bioresource Technology，2017，233：200-208.

[4] Ngwabie N M，Chungong B N，Yengong F L. Characterisation of pig manure for methane emission modelling in Sub-Saharan Africa [J]. Biosyst Engineering，2018，170：31-38.

[5] Dennehy C，Lawlor P G，Jiang Y，et al. Greenhouse gas emissions from different pig manure management techniques：A critical analysis [J]. Frontiers of Environmental Science & Engineering，2017，11（3）：13-18.

[6] Petersen S O，Olsen A B，Elsgaard L，et al. Estimation of methane emissions from slurry pits below pig and cattle confinements [J]. PLoS One，2016，11（8）：1-16.

[7] 许稳，刘学军，孟令敏，等. 不同养殖阶段猪舍氨气和颗粒物污染特征及其动态 [J]. 农业环境科学学报，2018，37（6）：1248-1254.

[8] 李厅厅，阮蓉丹，蒲施桦，等. 猪舍氨气防控措施研究进展 [J]. 中国畜牧杂志，2019，55（9）：5-10.

[9] 朱志平，董红敏，尚斌，等. 育肥猪舍氨气浓度测定与排放通量的估算 [J]. 农业环境科学学报，2006，25（4）：1076-1080.

[10] 石志芳，姬真真，席磊. 规模化猪场 NH_3 排放特征及影响因素研究 [J]. 中国畜牧杂志，2017，53（8）：100-104.

[11] 汪开英，代小蓉，李震宇，等. 不同地面结构的育肥猪舍 NH_3 排放系数 [J]. 农业机械学报，2010，41（1）：163-166.

［12］刘安芳，阮蓉丹，李厅厅，等．猪舍内粪污废弃物和有害气体减量化工程技术研究［J］．农业工程学报，2019，35（15）：200-210．

［13］纪英杰，沈根祥，徐昶，等．典型季节规模化猪场氨排放特征研究［J］．农业环境科学学报，2006，4：1076-1080．

［14］郭建斌，牛红林，韩玉花，等．基于PLC的养殖场氨气生物氧化装置设计与试验［J］．农业机械学报，2017，48（3）：310-316．

［15］郭军蕊，刘国华，杨斌张，等．畜禽养殖场除臭技术研究进展［J］．动物营养学报，2013，8：1708-1714．

［16］Huaitalla M R，Gallmann E，Liu X J，et al. Aerial pollutants on a pig farm in peri-urban Beijing, China［J］. International Journal of Agricultural and Biological Engineering，2013，6（1）：36-47．

［17］许稳，刘学军，孟令敏，等．不同养殖阶段猪舍氨气和颗粒物污染特征及其动态［J］．农业环境科学学报，2018，37（6）：1248-1254．

［18］黄健，王胜，李其松，等．养殖臭气污染及综合治理探讨［J］．畜牧业环境，2018，10：34-36．

［19］梁宏志．北方地区猪舍内引发生猪疫病的有害气体［J］．当代畜牧，2015，9：46-47．

［20］Van Ransbeeck N，Van Langenhove H，Michiels A，et al. Exposure levels of farmers and veterinarians to particulate matter and gases during operational tasks in pig-fattening houses［J］. Annals of Agricultural Environmental Medicine，2014，21（3）：472-478．

［21］黄凯，唐倩，沈丹，等．冬季不同类型猪舍内颗粒物与微生物气溶胶浓度分布规律研究［J］．农业环境科学学报，2019，38（7）：1616-1623．

低温等离子协同催化在聚落化猪场氨气减排中的应用[*]

张秀之　周思邈　寿冬金　侯景宇　李　同　张仲飞　闫之春

摘要： 随着规模化养猪场的臭气问题日益严峻，亟须探寻有效的臭气治理方法。本试验在山东省德州市聚落化猪场选择一猪舍排风口安装中试低温等离子设备，探究了低温等离子技术运用于聚落化猪场排风口的除 NH_3 效果。通过四种不同类型（高能1、高能2、低能1、低能2）的中试低温等离子设备进行除氨试验。结果表明，在猪舍高通风量的条件下，高能1试验组效果最好，除氨效率达到了 70.03%，并通过催化剂的添加使得除氨效果增加了 4.64%，猪舍风机口 NH_3 初始浓度存在变化（1.97～5.89 mg/m^3），经设备处理后 NH_3 浓度降低至 0.50～0.92 mg/m^3，并对本试验所利用的4台中试低温等离子设备进行了运行成本分析。

关键词： 聚落化猪场；氨气去除；低温等离子

近年来，随着各大养猪集团兴起，养猪场由原来的小户养殖转入规模化养殖，聚落化猪场养殖模式在维持高效率生产的同时，导致污染物的增加，给局部区域带来了臭气问题。臭气扩散会对周围环境产生一定程度的不良影响，其中 NH_3 会使人产生不愉快的感觉并容易引发一些呼吸道疾病，臭气的污染治理迫在眉睫[1-3]。据研究，年出栏量10.8万头的猪场，每小时可向大气中排放氨气 159.0 kg，硫化氢 14.5 kg，饲料粉尘 25.9 kg，养殖畜禽种类、生产管理方式、粪尿处理措施等因素都会影响养殖场臭气成分[3-5]。畜禽养殖场排泄物相关的臭气物质种类繁多，其中猪舍粪便中包括230种恶臭物质，主要包括 NH_3、H_2S、粪臭素、挥发性脂肪酸（VFA）硫醇类、醛类等，NH_3 大部分来源于排泄物中尿酸分解产生的具有强烈刺激性气味的气体[6-7]。

目前，畜禽养殖场臭气减量方法分为源头控制、养殖环节改进和末端处理，处理方法包括水洗法、药液清洗法、活性炭吸附法、一级生物除臭法等[5,8]。聚落化猪场大都采用纵向机械通风使得舍内气体相对集中且有组织的排放，从而使得在舍外处理臭气排放成为可能，但是大风量低浓度的持续排放又给臭气处理带来了技术挑战。低温等离子除臭技术作为一项新型的除臭技术，越来越广泛地运用于臭气处理，具有处理流程短、使用范围广、无二次污染的优点[9]。此外，低温等离子体可与催化剂氧化相结合，协同降低臭气浓度[10-12]。但是，目前对低温等离子技术运用至规模化养猪的研究较少，本试验将利用低温等离子与催化剂氧

---

* 该文章发表于《能源环境保护》2021年第35卷第6期。

化相结合技术作用于猪舍末端,探究其对猪舍排风口 NH_3 的处理效果,为聚落化猪场除臭治理提供技术参考。

1　试验材料与方法

1.1　试验设施

中试低温等离子设备系统示意图及实物图如图 1 所示,猪舍风机口与低温等离子设备首端利用密闭帆布连接,舍内气体从首端到末端依次经过除尘滤网、电场 1、催化剂填充层 1、电场 2、催化剂填充层 2、电场 3、催化剂填充层 3、尾端风机。设备安装地点及试验地点为山东省德州市夏津县夏庄养猪场,此猪舍共有 8 个风机口,依次选取 4 个风机口为试验对象安装低温等离子设备。

4 台中试低温等离子设备根据电流强弱和电极类型分为高能线筒式(高能 1)、高能线板式(高能 2)、低能线筒式(低能 1)、低能线板式(低能 2),其中高能电场运行时电流为 25～27A,低能电场运行电流为 4～6A。

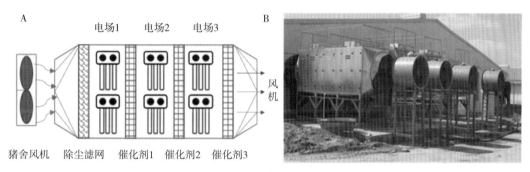

图 1　中试低温等离子除臭系统示意
A. 设备结构示意　B. 设备实景

1.2　试验设计

为验证低温等离子设备对猪场排风口臭气的去除效果,探究其最佳去除参数及成本估算,进行此试验。试验设备(图 1B)置于猪舍排风口处,其中每台设备尾部配有风机,试验时间为 2021 年 8 月 15 日至 9 月 15 日。

本试验进行了 3 种因素的对比试验,包括能量密度、内部结构、催化剂类型,其中电场类型和风机频率决定能量密度;电场类型包括高能电场、低能电场;设备风机运行频率分别为 20 Hz、30 Hz、40 Hz、50 Hz,对应风量为 23 200 m^3/h、34 800 m^3/h、46 400 m^3/h、58 000 m^3/h。内部结构分为线板式和线筒式两种类型。试验开始时 4 台等离子设备同时开启,探究设备的不同风机频率、不同电场类型对猪场排风口臭气中氨气的影响。在选取最优运行条件及设备后,通过填充更换 3 种不同类型的催化剂,即低锰系(催化剂 1)、铁系(催化剂 2)、高锰系(催化剂 3),保证催化剂的填充方式及数量相同,进而探究其对氨气去除率的影响。每组试验重复 3 次，试验结果取平均值。

1.3 样品采集与分析

氨气的监测方法为处理前后各 1 台在线检测器进行监测，在风机开启时，每组处理条件监测 15 min，其中每 1 min 记录 1 次数据，监测结果取平均值，每组处理重复 3 次，最终结果取平均值，每组处理条件的时间间隔为 10 min；臭气浓度采用《空气质量恶臭的测定三点比较式臭袋法》（GB/T 14675—1993）进行测定。NH_3 检测设备为便携式多功能二合一气体检测仪（Smart pro 10-C4-D-PID）。

1.4 统计分析

使用 Excel 进行数据处理，差异显著性用 LSD 法进行比较，数据通过 Origin 9.0 作图。能量密度计算公式为：

$$能量密度(J/m^3) = \frac{各电场单位时间功率总和 \times 3.6 \times 10^6}{空气流量}$$

式中，各电场单位时间功率总和（W/h）通过设备控制系统中所显示的各电场电流、电压进行计算；空气流量通过风机运行频率和最大出风量（58 000 m^3/h）值进行计算。

2 试验结果与分析

2.1 不同电场类型对于猪舍外氨气去除率的影响

如图 2 所示，低温等离子设备可降低猪舍排风口氨气浓度，原因在于低温等离子体可利用高能电子、自由基活性电子作用于废气中的污染物，使得污染物组分在极短的时间内分解，以达到降解污染物的目的[9]。不同类型的电场（高能 1、高能 2、低能 1、低能 2）对于氨气的去除率均随能量密度的增加而增加，其中能量密度由电场运行时的电流、电压及风机运行频率决定。能量密度越高，反应器中各种自由基等活性粒子（如羟基自由基、氧原子）的数量增多，更多的电子获得高能量，高能电子和 NH_3 分子（或其他恶臭分子）发生碰撞时更易打开其化学键使其分解，从而促进了 NH_3 分子的氧化分解[13,14]。所以，不同类型电场运行时可达到的能量密度决定了对臭气中氨气的去除率。

利用高能电场进行排风口除臭时，如表 1 所示，高能 1、高能 2 能量密度最高分别能达到 2 022.42 J/m^3、1 862.72 J/m^3，对应的氨气去除率分别为 70.03%、64.86%（图 2A、B），能量密度最低分别为 838.84 J/m^3、792.75 J/m^3，氨气去除率分别为 31.03%、29.33%（图 2A、B）。由图 2 可见，高能 1 运行时对于氨气的去除效果最大高于高能 2 设备 5.23%。低能电场在运行时所能达到的能量密度显著小于高能电场，导致氨气的去除率较低，低能 1、低能 2 运行的能量密度最高仅为 724.55 J/m^3、502.45 J/m^3，对应的氨气去除率分别为 30.55%、29.78%（图 2C、D）。

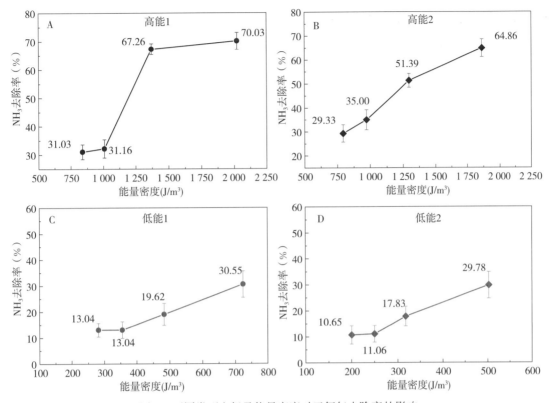

图 2　不同类型电场及能量密度对于氨气去除率的影响

表 1　不同风速条件下各低温等离子设备的能量密度

风速（m³/h）	能量密度（J/m³）			
	高能 1	高能 2	低能 1	低能 2
23 200	2 022.42	1 862.72	724.55	502.45
34 800	1 371.93	1 296.71	483.23	318.00
46 400	1 007.15	972.53	355.59	250.30
58 000	838.84	792.75	281.83	200.24

　　龚永骏等[9]将低温等离子除臭技术运用到了医疗废水处理中，臭气通风量为 1 800 m³/h，发现 NH_3 的去除率达到 90％以上。付丽丽等[15]利用雾化协同等离子体，对模拟氨气恶臭气体进行降解，研究发现，气体停留时间 10 s 时低温等离子对于氨气的去除率可达 80％以上。本试验将低温等离子技术利用到猪场风机口，高能 1 试验组 NH_3 去除率最高为 70％，原因在于试验猪场安装的低温等离子设备风机口最低通风量（23 200 m³/h）高于 1 800 m³/h，气体在设备中的滞留时间较短，臭气污染物未能充分分解。

　　在能量密度无明显差异时，线筒式结构的低温等离子设备对于氨气的去除率高于线板式。由图 2 可见，高能 1 运行时对于氨气的去除效果最大高于高能 2 设备 5.17％，低能 1 运行时对于氨气的去除效果最大高于低能 2 设备 2.39％，原因在于线筒式结构可使得反应器中的活性离子聚集密度增加，从而提高了氨气去除效果，但提高效果并不显著。

2.2　不同催化剂类型对于氨气去除效果的影响

已有研究表明，添加催化剂可增强活性粒子与污染物的碰撞概率，同时延长污染物的停留时间，提供氧活性点，催化剂中晶格氧易被低温等离子体激活，促进氧在催化剂表面的氧化反应，从而促进污染物降解[16-19]。基于之前的试验结果，本试验选择高能 1 等离子设备作为研究对象，通过向设备之内填充不同类型的催化剂，研究不同催化剂的填充对于猪舍排风口氨气去除效果的影响。

本试验所用工艺为催化剂后置二段式工艺[20]，包括三个电场及三段催化层，采用低锰系（催化剂 1）、铁系（催化剂 2）、高锰系（催化剂 3）填充至低温等离子设备时发现，3 种催化剂的添加并没有使得系统氨气去除率显著增加（图 3），低锰系催化剂的添加得到了最高 NH₃ 去除率。由图 3 可见，催化剂的添加对于猪场臭气中氨气的去除效果并没有显著性增加效果，其中催化剂 1 的填充在高能量密度运行时高于无添加催化剂试验组 4.58%，3 种催化剂添加后氨气去除率最高分别为 74.67%、65.23%、69.52%。

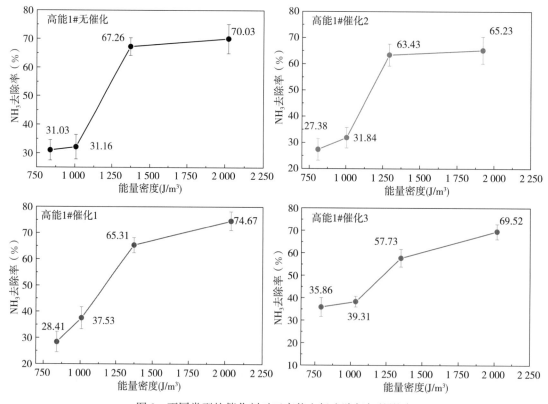

图 3　不同类型的催化剂对于高能电场去除氨气的影响

影响催化剂未能达到理想去除效果的原因在于，一是废气在等离子体区停留时间较低，导致污染物分子来不及反应就离开等离子体；二是粉尘的存在与污染物分子之间产生吸收高能粒子的竞争关系，而且，粉尘中的碱金属和痕量重金属均能影响催化剂活性组分，大量粉尘可堵塞催化剂的气体通道;三是从猪舍内部排出的气体中可能存在有机污染物,在等离子体区形成纳米气溶胶,覆盖在催化剂活性位,引起催化剂失活[20-24]。

2.3　不同电场处理前后猪舍外各指标的浓度变化

对于猪场臭气中氨气的浓度变化,最佳条件处理后臭气中氨气质量浓度如图 4 所示。本试验中各试验组初始浓度存在变化,原因在于猪场内部猪只活动、舍内外温度变化、微生物活性等原因,导致排风口的 NH_3 初始浓度发生变化[2,3]。根据本试验测试结果,在处理前猪舍外风机口的 NH_3 浓度为 $1.97 \sim 5.89\ mg/m^3$,这与梁晓飞等[5]对水泡粪妊娠舍风机口氨气浓度检测后的 NH_3 浓度值($1.63 \sim 5.34\ mg/m^3$)相近。

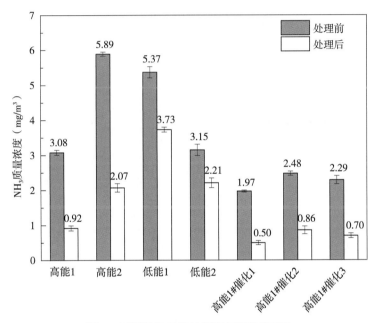

图 4　不同设备最佳处理后氨气浓度对比

高能 1 试验组获得了最佳的 NH_3 去除效果,去除后的 NH_3 浓度值为 $0.50 \sim 0.92\ mg/m^3$。根据陈杰等[13]的研究,通过低温等离子体去除氨吹脱技术中的氨气后发现去除率最高能达到 91%,但处理后的氨气浓度仍为 $7.88\ mg/m^3$。龚永骏等[9]通过低温等离子技术处理医学废水臭气中的 NH_3 时,从初始浓度 $1.93 \sim 5.28\ mg/m^3$ 降低至 $0.16 \sim 0.52\ mg/m^3$。由此说明,低温等离子体虽然有可观的 NH_3 去除率,但在 NH_3 高含量初始浓度的条件下,处理后的 NH_3 浓度仍旧很高。

将 3 种不同催化剂填充至高能 1 设备,并进行最佳去除条件运行时,猪舍排风口氨气浓度变化过程如图 5 所示。电场开启后,排风口的 NH_3 质量浓度陡崖式降低,其中高能 1 加 1 号催化剂时得到了最佳的氨气去除效果,处理后浓度在 $0.5\ mg/m^3$ 左右。

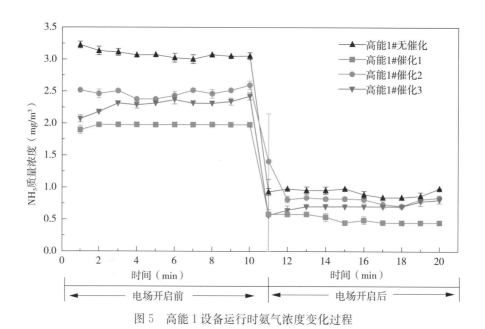

图5　高能1设备运行时氨气浓度变化过程

2.4　低温等离子设备运行成本估算

养殖场普遍反映除臭设备使用成本较高，大型养殖场安装设施设备比例明显比中小型养殖场高，需对所用设备进行运行成本分析，为低温等离子工艺运用于聚落化养殖提供经济性参考[5]。针对该聚落化猪场运营规律，猪舍通过控制排风口的风机来调节猪舍内部的温湿度，夏季平均通风时间为10 h，冬季平均通风时间为4 h。根据此运行条件，下文将对高能低温等离子除臭设备进行运行成本估算。

高能低温等离子除臭设备的运行成本，包括三个电场的能耗及风机运行能耗，其中，运行时三个电场的电流值平均为25.77 A、0.89 A、27.72 A，风机功率为22 kW，以系统最大能量密度，即20 Hz、臭气处理量为23 200 m³/h条件下进行成本估算，电费按照35 kW以上国家标准农业用电价格0.379元/kW·h进行计算，运行成本如表2所示。

表2　低温等离子设备夏季和冬季每天运行成本

季节	电场运行功率（kW）	风机运行功率（kW）	运行时间（h）	成本（元/d）
夏季	47.85	22	10	258.39
冬季	47.85	22	4	103.36

猪舍排风口风机开启的时间决定了设备的运行成本，以本试验4台低温等离子设备为例，在23 200 m³/h臭气处理量条件下，夏季每天运行成本为258.39元，冬季每天运行成本为103.36元。张钊彬[25]通过生物化学方式除臭、化学除臭处理10 000 m³/h污水厂臭气，成本分别为260.3元/d、232.88元/d，说明低温等离子除臭方式运行成本低于生物化学除臭。

通过对本试验所用 4 台低温等离子设备的运行成本分析后发现，该设备运用于聚落化猪场时，由于运行时间随猪场末端风机开启时间变化，而猪舍尾端风机开启时间决定于猪舍内部温度和湿度的变化，因此夏季气温较高猪舍末端风机开启时间长，低温等离子设备运行时间长、成本高；冬季末端风机开启时间短，设备运行成本降低。

3　结论

从本试验可以得出，四种不同类型的中式低温等离子设备均对聚落化猪场 NH_3 的排放具有去除效果。

（1）风机的通风量通过影响设备内部的能量密度而影响 NH_3 去除率，风机通风量越低，能量密度越高，NH_3 去除率越高。高能线筒式低温等离子设备对于猪舍末端排风口的 NH_3 去除效果最佳，去除率可达 70.03%；低能线板式的去除效果最低，去除率最高为 29.78%。内部结构为线筒式的低温等离子设备除氨效果最大高于线板式 5.17%。

（2）催化剂的添加对设备除氨效果并没有显著性的提高，通过低锰系催化剂的添加可提高设备的氨气去除率 5.17%。

（3）中试低温等离子设备运用于猪舍末端时，夏季运行成本为 258.39 元/d，冬季运行成本为 103.36 元/d。

参考文献

[1] 周思邈，柴小龙，梁晓飞，等．规模化养猪过程中臭气的减排措施研究进展 [J]．中国畜牧杂志，2020（12）：1-12．

[2] 楼芳芳，杜喜忠，胡旭进，等．金华市畜禽养殖场减臭技术现状调查及建议 [J]．湖北畜牧兽医，2020，41（7）：2．

[3] 李华琴，何觉聪，陈洲洋，等．低温等离子体-生物法处理硫化氢气体研究 [J]．环境科学，2014，035（004）：1256-1262．

[4] 杨柳，邱艳君．畜禽养殖场恶臭气体的生物控制研究 [J]．中国沼气，2013，02：30-33．

[5] 梁晓飞，柴小龙，周思邈，等．聚落化猪场不同猪舍及清粪工艺对臭气排放的影响 [J]．中国畜牧杂志，2020，12：1-9．

[6] 徐廷生，雷雪芹，赵芙蓉，等．养殖场粪污的恶臭成分及其产生机制 [J]．中国动物保健，2001，7：36-37．

[7] 郭军蕊，刘国华，杨斌，等．畜禽养殖场除臭技术研究进展 [J]．动物营养学报，2013，08：1708-1714．

[8] Ngwabie N M, Chungong B N, Yengong F L. Characterisation of pig manure for methane emission modelling in Sub-SaharanAfrica [J]. Biosyst Eng, 2018, 170: 31-38.

[9] 龚永骏，孙英战．低温等离子除臭技术在医疗废水处理中的应用 [J]．能源与环境，2017，1：81-82．

[10] Chung W C, Mei D H, Tu X, et al. Removal of VOCs from gas streams via plasma and catalysis [J]. Catalysis Reviews, 2018, 61 (2): 270-331.

［11］Nozaki T，OkazariK. Non-thermal plasma catalysis of methane：Principles，energy efficiency，and applications［J］. Catalysis Today，2013，211：29-38.

［12］李一倬. 低温等离子体耦合催化去除挥发性有机物的研究［D］. 上海：上海交通大学，2014：51-72.

［13］陈杰，王倩楠，叶志平，等. 低温等离子体结合吹脱法去除垃圾渗滤液恶臭［J］. 化工学报，2012，63（11）：3660-3665.

［14］Kim H. Nonthermal Plasma Processing for Air - Pollution Control：A Historical Review，Current Issues，and Future Prospects［J］. Plasma Processes and Polymers，2004，1（2）.

［15］付丽丽，刘天会，姜彬慧，等. 雾化协同低温等离子体去除氨气的实验研究［J］. 环境工程，2016，7：125-128.

［16］陈鹏，陶雷，谢怡冰，等. 低温等离子体协同催化降解挥发性有机物的研究进展［J］. 化工进展，2019，038（009）：4284-4294.

［17］竺新波. 等离子体协同催化脱除挥发性有机物（VOCs）的机理研究［D］. 杭州：浙江大学，2015.

［18］鲁美娟，汪怀建，黄荣，等. 催化剂对等离子体协同催化降解挥发性有机物影响的研究进展［J］. 环境污染与防治，2018，40（1）：88-94.

［19］梁煜，李茹，李青. 低温等离子体协同催化剂降解甲醛气体研究进展［J］. 当代化工，2020，49（2）：361-364.

［20］付鹏睿，范淑珍，张帅，等. 低温等离子体协同催化降解废气污染物的研究进展［J］. 能源环境保护，2020，34（2）：14-18.

［21］王鑫. 活性炭纤维协同等离子体治理恶臭废气技术研究［D］. 杭州：浙江大学，2006：36-37.

［22］章旭明. 低温等离子体净化处理挥发性有机气体技术研究［D］. 杭州：浙江大学，2011：80-103.

［23］赵业红. 直流电晕低温等离子体协同催化降解低浓度发性有机废气的研究［D］. 杭州：浙江大学，2016：35-37.

［24］何忠，徐明，李建军，等. 某兽药厂废气治理改造工程实例［J］. 环境科技，2015，02：37-39，43.

［25］张钊彬. 污水处理厂高标准除臭技术研究［D］. 哈尔滨：哈尔滨工业大学，2013.

不同施肥梯度对桃树果实品质的影响*

杜鑫宇　周思邈　梁晓飞　张哲栋　李　同　李青宜　闫之春

摘要： 为明确不同施肥梯度对桃树果实品质的影响，本试验通过有机、无机相结合的方式进行梯度试验，旨在确定有机肥对桃树的最佳用量。结果表明，与不施用有机肥比较，增施有机肥能有效提高"小红桃""锦绣黄桃"果实品质。其中，株施有机肥 34.1 kg＋复合肥 1.0 kg＋花后肥处理，"小红桃"果实可溶性固形物含量比对照提高 0.67％，总酸含量比对照低 0.07％，维生素 C 含量比对照提高 60.8％；株施有机肥 34.1 kg 的，"锦绣黄桃"可溶性固形物含量比对照提高 1.20％，总酸含量比对照低 0.05％，维生素 C 含量比对照提高 35.59％。2 个品种的适宜有机肥用量为 34.1 kg；化肥宜与有机肥同时施用，以提高其肥效。

关键词： 有机肥；单果重；果实内在品质；可溶性固形物；维生素 C

我国传统种植过程中，通常对果树大量使用化肥，导致土壤大面积板结[1]。畜禽粪污通过高温腐熟后变成可以还田的有机肥，而施用有机肥对于增加土壤养分、培肥地力、改善土壤结构有着不可替代的作用[2,3]。通过有机肥与无机肥配合施用，不仅可提供植物生长所需的各种营养元素及氨基酸、胡敏酸等营养物质，增加土壤生物活性，改善土壤微环境，增强机体免疫力，也可以通过施用有机肥提高化肥利用率，促进养分资源高效利用，从而提高果树产量和果实品质[4,5]。目前，替代性施肥对桃树果实品质的影响尚未明确，为了进一步了解有机-无机相结合的方式对桃果品质的影响，以发酵猪粪为主要肥料进行此次施肥试验，为化肥减量、有机肥与化肥搭配施用提供理论依据。

1　试验材料与方法

本试验于 2020 年 10 月至 2021 年 7 月在湖北省枣阳市古岭村桃园开展。当地年平均温度为 22 ℃，年平均降水量为 851 mm。供试果树为当地常种品种小红桃"锦绣黄桃"，株行距 4 m×4 m，均为成年树且生长良好，试验前评估当地土壤肥力状况，具体养分含量如表 1 所示。

---

* 该文章发表于《北方果树》2022 年第 2 期。

表 1 试验园土壤养分含量

项目	"小红桃"园	"锦绣黄桃"园
pH	5.03	5.28
有机质（g/kg）	12.73	10.14
全氮（g/kg）	0.63	0.51
全磷（g/kg）	3.63	0.23
全钾（g/kg）	3.63	4.53
水解氮（mg/kg）	35.80	30.13
有效磷（mg/kg）	5.93	3.43
速效钾（mg/kg）	51.00	36.67

1.1 试验设计

根据基肥中有机肥与无机肥的不同配比，以及是否施入花后肥，试验决定在"小红桃"园中设置 1 个当地常规施肥（只施化肥）为对照，6 个不同有机肥与化肥组合如表 2 所示，其中果实膨大肥各处理一致；不同的是基肥中有机肥与化肥的量，花后肥有施和不施之分。在"锦绣黄桃"园设置 1 个当地常规施肥量（只施化肥）为对照，4 个不同有机肥与化肥组合如表 3 所示。其中基肥中的复合肥、果实膨大肥各处理一致，不同的是基肥中有机肥的量不同，花后肥有施和不施之分。基肥施用时间选择在桃树落叶前 30 d，花后肥施用时间为桃树开花后的 7 d，果实膨大肥在采收前 30 d，所施用硫酸钾型复合肥含氮 18%、P_2O_5 6%、K_2O 24%，尿素含氮 46%，农业用硫酸钾含 K_2O 45%，猪粪有机质含量为 71.4 g/kg，总养分 9.92%（均为烘干基计），每个处理选取树势一致的 9 株树。各处理田间管理一致。

表 2 "小红桃"园施肥处理（kg/株）

处理	基肥		花后肥（尿素）	果实膨大肥	
	有机肥	硫酸钾型复合肥		硫酸钾型复合肥	硫酸钾
对照	0.00	1.00	0.00	0.15	0.30
有机肥低量 1	22.70	0.75	0.00	0.15	0.30
有机肥低量 2	22.70	1.00	0.00	0.15	0.30
有机肥低量 3	22.70	1.00	0.30	0.15	0.30
有机肥高量 1	34.10	1.00	0.30	0.15	0.30
有机肥高量 2	34.10	1.00	0.00	0.15	0.30
有机肥高量 3	34.10	0.75	0.00	0.15	0.30

表 3 "锦绣黄桃"园施肥处理（kg/株）

处理	基肥		花后肥（尿素）	果实膨大肥	
	有机肥	硫酸钾型复合肥		硫酸钾型复合肥	硫酸钾
对照		1.00		0.15	0.30
处理 1	15.10	1.00	0.30	0.15	0.30

（续）

处理	基肥		花后肥 （尿素）	果实膨大肥	
	有机肥	硫酸钾型复合肥		硫酸钾型复合肥	硫酸钾
处理 2	22.70	1.00		0.15	0.30
处理 3	34.10	1.00		0.15	0.30
处理 4	45.50	1.00	0.30	0.15	0.30

1.2　测定项目与方法

果实成熟后，每个小区按树的东、南、西、北 4 个方向共采集 30 个果实进行测定。

单果重使用电子天平进行测定；果实纵径/横径/侧径使用 DL91150 游标卡尺测定；可溶性固形物以折射仪法测定（NY/T 2637—2014）；总酸以 pH 计点位滴定法测定（GB 12456—2021）；可溶性糖以 3,5-二硝基水杨酸比色法测定（NY/T 2742—2015）；维生素 C 以高效液相色谱法测定（GB 5009.86—2016）。

1.3　数据分析

采用 Excel 进行数据处理，通过 SPSS 23.0 统计分析数据，利用 Duncan 进行差异显著性分析，利用 Origin 对数据进行图表绘制。

2　试验结果与分析

2.1　不同施肥处理对"小红桃"品质的影响

2.1.1　不同施肥处理对"小红桃"果型指数和单果重的影响　由表 4 可见，各处理果型指数基本一致（差异不显著），说明施肥种类和数量与果型指数无关；株施有机肥 22.7 kg＋复合肥 1.0 kg、株施有机肥 22.7 kg＋复合肥 1.0 kg＋花后肥、株施有机肥 34.1 kg＋复合肥 1.0 kg＋花后肥 3 个处理的侧径最大，为 7 cm，显著高于对照和有机肥 34.1 kg＋复合肥 0.75 kg 的处理，与其他处理差异不显著。这说明施有机肥与果实膨大有关，但与有机肥施用量无关，与基肥中复合肥有关，与花后肥无关。单果重高的也是上述 3 个处理，依次为 162.86 g、162.91 g、163.57 g，显著高于对照 10.97%，高于有机肥 34.1 kg＋硫酸钾复合肥 0.75 kg 的处理。说明单果重与施有机肥有关，但与有机肥施用量无关，与基肥中的复合肥施用量有关，与花后肥无关。

表 4　不同施肥处理对"小红桃"果形、单果重的影响

处理	果实纵径（cm）	果实侧径（cm）	果实横径（cm）	果形指数（%）	平均单果质量（g）
对照	6.48±0.31[b]	6.77±0.37[b]	6.59±0.25[a]	0.98±0.01	146.76±3.20[c]
有机肥低量 1	6.45±0.33[b]	6.94±0.38[ab]	6.67±0.31[a]	0.97±0.01	154.14±4.20[abc]
有机肥低量 2	6.70±0.23[a]	7.02±0.30[a]	6.70±0.39[a]	1.00±0.02	162.86±5.21[a]

（续）

处理	果实纵径（cm）	果实侧径（cm）	果实横径（cm）	果形指数（%）	平均单果质量（g）
有机肥低量 3	6.69±0.29ᵃ	7.03±0.33ᵃ	6.72±0.28ᵃ	1.00±0.01	162.91±2.50ᵃ
有机肥高量 1	6.73±0.33ᵃ	7.12±0.35ᵃ	6.72±0.28ᵃ	1.00±0.01	163.57±3.82ᵃ
有机肥高量 2	6.69±0.28ᵃ	6.90±0.28ᵃᵇ	6.64±0.25ᵃ	1.01±0.01	157.81±3.13ᵃᵇ
有机肥高量 3	6.61±0.29ᵃᵇ	6.77±0.26ᵇ	6.62±0.29ᵃ	1.00±0.02	149.10±5.23ᵇᶜ

注：同列数据肩标不同字母代表存在显著性差异（$P<0.05$），下同。

2.1.2　不同施肥处理对"小红桃"内在品质的影响　由表 5 可见，株施有机肥 34.1 kg＋复合肥 1.0 kg＋花后肥和株施有机肥 22.7 kg＋复合肥 1.0 kg 处理的还原糖含量最高和次高，分别为 3.53% 和 3.33%，二者之间差异不显著，显著高于对照 0.8%、0.6%，与其他处理差异不显著，说明还原糖与施有机肥有关，但与有机肥施用量无关，与花后肥无关。株施有机肥 34.1 kg＋复合肥 1.0 kg＋花后肥和株施有机肥 34.1 kg＋复合肥 1.0 kg 处理的可溶性固形物含量最高和次高，分别为 10.07%、9.97%，二者之间差异不显著，显著高于对照 0.67%、0.57%，与其他处理差异不显著，说明可溶性固形物含量与施有机肥有关，与有机肥施用量关系不大，与花后肥无关。株施有机肥 34.1 kg＋复合肥 1.0 kg＋花后肥、株施有机肥 34.1 kg＋复合肥 1.0 kg 处理的总酸含量最低，均为 0.38%，显著低于对照 0.07%、低于株施有机肥 22.7 kg＋复合肥 0.75 kg 处理 0.13%、低于株施有机肥 22.7 kg＋复合肥 1.0 kg 处理 0.06%，与其他处理差异不显著，说明总酸含量与施用有机肥有关，与有机肥施用量有关，与复合肥施用量无关，与花后肥无关。株施有机肥 34.1 kg＋复合肥 10 kg＋花后肥处理的维生素 C 含量最高，为每 100 g 原料 1.56 mg，显著高于对照 60.8%、高于株施有机肥 22.7 kg＋复合肥 0.75 kg 处理 73.3%，与其他处理差异不显著。说明维生素 C 含量与施用有机肥有关，与有机肥施用量有关，与复合肥施用量有关，与花后肥似有关。

表 5　不同施肥处理对"小红桃"内在品质的影响

处理	还原糖 （g, 按 100 g 原料计）	可溶性固形物 （%）	总酸 （g/kg）	维生素 C （mg, 按 100 g 原料计）
对照	2.73±0.05ᶜ	9.40±0.43ᵇ	4.54±0.09ᵃᵇ	0.97±0.21ᵇ
处理 1	3.00±0.16ᵇᶜ	9.23±0.17ᵃᵇ	5.14±0.17ᵃ	0.90±0.06ᵇ
处理 2	3.33±0.40ᵃᵇ	9.6±0.21ᵃᵇ	4.36±0.28ᵃᵇ	1.13±0.25ᵃᵇ
处理 3	3.27±0.21ᵃᵇ	9.73±0.37ᵃ	4.00±0.16ᵇᶜ	1.11±0.10ᵃᵇ
处理 4	3.53±0.23ᵃ	10.07±0.19ᵃ	3.78±0.10ᶜ	1.56±0.24ᵃ
处理 5	2.97±0.12ᵇᶜ	9.97±0.29ᵃ	3.79±0.19ᵇᶜ	1.26±0.38ᵃᵇ
处理 6	3.13±0.05ᵃᵇᶜ	9.80±0.28ᵃᵇ	3.93±0.19ᵇᶜ	1.19±0.18ᵃᵇ

2.2　不同施肥处理对"锦绣黄桃"品质的影响

2.2.1　不同施肥处理对"锦绣黄桃"果形的影响　由表 6 可见，果径变化与施用有机肥有关，随着有机肥施用量的增加，果实纵径呈逐渐增加的趋势。株施有机肥 45.5 kg＋花后肥处理的纵径最长，为 8.11 cm，显著高于对照、株施有机肥 15.1 kg＋花后肥和株施有机肥 22.7 kg 的处理，与株施有机肥 34.1 kg 的差异不显著；其次为株施有机肥 34.1 kg 处理的纵径为 8.01cm，显著高于对照和株施有机肥 15.1 kg＋花后肥的处理。横径与果型指数和侧径的变化趋势与纵径一致，不同处理单果重的变化趋势与果径变化一致。这说明果径和果重与有机肥施用量有关（但不是有机肥越多越好，从经济效益考虑，宜施有机肥 34.1 kg），与花后施肥无关。

表6　不同施肥处理对"锦绣黄桃"果形、单果质量的影响

处理	果实纵径（cm）	果实侧径（cm）	果实横径（cm）	果形指数（%）	平均单果质量（g）
对照	7.51±0.46c	7.47±0.28d	7.79±0.32c	0.96±0.01	146.76±3.20c
有机肥 15.1 kg	7.58±0.31c	7.56±0.29cd	7.82±0.38c	0.97±0.05	154.14±4.20c
有机肥 22.7 kg	7.84±0.35b	7.73±0.45bc	7.98±0.40bc	0.98±0.02	162.86±5.21b
有机肥 34.1 kg	8.01±0.32ab	8.02±0.31a	8.17±0.38ab	0.98±0.03	162.91±2.50a
有机肥 45.5 kg	8.11±0.34a	7.87±0.24ab	8.26±0.40a	0.98±0.04	163.57±3.82a

2.2.2　不同施肥处理对"锦绣黄桃"内在品质的影响　由表7可见，株施有机肥 22.7～45.5 kg 的果实还原糖含量依次为 4.70%、4.71%、4.71%，显著高于对照 0.53%、0.54%、0.60%，高于株施 15.1 kg＋花后肥 0.77%、0.78%、0.84%；维生素 C 含量依次为每 100 g 原料 5.31 mg、5.41 mg、5.63 mg，显著高于对照 33.08%、35.59%、41.10%。株施有机肥 34.1 kg 和 45.5 kg 的果实可溶性固形物含量达到 12.13%、12.37%，显著高于对照 1.20%、1.44%；总酸含量均最低，为 0.25%，显著低于对照 0.05%，与其他处理差异不显著。说明果实内在品质与有机肥施用量有关，但不是越多越好，以株施有机肥 34.1 kg 为宜。

表7　不同施肥处理"锦绣黄桃"内在品质的影响

处理	还原糖（g，按 100 g 原料计）	可溶性固形物（%）	总酸（g/kg）	维生素 C（mg，按 100 g 原料计）
对照	4.17±0.15bc	10.93±0.97b	2.97±0.11a	3.99±0.94b
有机肥 15.1 kg	3.93±0.23c	11.00±0.52b	2.82±0.14ab	4.50±0.45ab
有机肥 22.7 kg	4.70±0.30ab	11.53±0.45ab	2.79±0.47ab	5.31±0.19a
有机肥 34.1 kg	4.71±0.31ab	12.13±0.15a	2.47±0.05b	5.41±0.22a
有机肥 45.5 kg	4.77±0.17a	12.37±0.40a	2.51±0.17b	5.63±0.36a

3　讨论与结论

有机肥含有作物生长所需的各种营养元素[6]，施用有机肥有利于缓解单一施用化肥的不良影响，并能增强作物抗逆性，提高果树产量和品质[7]。有研究表明，施用有机肥后枸杞的千粒重提高 33.2 g，增产 1 341 kg/hm²，果实内在品质也均有所提高。臧小平[9]以芒果为

试验材料，将化肥设置为固定基础，通过施用不同量的有机肥。芒果果实可溶性固形物呈现上升趋势，可滴定酸呈下降趋势，固酸比大。本试验结果与此相似。本试验中，"小红桃"株施有机肥 22.7 kg＋复合肥 10 kg，单果重达到 162.86 g 以上，比对照增加 10.97％，比株施有机肥 34.1 kg＋复合肥 0.75 kg 的处理增加 9.23％；株施有机肥 34.1 kg＋复合肥 1.0 kg＋花后肥的可溶性固形物含量为 10.07％，比对照提高 0.67％；总酸含量为 0.38％，比对照低 0.07％；维生素 C 含量为每 100 g 原料 1.56 mg，比对照提高 60.8％。这说明从提高果实内在品质考虑，"小红桃"的有机肥适宜用量为株施 34.1 kg。"锦绣黄桃"株施有机肥 22.7～45.5 kg 的单果重均达到 162 g 以上，比对照增加 10.97％；维生素 C 含量达每 100 g 原料 5.31 mg，比对照提高 33％以上；株施有机肥 34.1 kg 和 45.5 kg 的可溶性固形物含量达到 12％以上，比对照高 1.2％以上；总酸含量为 0.25％，比对照降低 0.05％。这说明从提高果实内在品质考虑，"锦绣黄桃"的有机肥适宜用量为株施 34.1 kg。复合肥与有机肥一起作基肥施用，有利于化肥肥效的发挥。本试验中的花后肥无实际意义。

参考文献

[1] 李先，刘强，荣湘民，等 . 有机肥对水稻产量和品质及氮肥利用率的影响 [J]. 湖南农业大学学报（自然科学版），2010，36（3）：258-262.

[2] 张建军，党翼，赵刚，等 . 不同用量有机肥对陇东旱塬黑垆土磷素形态转化及有效性的影响 [J]. 中国土壤与肥料，2016，(2)：32-38.

[3] 许俊香，邹国元，孙钦平，等 . 施用有机肥对蔬菜生长和土壤磷素累积的影响 [J]. 核农学报，2016，30（9）：1824-1832.

[4] 杨旭初，龙莉，屠乃美，等 . 微生物肥料在不同节肥水平下对烟草根际微生物碳代谢指纹的影响[J]. 核农学报，2017，31（10）：2016-2022.

[5] 杨蕾 . 有机肥在果树栽培中的施用技术探究 [J]. 南方农业，2017，11（20）：11-12.

[6] 何文寿 . 植物营养学通论 [M]. 银川：宁夏人民出版社，2004：109-189.

[7] 姜瑞波，张晓霞，吴胜军 . 生物有机肥及其应用前景 [J]. 磷肥与复肥，2003，18（4）：59.

[8] 高亮，丁春明，史卓强，等 . 生物有机肥对枸杞的增产效应 [J]. 山西农业科学，2010，38（8）：45-49.

[9] 臧小平，周兆禧，林兴娥，等 . 不同用量有机肥对芒果果实品质及土壤肥力的影响 [J]. 中国土壤与肥料，2016，01：98-101.

不同固液分离技术对聚落化猪场废水指标的影响

张秀之　周思邈　张　绪　梁晓飞　闫之春*

摘要： 本试验旨在研究不同固液分离方式对于聚落化猪场废水固体的脱除效果，并进行经济性分析。选用组合气浮、叠片螺旋挤压固液分离（简称叠螺固液分离）、卧螺离心固液分离三种不同方式，对聚落化猪场粪水进行中式固液分离试验。结果表明，组合气浮、叠螺固液分离、卧螺离心固液分离均对猪场粪水的固体含量化学需氧量（COD）、总氮（TN）、总磷（TP）起到了不同程度的脱除作用。其中，卧螺离心固液分离对于固体脱除的效果最佳，去除率为92.88%；叠螺固液分离对于粪水的固体去除效果最差，仅为28.46%；组合气浮对COD去除率最高67.8%；三种固液分离方式均对粪水中TN的去除效果最差，最高为49.77%；卧螺离心固液分离设备运行成本最低，吨水处理成本为2元，是本试验条件下最为经济的固液分离方式。

关键词： 猪场粪水；固液分离；悬浮物去除；经济测算

随着集约化养猪场的快速发展，养猪场高浓度排泄物的处理压力越来越大。养殖废水的处理一般是以城市生活污水厂工艺为基础进行开发的，新希望六和集团有限公司多以UASB＋2A/O生化处理方式为主，通过厌氧与好氧或缺氧过程相结合的方式，达到去除水中有机污染物及氮、磷等元素的目的。此外还采用沼气工程＋沼液还田处理工艺。无论采取哪种处理方式，其处理成本主要受污水中污染物浓度的影响，污水中污染物浓度越高，就需要越长的停留时间[1]。因此，在进行生化处理前，尽可能多地去除水中的总固体含量，可以有效降低污水生化处理的时间和成本[2-4]。

开展本试验的养殖场区已配置的固液分离设备包括斜筛式螺旋挤压设备、组合气浮设备、叠片螺旋挤压设备，在生化前端的固液分离方式以斜筛式螺旋挤压设备＋组合气浮设备工艺为主，但是并没有达到理想的固液分离效果，导致后续生化处理压力较大、水处理成本较高，带来严重的投诉和政府惩罚压力。王明等[5]的研究提出，生化前端高效固液分离较直接生化处理可节约45%的废水处理成本，说明粪水的高效固液分离极其必要。针对养殖产业粪水固液分离技术，研究者们已证实了螺旋挤压式固液分离、滚轴过滤固液分离、卧螺离心固液分离、溶气气浮固液分离技术的可行性[6-8]。高其双等[9]通过螺旋挤压式固液分离滚轴过滤固液分离卧螺离心固液分离综合分析后认为，用卧式螺旋沉降离心机对猪场粪水进行初步处理较为合适。但是，江滔等[10]认为，离心技术因运营成本太高，在中国推广难度大。

本试验利用组合气浮设备、叠螺固液分离设备和卧螺离心固液分离设备,对不同固液分离技术进行选型,并验证固液分离效果,为猪产业粪水固液分离提供参考。

1 试验材料与方法

1.1 试验材料

1.1.1 叠螺固液分离工艺对猪场水质的影响 本试验选择东营新好现代农牧有限公司方家一期项目作为试验点,该厂区原工艺为 UASB+沼液还田。本试验探究叠螺固液分离机用于废水处理时,对于猪场废水各指标的影响,为固液分离的选型和使用提供参考,所用设备为叠螺固液分离机,运行功率为 3.75 kW·h。

试验地点:东营新好现代农牧有限公司方家一期项目,72 000 头规模育肥厂区,标准厂,原有污水处理工艺为 UASB+沼液还田。

1.1.2 组合气浮工艺对猪场水质的影响 本试验选择五河新希望六和乔张项目作为试验点,该猪场污水处理工艺为黑膜沼气加沼液还田技术,黑膜沼气所产生的沼液在还田后存在气味大及烧苗的问题。目前氧化塘中的沼液为外运消纳处理,处理成本较高,所以对该猪场工艺进行改造,目标是保证现有沼液资源化处理的同时,实现猪舍进猪后的原水处理,出水悬浮物(SS) <100 mg/L,组合气浮机由山东智博环境工程有限公司提供,总功率为 4.5 kW,处理量为 10 m³/h。组合气浮机试验流程示意见图 1。

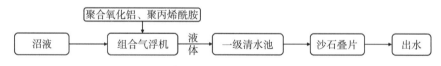

图 1 组合气浮机试验流程示意

1.1.3 卧螺离心机对猪场水质的影响 试验场地为浙江省丽水市中圣环保科技有限公司。试验取水为衢州一海项目集水池污水;卧螺离心机为 LW350-1600 型,功率为 22.5 kW·h,处理量为 5~10 m³/h;聚丙烯酰胺(PAM)阳离子。从衢州一海项目运送 3 000 L 污水至设备厂家进行现场试验。

1.2 试验设计

1.2.1 叠螺固液分离工艺对于猪场水质的影响

(1)分别将 PAM 阴离子、PAM 阳离子、聚合氧化铝(PAC)阳离子作为絮凝剂进行试验(表 1),分别按 0.2%、0.2%、10% 比例稀释,原水取自经格栅和斜筛固液分离后的猪场水泡粪污水,试验每组处理量为 400 mL。

表 1 不同加药量试验设计 (mL)

絮凝剂	处理 1	处理 2	处理 3	处理 4
PAM	10	10	10	15

（续）

絮凝剂	处理 1	处理 2	处理 3	处理 4
PAC	12	15mL	18	15

（2）猪场废水依次经过斜筛固液分离和叠螺式固液分离设备，经斜筛前后取样检测，经叠螺固液分离机时，设置不同进水流速为 3 m³/h、5 m³/h、7 m³/h，通过试验确定最佳加药量，探究叠螺固液分离机，对于猪场废水指标的影响，优化最佳运行条件。

检测指标：悬浮物浓度（SS）、COD、氨氮、TN、TP。

1.2.2　组合气浮机工艺对猪场水质的影响　为保证后续中式气浮设备的运行，选择加药量并提供技术参考，该试验以黑膜沼气工程所产沼液为处理对象，利用 PAM 作为絮凝剂 含铁聚合氯化铝溶液（PFC）为助凝剂。先将 PAM 以 0.3% 的比例溶于水中，试验设计如表 2 和表 3 所示，选择最佳加药量后，进行组合气浮机中试试验。

表 2　不同 PAM 剂量絮凝试验（mL/L）

絮凝剂	处理 1	处理 2	处理 3	处理 4
PAM	50	40	30	20
PFC	3	3	3	3

表 3　不同 PFC 剂量絮凝试验（mL/L）

絮凝剂	处理 1	处理 2	处理 3	处理 4
PAM		根据表 1 实验得出		
PFC	3	3	3	3

检测指标：悬浮物浓度（SS）、COD、氨氮、TN、TP。

1.2.3　卧螺离心机对猪场水质的影响　试验设计如表 4 所示，设置不同进水量为 4 m³/h、7 m³/h、8 m³/h、9 m³/h，对照为未处理水样，探究不同进水量对卧螺离心效率的影响，每组试验重复 3 次，检测指标为 COD、NH₄-N、TN、TP 和 SS 质量浓度。

表 4　不同进水量卧螺离心试验

项目	处理 1	处理 2	处理 3	处理 4
进水量（m³/h）	4	7	8	9
加药比例（%）	5	10	7	5

检测指标：悬浮物浓度（SS）、COD、氨氮、TN、TP。

2　试验结果

2.1　叠螺固液分离工艺对猪场水质的影响

从猪舍被排放出的污水经格栅和斜筛固液分离后，利用 PAC＋PAM 阳离子没有明显的絮凝效果，表观现象如图 2 所示，只添加 PAM 阳离子效果不佳，添加 PAM 阳离子＋PAC 时具有明显絮凝效果，在 PAM 添加量为 25 L/m³、PAC 添加量为 37.5 L/m³时，处理效果最好。

图 2   絮凝试验表观现象

确定最佳加药量后，进行中试试验，5 m³/h、3 m³/h 进水条件下，叠螺固液分离机对于猪场废水指标的影响如图 3 所示。经斜筛固液分离后，各指标实现了第一阶段的去除，COD、NH₄-N、TN、TP、SS 去除率分别为 35.09%、21.01%、21.07%、11.66%、25.93%，经 UASB 进一步降解后，COD 去除率达到了 77.60%，NH₄-N、TN、TP 去除率分别为 58.75%、49.77%、9.14%。原污水处理工艺对于 COD 具有较好的去除效果，但是对于 TP、SS 去除效果并不显著。经叠螺式固液分离工艺去除后，SS 去除率最高为 28.46%，与原工艺无显著差异，絮凝剂＋叠螺式固液分离工艺对于 TP 的去除效果优于原工艺，去除率达到 97.41%、92.88%，且低流量进水有利于叠螺机的固液分离效果。

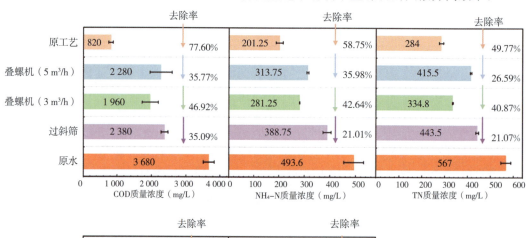

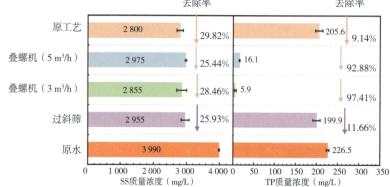

图 3   叠螺固液分离工艺对猪场废水各指标的影响

　　叠螺固液分离机处理 SS 的去除效果并未达到预期，处理后 SS 质量浓度最低为2 855 mg/L，原因在于叠螺固液分离机本身结构适用于 SS 质量浓度较高的污泥脱水，并不适合直接运用至水泡粪处理，用药量高对各指标去除效果并不理想，分离后仍有部分絮凝体存在于水中，无法保证后续雾化喷灌的顺利进行。

2.2　组合气浮机工艺对猪场水质的影响

　　通过图 4 可以看出，PAM、PFC 添加量越高，絮凝效果越好，考虑到絮凝效果和最低加药量，最终选择 40 mL/L PAM＋3 mL/L PFC 添加量进行中试试验。沼液经气浮机处理后进入沙石叠片过滤系统，在进入气浮机前向管道中加入絮凝剂，试验结果如图 5 所示。沼液经絮凝＋气浮处理后，COD 、SS、TP 质量浓度分别降低至值 673 mg/L、95 mg/L、1.84 mg/L，去除率分别为 67.8％、88.6％、95.9％，对于氨氮、TN 没有显著性影响，组合气浮＋沙石叠片系统，对于各指标的去除效果 TP＞SS＞COD≈BOD＞NH$_4$-N＞TN，其中沼液经沙石叠片过滤后与一级清水池指标没有显著性差异。

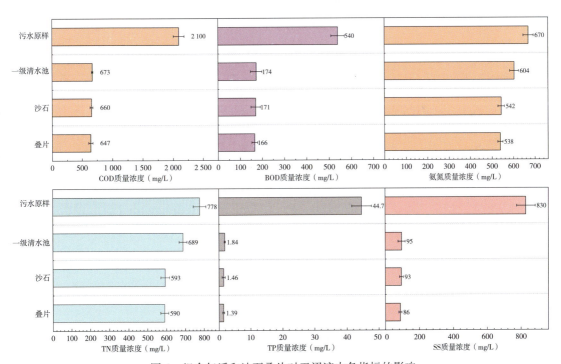

图 4　组合气浮和沙石叠片对于沼液中各指标的影响

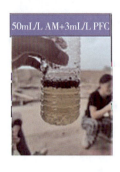
50mL/L AM+3mL/L PFC

40mL/L PAM+3mL/L PFC

30mL/L PAM+3mL/L PFC

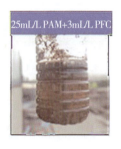
25mL/L PAM+3mL/L PFC

40mL/L PAM+3mL/L PFC

40mL/L PAM+2.5mL/L PFC

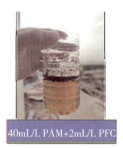
40mL/L PAM+2mL/L PFC

图 5   絮凝试验表观现象（组合气浮工艺）

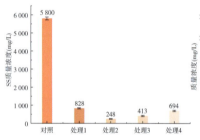
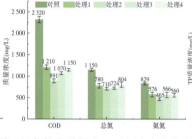
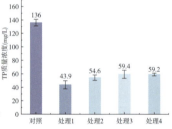

图 6   不同运行条件的卧螺离心对污水各指标的影响

2.3   卧螺离心机对猪场水质的影响

试验结果如图 6 所示，卧螺离心机能够显著降低污水中的 SS、COD、TP 质量浓度，SS 由污水的 5 800 mg/L 降低到 248 mg/L，去除率 95.72%；COD 由污水的 2 320 mg/L 降低到 891 mg/L，去除率 61.69%；TP 去除率 59.82%（表 5）。

表 5   不同条件处理对污水各指标的去除率

处理	进水量（m³/h）	PAM 加药比例（%）	SS 去除率	COD 去除率	氨氮去除率	TN 去除率	TP 去除率
处理 1	4	5	85.72	47.85	30.16	32.17	67.72
处理 2	8	7	92.88	53.88	31.72	37.04	56.32
处理 3	9	5	88.03	50.43	32.45	30.09	56.47
处理 4	7	10	95.72	61.59	43.91	38.26	59.82

3　讨论

3.1　不同固液分离方式效果对比

养殖废水中高浓度污染物质主要来自固态粪污的溶解或微生物的分解作用，在废水产生后立即进行固液分离，可以有效将废水中还未溶解的固态物质分离出去，从而降低废水中污染物的含量和减轻后续生化处理的压力，絮凝剂能和大多数的分离技术结合使用从而提高污染物分离效率[11-14]。本试验利用絮凝剂和固液分离设备相结合，探究了不同固液分离方式对于粪水的影响。

根据试验结果（表6），利用组合气浮机可实现COD最高去除率（68.57%），TP去除率96.73%。刘建生等[15]利用加压气浮可使得COD去除率提高至71%，仅高于本试验2.43%。三种固液分离工艺对于氮元素的去除效果均不理想，其中气浮机对于TN的去除率最低，为23.78%，原因在于，畜禽粪便中85.3%～89.3%的氮素存在于粒径<0.15 mm的细小颗粒中，通过重力沉降很难达到氮素的去除[16]。对于固体脱除，卧螺离心机达到了最高的92.88%，组合气浮机和卧螺离心机达到了88.8%，均显著高于叠片螺旋固液分离机（28.46%）。与本试验相符，Moller等[6]利用3台不同类型的螺旋压缩机对于固体分离的效率为19.2%～49.4%。高其双等[9]研究也表明，卧螺离心机可使高浓度猪粪水中总固体含量降至0.5 g/L，高于螺旋挤压式分离机（总固体质量浓度则高达40.2 g/L）。

综上所述，组合气浮固液分离技术和卧螺离心固液分离技术对于猪场粪水中的固体去除效果优于叠螺固液分离机。

表6　不同固液分离方式处理前后各指标浓度及其去除率

项目	组合气浮机		叠螺固液分离机		卧螺分离机	
	质量浓度（mg/L）	去除率（%）	质量浓度（mg/L）	去除率（%）	质量浓度（mg/L）	去除率（%）
COD	B：2 100±32 A：673±21	68.57±3.1	B：3 680±160 A：1 960±240	46.92±4.21	B：2 320 A：1 070	53.88
NH₄-H	B：670±26 A：604±21	19.1±2.2	B：493.6±38.9 A：281.5±1.25	42.64±4.77	B：829 A：566	31.72
TN	B：778±35 A：689±21	23.78±3.2	B：567±22 A：334.8±1	40.87±2.12	B：1 150 A：724	37.04
TP	B：44.7±8.1 A：1.84±0.2	96.73±2.8	B：226.5±5.5 A：5.9±2.5	97.41±1.06	B：136 A：59.4	56.32
SS	B：830±7 A：95±5	88.8±1.9	B：3 990±10 A：2 855±155	28.46±3.71	B：5 800 A：413	92.88

注：B表示固液分离后，A表示固液分离前。

3.2　以卧螺离心机及组合气浮机为核心的水处理工艺经济性讨论

如图7所示，两种沼液处理方式的固液分离方式不同。以650 m³/d的污水处理量进行设备投入和运行费用计算，如表7所示，组合气浮工艺药费占92%，处理成本为

3.09 元/m^3；卧螺离心工艺处理成本最低，为 2 元/m^3，虽然卧螺离心机的运行功率最高，但是组合气浮设备须配合叠螺固液分离机才能发挥效果，而且，卧螺离心机相对于絮凝剂用量较低。有研究表明，离心机在不加任何絮凝剂的条件下可以去除 60％ 的磷[14]。卧螺离心固液分离工艺是三种固液分离工艺中去除效果好，且运行成本最低的固液分离工艺。

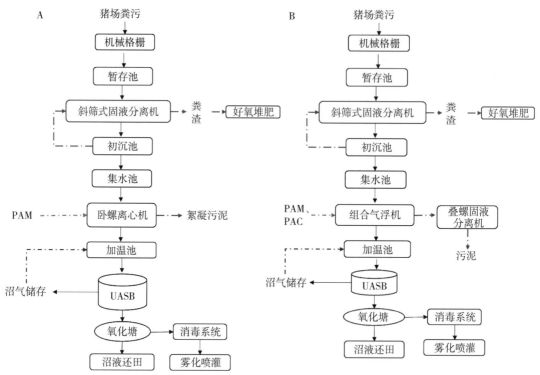

图 7　不同固液分离方式的水处理工艺流程
A. 以卧螺离心机为核心的水处理工艺　B. 以组合气浮机为核心的水处理工艺

表 7　不同固液分离方式的水处理工艺单元成本分析 650 m^3/d 污水处理规模

处理工艺	设备名称	运行功率(kW·h)	数量（个）	购置价格（万元）	运行时间（h）	运行成本（元）
组合气浮机工艺	组合气浮机	13.95	1	13.85	10	83.7
	叠螺固液分离机	3	2	29.3	6	21.6
	配药搅拌器	1.5	3	2.67	10	27
	加药离心泵	0.55	3	1.65	10	9.9
	压泥泵	1.5	3	1.38	6	16.2
	电费合计					158.4
	药剂费用		PFC：3 L/m^3，PAM：0.06 kg/m^3			1 852
	合计			48.85		1 783.4
	吨水成本＝电费＋药费＝3.09（元/t）					

（续）

处理工艺	设备名称	运行功率(kW·h)	数量（个）	购置价格（万元）	运行时间（h）	运行成本（元）
卧螺离心机工艺	卧螺离心机	63.5	1	40	10	381
	配药搅拌器	1.5	1	0.89	10	9
	加药离心泵	0.55	1	0.55	10	3.3
	电费合计					393.3
	药剂费用		PAM：0.07 kg/m³			910
	合计			41.44		1 303.3
	吨水成本＝电费＋药费＝0.61+1.4＝2.00（元/t）					

4　结论

固液分离方式对于聚落化猪场粪水中的固体均起到了不同程度的脱除作用，卧螺离心机去除效果最好，去除率为 92.88%；叠螺固液分离机对于固体脱除效果最差，去除率仅为 28.46%；组合气浮设备对于固体去除率达到了 88.8%。

卧螺离心机是三种固液分离设备中综合去除效果最好，且运行成本最低的固液分离方式，按 650 m³/d 污水处理规模，处理成本为 2 元/m³。

参考文献

[1] Hjorth M，Christensen K V，Christensen M L，et al. Solid-liquid separation of animal slurry in theory and practice. A review [J]. Agronomy for Sustainable Development，2010，30 (1)：153-180.

[2] Bai C，Park H，Wang L. Modelling solid-liquid separation and particle size classification in decanter centrifuges [J]. Separation and Purification Technology，2021.

[3] 郑立忠. 固液分离-UASB-A~2/O混凝沉淀-消毒工艺处理畜禽养殖废水及工程应用 [D]. 武汉：武汉工程大学，2015.

[4] 刘文权，吕小琳，肖锋，等. 螺旋挤压式固液分离机出料口优化设计 [J]. 机械工程与自动化，2019，4：3.

[5] 王明，孔威，晏水平，等. 猪场废水厌氧发酵前固液分离对总固体及污染物的去除效果 [J]. 农业工程学报，2018，34 (17)：235-240.

[6] Moller H B，Lund I，Sommer S G. Solid-liquid separation of livestock slurry：efficiency and cost [J]. Bioresource Technology，2000，74：223-229.

[7] 姚惠娇，董红敏，陶秀萍，等. 浸没式膜生物反应器处理猪场污水运行参数优化 [J]. 农业工程学报，2015，31 (15)：223-230.

[8] Santos A C D，Oliveira R A D. Swine waste water treatment in horizontal anaerobic reactor followed by aerobic sequencing batch reactor [J]. Engenharia Agrícola，2011，31 (4)：781-794.

[9] 高其双，彭霞，卢顺，等. 三种固液分离设备处理猪场粪污的效果及成本比较 [J]. 湖北农业科学，2016，55 (22)：5879-5881.

［10］江滔，温志国，马旭光，等. 畜禽粪便固液分离技术特点及效率评估［J］. 农业工程学报，2016，32（增刊 2）：218-225.

［11］何立红. 畜禽养殖废水处理技术应用及研究进展［J］. 节能与环保，2021，6：84-85.

［12］R Wnetrzak，W Kwapinski，K Peters，et al. The influence of the pig manure separation system on the energy production potentials［J］. Bioresource Technology，2013，136（1）：502-508.

［13］吴军伟. 畜禽粪便固液分离技术研究［D］. 南京：南京农业大学，2009.

［14］孔凡克，邵蕾，杨守军，等. 固液分离技术在畜禽养殖粪水处理与资源化利用中的应用［J］. 猪业科学，2017，34（4）：96-98.

［15］刘建生. 加压溶气气浮机对养猪废水预处理的应用研究［J］. 资源节约与环保，2019，3：3.

［16］Chastain J P. Removal of solids and major plant nutrients from swine manure using a screw press separator［J］. Applied Engineering in Agriculture，2001，17（3）：355-363.

Chapter

5

设备及智能化篇

A Tristimulus-formant Model for Automatic Recognition of Call Types of Laying Hens[*]
（一种用于蛋鸡发声识别的三色共振峰模型）

Xiaodong Du   Guanghui Teng   Chaoyuan Wang   Lenn Carpentier   Tomas Norton

Abstract：An essential objective of Precision Livestock Farming (PLF) is to use sensors that monitor bio-responses that contain important information on the health，well-being and productivity of farmed animals. In the literature，vocalisations of animals have been shown to contain information that can enable farmers to improve their animal husbandry practices. In this study，we focus on the vocalisation bio-responses of birds and specifically develop a sound recognition technique for continuous and automatic assessment of laying hen vocalisations. This study introduces a novel feature called the "tristimulus-formant" for the recognition of call types of laying hens (i. e.，vocalisation types). Tristimulus is considered to be a timbre that is equivalent to the colour attributes of vision. Tristimulus measures the mixture of harmonics in a given sound，which grouped into 3 sections according to the relative weights of the harmonics in the signal. Experiments were designed in which calls from 11 Hy-Line brown hens were recorded in a cage-free setting (4 303 vocalisations were labelled from 168 h of sound recordings). Then，sound processing techniques were used to extract the features of each call type and to classify the vocalisations using the LabVIEW® software. For feature extraction，we focused on extracting the Mel frequency cepstral coefficients (MFCCs) and tristimulus-formant (TF) features. Then，two different classifiers，the backpropagation neural network (BPNN) and Gaussian mixture model (GMM)，were applied to recognise different call types. Finally，comparative trials were designed to test the different recognition models. The results show that the MFCCs-12 + BPNN model (12 variables) had the highest average accuracy of (94. 9 ± 1. 6)% but had the highest model training time (3 201 ± 119) ms. At the same time，the MFCCs-3 + TF + BPNN model had fewer feature dimensionalities (6 variables) and required less training time (2 633 ± 54) ms than the MFCCs-12 + BPNN model and could classify well without compromising accuracy (91. 4 ± 1. 4)%. Additionally，the BPNN classifier was better than

---

* 该文章发表于 *Computers and Electronics in Agriculture*，2021，187：106221。

the GMM classifier in recognising laying hens' calls. The novel model can classify chicken sounds effectively at a low computational cost, giving it considerable potential for large data analysis and online monitoring systems.

Keywords: animal vocalization; sound recognition; chicken; MFCC; tristimulus-formant

1　Introduction

In the past, decisions on the management of farm animals have traditionally been based on the observation, judgment, and experience of farmers. However, the consolidation of livestock production throughout the world has made it increasingly difficult forfarmers to monitor and manage their animals at the level of detail that was once possible. Currently, it has become possible for cameras, microphones, and sensors to take the place of farmers' eyes and ears to monitor animal houses effectively[1,2]. Moreover, such technology can provide further benefits by monitoring animals continuously for 24 h per day and 365 days per year to provide more complete information on livestock [3].

Automatic and continuous sound analyses can provide more detailed information about the state of farm animals. Sound analysis has been used to estimate the thermal comfort of chicks in different thermal environments [4,5]. Sound analysis has also been used to monitor drinking behaviour by analysing pecking sounds [2,6]. Similarly, sound analysis has been used to predict broiler feed intake by acquiring pecking sounds during feeding [7,8]. Moreover, sound analysis has recently formed the basis of a pig cough monitoring system based on the fact that coughing can serve as a biomarker for respiratory disease and aerial pollution in livestock houses[9-11]. Extensions of sound analysis include sound source localisation analysis, which has also been applied to detect the source of pig respiratory diseases and chicken nocturnal vocalisations[12,13].

Animal vocalisations are a fundamental component of animal behaviour, and they can be used to provide information on animal health and welfare[14]. Capturing animal sounds precisely and quickly on the farm has the capacity to help farmers improve their husbandry practices. In animal vocalisation studies, information about a specific vocalisation is often extracted manually from its spectrogram, while the choice of parameters is often driven by the intuition of the researcher. This manual extraction makes the process unsuitable for online and real-time large data analysis[15]. Moreover, given the high level of mechanisation and the large population of animals in typical livestock buildings, a large quantity of different sounds can be heard throughout the day and night. The goal of designing a suitable classifier is a particular challenge in the case of poultry buildings (containing approximately 50 000 broilers or 80 000 laying hens) because it is difficult to realise accurate sound detection algorithms that can be implemented in such buildings[16]. Therefore, to date, it has been almost impossible to realise accurate algorithms for monitoring laying hen sounds and vocalisations. To improve the current state-of-the-art in this field, two key improvements are required for accurate animal vocalisation detection, namely, feature extraction and classification.

For feature extraction, the source-filter theory of vocalproduction is a robust framework for studying animal vocal communication[17]. According to this theory, calls are produced through the syrinx, which is regarded as a 'source' and located at the base of the trachea. Syringeal constriction functionally overlaps the role of the larynx in mammalian phonation, and the trachea acts as a 'filter' to remove certain frequencies or leave others unchanged[18-20]. Source-related vocal features (the fundamental frequency, F0), which are related to the vibrating mass in the syrinx, are stable (laying hens, F0: 400-2 500 Hz)[16], while filter-related features (formants), which are related to the supra-syringeal vocal tract, are dependent on different vocalisations. Specifically for the latter, the first three formants of each vocalisation and the tristimulus values of these formants contain the most energy and variance[21,22]. These tristimulus values were first introduced as a timbre equivalent to the colour attributes in vision analysis, and they represent three different types of energy ratio, which allow a fine description of the first harmonic of the spectrum[23]. Various studies have also shown that the widely used features of Mel frequency cepstral coefficients, MFCCs, can be employed when classifying animal sounds with good effect[24-27]. However, MFCCs often have more feature dimensionalities than other features, which can slow the computational rate. For this reason, an optimal feature combination from tristimulus values and MFCCs with fewer feature dimensionalities can perform better than individual features.

The classification algorithm is the second key component of sound recognition algorithms that operates on the feature output. Researchers have mainly used classifiers such as the Decision Tree (DT)[28-30], Gaussian Mixture Model (GMM)[24,31,32], Neural Network (NN)[31-35], and Support Vector Machine (SVM)[25,27,36]. Although there has been no agreement on which classifier is the most suitable for poultry vocalisation classification, classifiers for chicken calls should be carefully considered[37]. The major challenge is that some laying hens' sounds overlap in the frequency domain. The GMM and NN methods can potentially solve this problem because they both have the ability to differentiate overlapping features, which are already well known in ASR (Automatic speech recognition) and animal vocalisation recognition.

Given the above rationale, this study aims to explore and develop an optimal recognition model for classifying hens' call types (including drinking, laying, twitter and grunt calls). The objectives of this study are as follows: (i) sound feature extraction, (ii) sound classification, and (iii) modelling analysis and comparison.

2　Materials and methods

2.1　Animal and housing

Experiments were conducted on a pilot farm (Shangzhuang Experimental Station of China Agricultural University，Beijing，China). Eleven Hy-Line brown hens were reared to an age of 35-36 weeks. The floor-rearing area was 1.5 m (L) × 1.35 m (W) × 1.8 m (H) (Fig.1). The birds were given ad libitum access to food and water，and a timer-controlled light schedule (light period：6：00 a.m. to 10：00 p.m.) was applied during the experimental period (35-36 weeks). The room environment was suitably controlled to maintain a good level of thermal comfort.

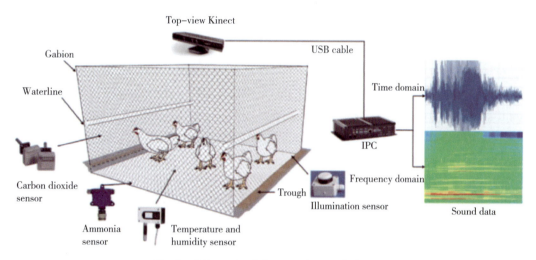

Fig. 1　Schematic of the experiment platform.

2.2　Data collection

A top-view Kinect camera for Windows V1 (Microsoft Corp.，Redmond，WA，USA) was installed at a height of 1.8 m above the ground and used to continuously acquire sound data in WAV format (1 channel，32 bit，16 000 Hz，recording at approximately 55 s of each file) (Fig. 1). The Kinect was connected to a mini industrial personal computer (IPC) via a USB cable. A 2 TB storage USB 3.0 mobile hard disk drive (HDD) was used to store recorded sound data. Data were recorded for approximately 24-h a day for seven days (a total of 168 h). NI LabVIEW 2015 (American National Instrument Corp.，Austin，TX，USA) and were used to pre-process and extract the sound features with the toolkits SVM (sound and vibration module) and MLT (machine learning toolkit)，and the implementation was developed as part of the vocalisation classifier.

2.3　Sound signal pre-processing

An improved spectrum subtraction algorithm was used to pre-process raw sound data by filtering background noise. The algorithm transformed the sound signal from the time domain to the frequency domain, and squared difference values between the sound signal and noise were regarded as the estimated signal power spectrum (Fig. 2). Then, the results were retransformed into time-domain signals using the Fourier inverse transformation for the subsequent processing operation[38,39]. This method has been proven to be a suitable de-noising approach with a low computational cost in practical situations[40]. The improved spectrum subtraction method was calculated following Berouti et al. (1979) and Upadhyay and Karmakar (2013):

$$P(w) = P_s(w) - \alpha P_n(w) \qquad (1)$$

$$P(w) = P_s(w) - \alpha P_n(w) \qquad (2)$$

$$P_s(w) = \begin{cases} P(w), & P(w) > \beta P_n(w) \\ \beta P_n(w), & P(w) \leqslant \beta P_n(w) \end{cases} \qquad (3)$$

where: $P_s(w)$ is the amplitude of the noisy signal power spectrum; $P_n(w)$ is the amplitude of the noise power spectrum; α is the subtraction factor; and β is the spectral floor parameter ($\alpha \geqslant 1$, $0 < \beta \leqslant 1$).

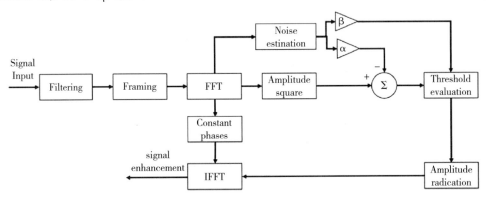

Fig. 2　Flowchart of the imporved spectral subtration algorthm.

2.4　Labelling

After filtering, the sound data were labelled by manual audio-visual inspection[41] performed by the first author, who is an experienced researcher in animal sound analysis. Audacity® software version 2.3.0 was used to label the data by human observers to inspect each recording and annotate the start and end time of the call events. In the process of replaying and visualising the sound recordings using spectrograms, overlapped sounds were not selected as testing samples for further analysis due to their complex acoustic features and the current limitations in sound source separation technology. Finally, there were approximately 15 min and 22 s of data being labelled for subsequent analysis (0. 15% of the original data).

2.5 Feature extraction

2.5.1 Mel frequency cepstral coefficients One of the most popular sources of features that is widely used in animal vocalisation recognition is Mel frequency cepstral coefficients (MFCCs), which are short-term spectrum-based features. The extraction of MFCCs includes the following steps:

(1) Pre-emphasis Usually, the system function is given by H (z) = 1—az^{-1}, where a ∈ [0. 95, 0. 98].

(2) Framing Overlapping frames with a 50% overlap were recommended in each 0. 2 s sound clip to avoid losing information, and the Hamming window was used to reduce the edge effects and spectral leakage in each frame.

(3) Discrete Fourier transform (DFT) Every frame passed through a DFT, and the frequency band was filtered using a filter-bank of triangular filters spaced on the Mel-scale (approximately linear below 1 kHz and logarithmic above 1 kHz) [26]:

$$Mel \ (f) = 2\ 595\log\left(1+\frac{f}{700}\right) \tag{4}$$

(4) Discrete cosine transformation (DCT) The spectral envelope in the decibel unit was obtained by applying a logarithm to the amplitude spectrum. Then, the signal was processed via DCT. The zeroth coefficient of the MFCCs is usually dropped because its value is the average log-energy. The first and second order coefficients of the MFCCs are often used as feature parameters. At the same time, this study only chose a 12-dimensional MFCC (12 vectors) because of the smaller number of feature dimensionalities.

2.5.2 Tristimulus-formant feature The first three formants (F_1-F_3) of each call can represent the main timbre information (Table 1). A popular autocorrelation approach was used for tracking formants [42]. Fig. 3 shows an example of how to locate and label the formants (F_1-F_3). In the spectrogram setting, each frame consisted of N = 512 samples to comply with the size of the short-time Fourier transform (STFT). Overlapping frames with a 50% overlap were recommended to avoid losing information, and the Hamming window was used to reduce the edge effects and spectral leakage in each frame. To highlight the variability of the dominant formants, the tristimulus values of each of the formants were transformed into energy ratios of the first three formants as new tristimulus values (TF_1, TF_2, and TF_3):

$$TF_i = \frac{F_i}{F_1+F_2+F_3}, \ i=1,\ 2,\ 3 \tag{5}$$

Table1   Deacription of feature parameters.

Feature parameter	Description
F_1 （Hz）	The lowest frequency band with substantial enerey is regarded as the frist formant. This is the first harmonic of resonance
F_2 （Hz）	The second harmonic of resonance
F_3 （Hz）	The third harmonic of resonance
TF_1 （%）	F_1 energy ratio derives form the trismulus values
TF_2 （%）	F_2 energy ratio derives form the trismulus values
TF_3 （%）	F_3 energy ratio derives form the trismulus values
MFCCs-12	12-dimensional MFCCs feature

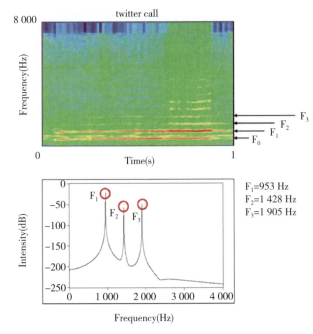

Fig. 3   Excmple of the extraction of the loction of the formants （twitter calls）.

2.6 Classification

Two classifiers，BPNN and GMM，were chosen to recognise different call types， and a total of 4 304 samples were labelled as training data and testing data. To avoid overfitting， k-fold cross-validation was used to estimate the accurate performance of the algorithm[43]. Due to the limitations in the data size， the sound data were split into five smaller sets， and the algorithm was trained using four sets （80% of the data） and validated on the remaining set （20% of the data） .The average of the five validation sets was used to measure the performance of the algorithm as demonstrated by Carpentier et al. （2018）[44].

2.6.1   BPNN   The network selected for this study was a variation of a multilayer, backpropagation neural network, which is a commonly used NN for vocalisation recognition. The network consists of three parts: (1) the input layer; (2) hidden layer (s); and (3) the output layer[15]. The basic principle of the BPNN algorithm is that the learning process consists of two processes-information forward propagation and error back propagation. When information is propagated forward, the input sample is passed into the input layer and then transmitted to the output layer after processing in each hidden layer. If the actual output does not match the expected output, there is back propagation of the errors. In the back-propagation phase, the output is transmitted backward step by step through the hidden layers in a certain form, and the errors are distributed to all the elements of each layer to correct the weights according to the error signal. The definition of the error function is the sum of the square of the difference between the expected output and the actual output[45].

$$e = \frac{1}{2} \sum_{p=1}^{m} (y_p - q_p)^2 \tag{6}$$

Where: e is the error function; y_p is the actual output; q_p is the expected output; p is the index of the output vectors; and m is the number of output vectors. The training data set was used to minimise the error between the predicted call type (the output of the network) and the actual call type (the known sounds in the training set). The weights were adjusted using a gradient descent function with momentum and an adaptive learning rate[33]. The maximum iteration and tolerance were set at 1 000 and 0. 000 1, respectively. Moreover, the accuracy, sensitivity and precision rates were chosen to assess the performance of the BPNN classifier[44]:

$$accuracy = \frac{number\ of\ ture\ positives + number\ of\ ture\ negatives}{number\ of\ total\ samples} \times 100\% \tag{7}$$

$$sensitivtiy = \frac{number\ of\ ture\ positives}{number\ of\ ture\ positives + number\ of\ flase\ positives} \times 100\% \tag{8}$$

$$precision = \frac{number\ of\ ture\ positives}{number\ of\ ture\ positives + number\ of\ flase\ positives} \times 100\% \tag{9}$$

2. 6. 2 GMM Feature classification methods developed for human speech recognition have been applied to species, individual and call type recognition in animals[15,32]. GMM is also widely used because any probability distribution model can be approximated by a weighted combination of multiple Gaussian distributions. The parameters of the Gaussian Mixture Model were calculated by maximising the likelihood function and iteratively using the expectation-maximisation algorithm[31]. The mathematical expression of the GMM for the probability density function is shown as follows[46]:

$$p(x \mid \lambda) = \sum_{i=1}^{m} w_i p_i(x) \tag{10}$$

Where: x is a d-dimensional random vector, p_i (x), $i = 1$, 2, $\cdots$, M, is the component density and w_i, $i = 1$, 2, $\cdots$, M, is the mixture weight. The component densities are d-variate Gaussian functions given by[46]:

$$p_i \ (x) = \frac{1}{\sqrt{(2\pi)^d det \ (\sum_i)}} \exp\left(-\frac{1}{2} \ (x - \mu_i)^T \sum_i^{-1} \ (x - \mu_i)\right) \tag{11}$$

Where: μ_i is the mean, Σ_i is the covariance matrix, and d is the number of features incorporated into every feature vector. The weights w_i must satisfy the following relation[46]:

$$\sum_{i=1}^{M} w_i = 1 \tag{12}$$

Each model can be expressed as a function of the following parameters: $\lambda = (w_i, \mu_i, \Sigma_i)$, $i = 1$, 2, $\cdots$, M. For the Gaussian Mixture Model, different numbers of Gaussian components were selected. The EM algorithm was implemented with a maximum of 1 000 iterations, and the value for tolerance was set to 0.000 1. The Rand index was used to assess the clustering effect, which is expressed as [45]:

$$RI \ \frac{a + b}{C_2^{n \ samples}} \times 100\% \tag{13}$$

Where: C represents the actual call category, K represents the clustering result, a means that both C and K are elements of the same call type, b means that both of C and K are in different categories, and $C_2^{n \ samples}$ represents the number of coupled samples from the data set. The value range of RI in Eq. (13) was 0-100%, which indicates the performance of the clustering effect of GMM.

3　Results and discussion

Comparative trials were conducted to determine the best features and classifier suitable for recognisingthe call types of laying hens. Table 2 and Fig. 4 present the descriptions of the different sound types and their spectrograms, respectively.

Table 2　Description of different sound types.

Sound type	Description	Number of sound clips
Drinking	Pecking sound for water, contains a short duration ($<$0.3 s) and a wide range of frequence bsnd (1-8 kHz)	353
Twitter	Noraml chirp call, contains a long duration ($<$1.0 s) and a distinst harmonic structure (0.1-2 kHz)	744
Laying	Sound in the process of egg laying, contain a succession of shorter notes ($>$1.0 s) and a distinct harmonic structure	1 984

(continued)

Sound type	Description	Number of sound clips
Grunt	Sound of hens snoring at night，contain a long duration (>0.5 s) and a narrow range of frequency band (0.1-5 kHz，concentraed energy wihtin 0.1-2 kHz)	
Fans	Sounds of mechanical noise，contain a random signal and a stable and narrow fondamental frequency band (<1 000 Hz)	330
Total		4 304

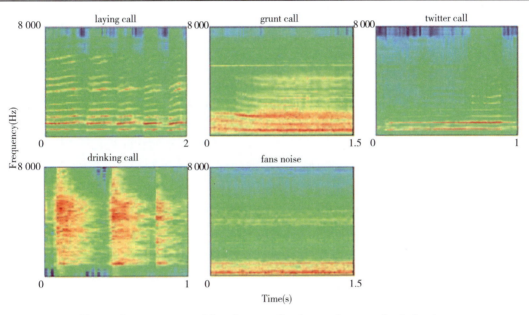

Fig. 4　Spectrogamans of four hens'vocalisations and one mechanical noise.

3.1　Using BPNN with different features

Comparative trials were designed，such as MFCCs-12 ＋ BPNN，MFCCs-3 ＋ TF ＋ BPNN，Formants＋TF＋BPNN，and MFCCs-3＋BPNN. As a result，MFCCs-12 had the highest accuracy，which was (94.9 ± 1.6)% (Table 3). However，MFCCs-12 also had the longest training time. For this reason，dimensionality reduction is a must for practical online identification for analysing large data sets. After a series of tests，the 1st，2nd and 5th dominant vectors of the MFCCs were extracted and reassembled into one 3-dimensional vector (MFCCs-3)，which gave an acceptable recognition rate of (87.3 ± 3.3)% (Table 3 and Table 4). Next，a combination of multiple features was explored to determine whether the recognition rates might be improved，as suggested in the literature [47,48]. Two combined features，formants-TF and MFCCs-3＋TF，were selected for training and testing. Although formants-TF exhibited a non-ideal recognition rate，the joint-feature (MFCCs-3 ＋ TF) approach worked well. Both Fig. 5 and Fig. 6 give information about the visual classification results, which were evaluated by observing a distinct block class of each call type. As shown in the two figures, the feature points of MFCCs-12 are much closer than those of MFCCs-3＋TF and the latter testing points have a more dispersed distribution, which means that MFCCs-3＋TF can better disperse the feature points of different animal call types. In summary, compared with using MFCCs-3 or MFCCs-12 as single, independent features, the combined MFCCs-3＋TF feature can not only increase the recognition effect of MFCCs-3 [from (87.3 ± 3.3)% to (91.4 ± 1.4)%] but also significantly reduce the model training time [from (3 201 ± 119) ms to (2 633 ± 54) ms].

Table 3 Classification performance using the BPNN classifier.

Feature category	Model training time ±SD (ms)	Call type	Classification performance		
			Senstivity±SD (%)	Precision±SD (%)	Accuracy ±SD (%)
MFCCs-12-12D	3 201±119	Drinking	84.2±4.9	83.3±4.3	—
		Twitter	90.1±4.1	92.1±2.0	—
		Laying	97.5±2.9	97.3±1.9	—
		Grunt	95.4±5.5	95.2±5.4	—
		Fans	100.0±0.0	100.0±0.0	
		Total	93.4±1.4	93.6±1.7	94.9±1.6
MFCCs-3＋TF-6D	2 633±54	Drinking	84.3±3.7	89.3±2.3	—
		Twitter	93.8±1.1	91.6±1.5	—
		Laying	92.3±0.6	94.5±0.2	—
		Grunt	87.3±2.4	81.1±3.3	—
		Fans	97.2±0.9	100.0±0.0	—
		Total	91.0±1.5	91.3±1.7	91.4±1.4
Formants＋TF-6D	2 667±28	Drinking	77.6±12.1	79.1±5.6	—
		Twitter	18.7±3.9	59.0±8.5	—
		Laying	89.6±4.8	67.3±1.9	—
		Grunt	61.5±4.8	67.5±5.8	—
		Fans	61.2±9.6	82.5±16.1	—
		Total	61.7±1.5	71.7±4.5	68.4±1.6
MFCCs-3-3D	2 395±38	Drinking	78.7±7.4	75.6±8.2	—
		Twitter	83.5±4.1	86.8±4.5	—
		Laying	89.5±8.0	92.0±5.6	—
		Grunt	84.5±12.6	85.3±13.4	—
		Fans	100.0±0.0	97.6±5.4	—
		Total	87.2±2.3	86.8±3.1	87.3±3.3

Note: D is an abbreviation for dimencison. —means null value.

Table 4　Accuracy rate of each MFCC ventor using the BPNN classifer.

Vector	1	2	3	4	5	6
Accuracy±SD（%）	75.4±0.6	62.8±1.4	54.6±2.0	56.5±1.1	66.3±0.8	58.4±1.0
Vector	7	8	9	10	11	12
Accuracy±SD（%）	60.7±0.7	54.2±0.9	54.1±1.2	61.6±0.3	56.5±2.1	48.0±1.0

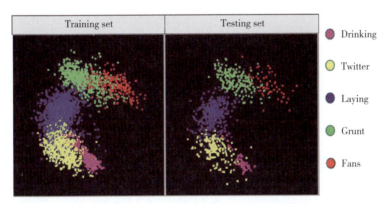

Fig. 5　Classfication results using the BPNN classifer（MFCCs-3＋TF feature）.

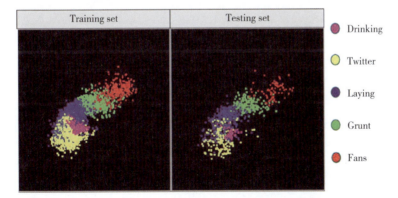

Fig. 6　Classfication results using the BPNN classifer（MFCCs-12 feature）.

Two confusion matrices were created to analyse the classification performance and difference between MFCCs-3＋TF and MFCCs-12（Table 5 and Table 6）. As shown in Table 3，fan noise was easily distinguished from the total 4 304 sound clips because of the difference between the signal of a dynamic sound system（animal）and a static sound system（machine）[5]. In terms of the classification rates of drinking vs. a twitter call，MFCCs-3＋TF outperforms MFCCs-12. The reason for the difference in performance could be due to the different vocal productions of the two call types because MFCCs-3＋TF can better differentiate based on the TF feature. In contrast，MFCCs-12 is superior to MFCCs-3＋TF in the recognition of laying vs. grunt calls. The main reason for the low precision and sensitivity of the grunt call might be due to the similarity of the first three formants in the laying call. Another reason might be the difficulty of tracking accurate formants in the grunt call.

Table 5   Confusion matrix of one vaildation set（MFCCs-3＋TF＋BPNN model）.

Actual call type	Classified by MFCCs-3＋TF feature						
	Drinking	Twitter	Laying	Grunt	Fans	Total	Sensitivity（%）
Drinking	65	12	0	0	0	77	84.4
Twitter	5	158	2	0	0	165	95.8
Laying	0	2	382	30	0	414	92.3
Grunt	0	0	20	128	0	148	86.5
Fans	0	0	0	1	56	57	98.2
Total	70	172	404	159	56	861	—
Precision（%）	92.9	91.9	94.6	80.5	100.0	—	91.6*

Note: * means the accuracy rate. —means null value.

Table 6   Confusion matrix of one vaildation set（MFCCs-12＋BPNN model）.

Actual call type	Classified by MFCCs-12 feature						
	Drinking	Twitter	Laying	Grunt	Fans	Total	Sensitivity（%）
Drinking	57	13	0	0	0	70	81.4
Twitter	10	138	1	0	0	149	92.6
Laying	0	2	392	5	0	397	98.7
Grunt	0	0	5	174	0	179	97.2
Fans	0	0	0	0	66	66	100.0
Total	67	151	398	179	66	861	—
Precision（%）	85.1	91.4	98.5	97.2	100.0	—	94.0*

Note: * means the accuracy rate. —means null value.

3.2   Using GMM with different features

Comparative trials were also designed based on the GMM model，such as MFCCs-12＋GMM，MFCCs-3＋TF＋GMM，Formants＋TF＋GMM，and MFCCs-3＋GMM. As shown in Table 7，MFCCs-12 still shares the highest RI index of（91.7 ± 5.3）%，with the longest model training time. Both MFCCs-3 and MFCCs-12 outperform MFCCs-3 ＋ TF in regard to the clustering effect，but MFCCs-3＋TF is superior to other features at the model training time. Both Fig. 7 and Fig. 8 show the visual clustering results that can be intuitively evaluated by observing the matching degree between the training set and testing set. Unfortunately，many feature points cannot be classified correctly by using MFCCs-3＋TF，but it can better disperse the feature points of different call types（Fig. 7）. In short， compared with MFCCs-3 and MFCCs-12 that take a single feature at a time，the combined feature MFCCs-3＋TF using GMM can decrease the model training time but has an inferior classification rate.

Table 7 Classification performance using the GMM classifier.

Performance parameters	MFCCs-12-12D	MFCCs-3＋TF-6D	Formants＋TF-6D	MFCCs-3-3D
RI（%）	91.5±5.3	73.0±2.1	62.1±0.7	81.9±0.5
Model training time(ms)	3 458±151	2 587±107	2 689±41	2 876±55

Note：D is an abbreviation for dimension.

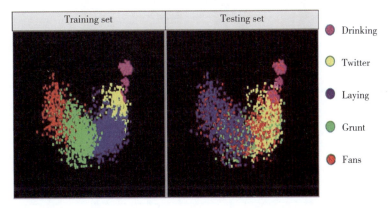

Fig. 7 Classification results using the GMM classifier（MFCCs-3＋TF feature）.

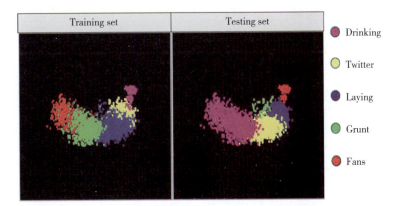

Fig. 8 Classification results using the GMM classifier（MFCCs-12 feature）.

3.3 Comparison of BPNN and GMM performances

To identify an optimal recognition model for recognising hen call types，the accuracy rate，*RI* and model training time of all of the call types were calculated. Fig. 9 and Fig. 10 show the differences in the performance between the BPNN and GMM classifiers. As shown in these two figures，the BPNN classifier obviously outperforms the GMM classifier，and the former also has a shorter model training time. Moreover，the MFCCs-12 feature shares the longest model training time in spite of its high classification rate，which is not suitable for big data analysis. In contrast，the novel MFCC-3＋TF feature is more competent for big data analysis as well as for real-time monitoring because it can effectively recognise hen call types at a low computational cost (a 12.8%-22.3% decrease in the execution time)．

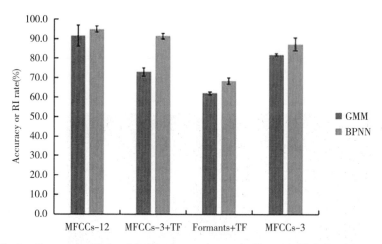

Fig. 9　Compaison of the classification performance between GMM and BPNN.

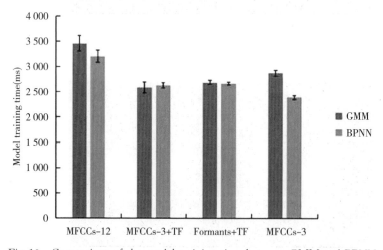

Fig. 10　Comparison of the model training time between GMM and BPNN.

Artificial Neural Networks (ANNs) were first introduced in animal behavioural studies in the early 1990s of the past century，and today，they have been widely used as a valuable acoustic classification tool [49,50]. Compared with traditional statistical approaches，the largest advantage of ANNs is their ability to model complex and non-linear relationships among acoustic parameters [20]. In this paper，the proposed method can be used to recognise the five call types of laying hens with a high accuracy of (94.9 ± 1.6)% (MFCCs-12＋ BPNN model) and (91.4 ± 1.4)% (MFCCs-3＋TF＋BPNN model)．The average precision rates are $(93.6 \pm 1.7)\%$ (MFCCs-12+BPNN model) and $(91.3 \pm 1.7)\%$ (MFCCs-3+ TF+BPNN model). Other similar animal sound recognition rates are the following: 98% for blue monkeys (2 call types: 'pyow' and 'hack' calls) [15], 92% for geese (an average accuracy for 3 behaviours) and 84% (an average precision for 3 behaviours) [36], 80.4%- 92.5% for birds [24], 90% for marine mammals (three call types: whistles, calls and squeaks) [35], 84% for cattle (three ingestive behaviours: chews, bites and composite chew- bites) [51] and 92.5%-95.6% for black lemurs[50]. Favaro demonstrated that ANNs are a powerful tool for studying goat kid contact calls. For each call, 27 spectral and temporal acoustic parameters (including formant parameters) were measured, and the accuracy rates were $(71.1 \pm 1.2)\%$ (vocal individuality, 10 goats), $(79.6 \pm 0.8)\%$ (3 social groups), $(91.4 \pm 0.8)\%$ (maturation, 2 classes) [34]. Similarly, the proposed Formants+TF (5 classes) also show a low accuracy of $(68.4 \pm 1.6)\%$. At the same time, the novel combined feature MFCCs-3+TF has a high accuracy of $(91.4 \pm 1.4)\%$, which overmatches the classification performance found in previous research with fewer feature dimensionalities. In previous research, fuzzy logic values and a feedforward network was used to classify alarm call barks, with 21 neurons in the input layer and 50 neurons in the hidden layer. For different predator species, the lowest accuracy of 79% was obtained when classifying all four species together [52]. Compared with previous studies, the proposed method can better classify 5 classes with a smaller number of neurons and features. It is very difficult to theoretically estimate the number of hidden layers due the possibility of overfitting and additional training time. Increasing this number can further enhance the risk of overfitting. Training time can be saved by avoiding overfitting [49]. The problem of overfitting the training set (overlearning) can be overcome using cross-validation sets (as employed in this study). The proposed method chose 5 hidden neurons after optimised tests[33].

The major disadvantage of machine learning algorithms is that they require large numbers of samples to train the model for high accuracy. Moreover, the training stage of most of the ML algorithms is computationally demanding due to the large number of features used as inputs[53]. To overcome this problem, an optimal feature combination of TF and MFCCs with fewer feature dimensions can classify well without compromising precision or accuracy [training time reducing from $(3\ 201 \pm 119)$ ms to $(2\ 633 \pm 54)$ ms]. In this paper, each call with only 6 feature variables is sufficient to obtain an acceptable classification.

Source-filter theory has been considered to be the commonly used theory for explaining the acoustic characteristics of bird vocalisations [20]. Williams concluded that the syrinx in birds can vary the harmonic amplitude output [54]. Moreover, formants are completely independent of the fundamental frequency (F0) [55]. In this paper, the chosen filter-related features (F1—F3, TF1—TF3) are different among the five call types of laying hens, which is helpful to disperse the feature points for better recognition performance. Additionally, the formant parameters can be used to estimate the biological information of mammals, such as the vocal tract length[56]. At the same time, the mammal vocal tract model might not be suitable for hens because the structure of their respiratory system is very different from that of mammals [17]. Unfortunately, we did not perform a physiological autopsy on the test chicken and were unable to verify the true vocal tract length. This matter remains to be fully investigated in future research.

Moreover, on-site machinery noise is an influencing factor that can reduce classification rates. Other researchers have suggested that ANNs may be very helpful to assign calls with high background noise [50]. To date, it is still a challenge for researchers to implement sound algorithms in a commercial henhouse that stocks a large population of animals (approximately 50 000 broilers, 80 000 laying hens) because of the large quantity of sounds produced during the daytime. At the same time, it was found that hens' vocalisations during the night were less than those during the daytime and that most of the vocalisations were sounds that indicated animal health and production performance, such as the sound of egg laying, grunting and coughing. The application of sound source localisation (SSL) algorithms makes it possible to detect anomalous animal vocalisations at night by monitoring the number of concerned vocalisations and the area distributions for precision analysis [13]. Here, it is recommended to use the tristimulus-formant model to monitor birds when there are fewer than one thousand in a subarea and the SNR (Signal-to-noise ratio) >5 dB [40]. Additionally, the performance of the algorithm might be lower than expected in a chicken barn because the distance between a sound source and a microphone is an affecting factor. A long distance can lead to an inadequate sound quality and a low sound intensity. These problems have not yet been solved completely and remain to be fully investigated. Further studies can explore the possibility of combining call type recognition and SSL algorithms for the automatic detection of specific sounds in a sub-area, which can be considered to be one of the potential applications.

4　Conclusions

In this study, we determined which acoustic features and classifiers have the potential to better recognise each call type of laying hens. The novel model "MFCCs-3+TF+BPNN" performs well without compromising accuracy in recognising hen vocalisations. This model also has less training time and fewer feature dimensions (6 variables) than those of other models. Compared with other animal sound recognition approaches, the proposed model shows considerable potential for online identification and for large data analysis. Further research could be performed to study the relationship between animal behaviour recognition and animal sound recognition by using a multi-modality of video and sound streaming technology.

References

[1] Vandermeulen J , Kashiha M, Ott S , et al. Combination of image and sound analysis for behaviour monitoring in pigs [C] . In: Proceedings of the 6th European conference on Precision Livestock Farming, Leuven, Belgium, 2013, 9 (10-12) : 62-67.

[2] Kashiha M , Bahr C , Haredasht S A , et al. The automatic monitoring of pigs water use by cameras [J] . Computers and Electronics in Agriculture. 2013, 90: 164-169.

[3] Guarino M , Norton T , Berckmans D , et al. A blueprint for developing and applying precision livestock farming tools: a key output of the EU-PLF project [J] . Animal Frontiers, 2017, 7: 12.

[4] De Moura D J. , Naeaes I D A , de Souza Alves E C , et al. Noise analysis to evaluate chick thermal comfort [J] . Scientia Agricola, 2008, 65: 438-443.

[5] Du X , Carpentier L , Teng G et al. Assessment of laying hens' thermal comfort using sound technology [J] . Sensors, 2020, 20 (2): 473.

[6] Pluk A , Cangar O , Bahr C , et al. Impact of process related problems on water intake pattern of broiler chicken [C] . In: Proceedings of the International Conference on Agricultural Engineering, Clermont-Ferrand, France, 2010, 9: (6-8) : 29.

[7] Aydin A , Bahr C , Viazzi S , et al. A novel method to automatically measure the feed intake of broiler chickens by sound technology [J] . Computers and Electronics in Agriculture, 2014, 101: 17-23.

[8] Aydin A , Berckmans D. Using sound technology to automatically detect the short-term feeding behaviours of broiler chickens [J] . Computers and Electronics in Agriculture, 2016, 121: 25-31.

[9] Van Hirtum A , Berckmans D. Objective recognition of cough sound as biomarker for aerial pollutants [J] . Indoor Air, 2004, 14: 10-15.

[10] Exadaktylos V , Silva M , Aerts J M , et al. Real-time recognition of sick pig cough sounds [J]. Computers and Electronics in Agriculture, 2008, 63: 207-214.

[11] Berckmans D , Hemeryck M , Berckmans D , et al. Animal sound... talks! real-time sound analysis for health monitoring in livestock [C] . In: Proceedings of Animal Environment and Welfare, Chongqing, China, 2015, 23-26 October: 215-222.

[12] Silva M , Ferrari S , Costa A , et al. Cough localization for the detection of respiratory diseases in pig houses [J] . Computers and Electronics in Agriculture, 2008, 64: 286-292.

[13] Du X , Lao F , Teng G. A sound source localisation analytical method for monitoring the abnormal night vocalisations of poultry [J] . Sensors, 2018, 18: 2906.

[14] Manteuffel G , Puppe B , Schön P C. Vocalization of farm animals as a measure of welfare [J]. Applied Animal Behaviour Science: 2004, 88: 163-182.

[15] Mielke A , Zuberbühler K. A method for automated individual, species and call type recognition in free-ranging animals [J] . Animal Behaviour, 2013, 86: 475-482.

[16] Cao Y , Yu L , Teng G , et al. Feature extraction and classification of laying hens' vocalization and mechanical noise [J] . Transactions of the Chinese Society of Agricultural Engineering, 2014, 18:

190-197 (in Chinese).

[17] Taylor A M，Reby D. The contribution of source-filter theory to mammal vocal communication research [J]. Journal of Zoology，2010，280：221-236.

[18] Fletcher N H. Bird song - a quantitative acoustic model [J]. Journal of Theoretical Biology，1988，135：455-481.

[19] Beckers G J，Suthers R A，Ten C C. Pure-tone birdsong by resonance filtering of harmonic overtones [J]. Proceedings of the National Academy of Sciences of the United States of America，2003，100：7372-7376.

[20] Favaro L，Gamba M，Alfieri C，et al. Vocal individuality cues in the African penguin (Spheniscus demersus)：a source-filter theory approach [J]. Scientific Reports，2015，5.

[21] Yeon S C，Jeon J H，Houpt K A，et al. Acoustic features of vocalizations of Korean native cows (Bos taurus coreanea) in two different conditions [J]. Applied Animal Behaviour Science，2006，101：1-9.

[22] Favaro L，Gamba M，Gili C，et al. Acoustic correlates of body size and individual identity in banded penguins [J]. PLoS One，2017，12：e0170001.

[23] Pollard H F，Jansson E V. A tristimulus method for the specification of musical timber [J]. Acustica，1982，51：162-171.

[24] Cheng J，Sun Y，Ji L. A call-independent and automatic acoustic system for the individual recognition of animals：a novel model using four passerines [J]. Pattern Recognition，2010，43：3846-3852.

[25] Chung Y，Oh S，Lee J，et al. Automatic detection and recognition of pig wasting diseases using sound data in audio surveillance systems [J]. Sensors，2013，13：12929-12942.

[26] Noda J，Travieso C，S'anchez-Rodríguez，D.. Automatic taxonomic classification of fish based on their acoustic signals [J]. Applied Sciences，2016，6：443.

[27] Bishop J C，Falzon G，Trotter M，et al. Livestock vocalisation classification in farm soundscapes [J]. Computers and Electronics in Agriculture，2019，162：531-542.

[28] Digby A，Towsey M，Bell B D，et al. A practical comparison of manual and autonomous methods for acoustic monitoring [J]. Methods in Ecology and Evolution，2013，4：675-683.

[29] Moi M，Naeaes I D A，Caldara F R，et al. Vocalization data mining for estimating swine stress conditions [J]. Engenharia Agricola，2014，34：445-450.

[30] Mcgrath N，Dunlop R，Dwyer C，et al. Hens vary their vocal repertoire and structure when anticipating different types of reward [J]. Animal Behaviour，2017，130：79-96.

[31] Alonso J B，Cabrera J，Shyamnani R，et al. Automatic anuran identification using noise removal and audio activity detection [J]. Expert Systems with Applications，2017，72：83-92.

[32] Jahn O，Ganchev T D，Marques M I，et al. Automated sound recognition provides insights into the behavioral ecology of a tropical bird [J]. PLoS One，2017，12：e0169041.

[33] Khunarsal P，Lursinsap C，Raicharoen T. Very short time environmental sound classification based on spectrogram pattern matching [J]. Informing Science，2013，243：57-74.

[34] Favaro L，Briefer E F，McElligott A G. Artificial neural network approach for revealing individuality，group membership and age information in goat kid contact calls [J]. Acta Acustica United with Acustica，2014，100：782-789.

［35］Gonz'alez-Hern'andez，F R，S'anchez-Fern'andez L P，Su'arez-Guerra S，et al. Marine mammal sound classification based on a parallel recognition model and octave analysis ［J］. Applied Acoustics，2017，119：17-28.

［36］Steen K A，Therkildsen O R，Karstoft H，et al. A vocal-based analytical method for goose behaviour recognition ［J］. Sensors，2012，12：3773-3788.

［37］Ramachandran R P，Ramachandran R，Farrell K R，et al. Speaker recognition—general classifier approaches and data fusion methods ［J］. Pattern Recognition，2002，35：2801-2821.

［38］Berouti M，Schwartz R，Makhoul, J. Enhancement of speech corrupted by acoustic noise ［C］. ICASSP 79. In：IEEE International Conference on Acoustics，Speech and Signal Processing，1979：208-211.

［39］Upadhyay N，Karmakar A. Spectral subtractive-type algorithms for enhancement of noisy speech：an integrative review ［J］. International Journal of Image and Graphics，2013，5 (11)：13-22.

［40］Du X，Teng G. Research on an improved de-noising method of laying hens' vocalization ［J］. Transactions of the Chinese Society of Agricultural Machinery，2017，48 (12)：327-333 (in Chinese).

［41］Tullo E，Fontana I，Diana A，et al. Application note：Labelling, a methodology to develop reliable algorithm in PLF ［J］. Computers and Electronics in Agriculture，2017，142：424-428.

［42］Rabiner R L，Schafer W R. Introduction to digital speech processing ［J］. Foundations and Trends in Signal Processing，2007，1 (1-2)：1-194.

［43］Refaeilzadeh P，Tang L，Liu H. Cross-validation. In：Liu, L.，Ozsu, M. T. (Eds.)，Encyclopedia of database systems ［M］. Springer US，Boston，MA，2009：532-538.

［44］Carpentier L，Berckmans D，Youssef A，et al. Automatic cough detection for bovine respiratory disease in a calf house ［J］. Biosystems Engineering，2018，173：45-56.

［45］Theodoridis S. Pattern Recognition，fourth edition ［M］. Publishing house of electronics industry，Beijing，2010.

［46］Reynolds D A，Rose R C. Robust text-independent speaker identification using Gaussian mixture speaker models ［J］. IEEE Transactions on Speech & Audio Processing，1995，3 (1)：72-83.

［47］Scheumann M，Roser A E，Konerding W，et al. Vocal correlates of sender-identity and arousal in the isolation calls of domestic kitten (Felis silvestris catus) ［J］. Frontiers in Zoology，2012，9：36.

［48］Fukushima M，Doyle A M，Mullarkey M P，et al. Distributed acoustic cues for caller identity in macaque vocalization ［J］. Royal Society Open Science，2015，2：150432.

［49］Reby D，Lek S，Dimopoulos I，et al. Artificial neural networks as a classification method in the behavioural sciences ［J］. Behavioural Processes，1997，40：35-43.

［50］Pozzi L，Gamba M，Giacoma C. The use of artificial neural networks to classify primate vocalizations：a pilot study on black lemurs ［J］. American Journal of Primatology，2009，72：337-348.

［51］Chelotti J O，Vanrell S R，Milone D H，et al. A real-time algorithm for acoustic monitoring of ingestive behavior of grazing cattle ［J］. Computers and Electronics in Agriculture，2016，127：64-75.

［52］Placer J，Slobodchikoff C N. A fuzzy-neural system for identification of species-specific alarm calls of Gunnison's prairie dogs ［J］. Behavioural Processes，2000，52：1-9.

［53］Acevedo M A，Corrada-Bravo C J，Corrada-Bravo H，et al. Automated classification of bird and amphibian calls using machine learning: a comparison of methods [J] . Ecological Informatics, 2009, 4: 206-214.

[54] Williams H , Cynx J , Nottebohm F . Timbre control in zebra finch (Taeniopygia guttata) song syllables [J] . Journal of Comparative Psychology, 1989, 103: 366-380.

[55] Fitch W T , Kelley J P . Perception of vocal tract resonances by whooping cranes grus americana [J]. Ethology, 2000, 106: 559-574.

[56] Reby D , McComb K . Anatomical constraints generate honesty: acoustic cues to age and weight in the roars of red deer stags [J] . Animal Behaviour, 2003, 65: 519-530.

An Automatic Detection Method for Abnormal Laying Hen Activities Using a 3D Depth Camera[*]
(一种利用二维深度相机自动检测蛋鸡异常活动的方法)

Xiaodong Du   Guanghui Teng

Abstract: With the increasing scale of farms and the correspondingly higher number of laying hens, it is increasingly difficult for farmers to monitor their animals in a traditional way. Early warningof abnormal animal activities is helpful for farmers' fast response to the negative impact on animal health, animal welfare and daily management. This study introduces an automatic and non-invasive method for detecting abnormal poultry activities using a 3D depth camera. A typical region including eighteen Hy-line brownlaying hens was continuously monitored by a top-view Kinect during 49 continuous days. A mean prediction model (MPM), based onthe frame difference algorithm, was built to monitor animal activities and occupation zones. As a result, this method reported abnormal activities with an average accuracy of 84.2% and a rate of misclassifying abnormal events of 15.8% (P_{FPR}). Additionally, it was found that the flock showed a diurnal change pattern in the activity and occupation quantified index. They also presented a similar changing pattern each week.

Keywords: Activity; occupation index; 3D depth camera; MPM; laying hens

1 Introduction

Livestock management decisions are mostly based on the observation, judgement, and experience of farmers. However, with the increasing scale of farms and the correspondingly higher number of animals, it is increasingly difficult for farmers to monitor their animals in a traditional way. Moreover, it is impossible for farmers to monitor their animals continuously for a full 24 h. Modern technology now makes it possible to use cameras, microphones, and sensors sufficiently close to and sometimes on the animal so that they can, in effect, assist farmers' eyes and ears in everyday farming[1]. These techniques can

* 该文章发表于 *Engenharia Agrícola*，2021，41（3）：263-270。

facilitate the development of "early warning systems", which shorten the response time to individual animal needs [2]. Employing such a tool to monitor flocks can help farmers substantially manage their animals and houses more efficiently [3]. More detailed individual information can be perceived with image analysis techniques. For example, depth image processing can realize automatic detection of hens' behaviours [4]. By using modified cameras, researchers found that problematic events can be detected with an automatic method to predict the distribution index of broilers [5]. A computer vision-based system can be used for automatic detection of dairy cow lying behaviour in free-stall barns [6]. The health and welfare status of animals is often closely related to their active state and behavioural changes, so a better understanding of animal activities is of great help in the study of animal behaviour, animal welfare, and animal productivity [7].

Abnormal animal activity and declining production performance indicate problems in the chicken house. For example, sudden changes in animal activity may be related to animal heat stress, human disturbance, environmental control system failures, and feeding and drinking water system failures, all of which will have varying degrees of influence on production performance: egg production rate, rate of death and elimination, and drinking water and consumption [1]. If these abnormal events can be monitored and identified in real time and problems can be found quickly, it will be of great help to production management. Usually, in animal research and production, the activity index is an important animal activity parameter that is available. Optical flow analysis is a popular method for studying animal movement, which involves detecting the rate of changes in brightness in each area of an image frame [8]. In a large-scale henhouse, however, researchers found that the quality of colour images and videos are easily affected by light levels and the distance between cameras and animals [8]. Therefore, the optical flow method might not be a good choice for monitoring flocks in a low-light level environment. Moreover, 2D digital image segmentation and recognition can be problematic under real farm conditions due to dynamic background restrictions, such as dim or uneven light intensity in the house and varying floor status. These factors can affect the robustness of the algorithm for accurate classification [9]. While 3D depth cameras can solve this problem, information captured by a depth image sensor differs considerably from that of colour digital images in that each pixel in the depth data reflects the distance between the object and the depth image sensor [9]. Depth image analysis is a new method that helps detect not only horizontal but also vertical distribution attributes of animals [10] without restricting the light environment. This feature allows for continuous monitoring of animal behaviours throughout the day. Depth image analysis has been used in the automatic detection of animal lameness [11,12]. Additionally, this method is non-invasive and contactless, and it can measure animal movement in real time [13].

Given the above rationale, this study aims to explore whether the application of 3D depth sensors is competent in contactless continuous 24h monitoring of laying hen activities. The objectives are as follows: (i) development of the activity monitoring model and (ii) application and testing of the algorithm.

2　Material and methods

2.1　Animals and house

Experiments were conducted on a small-scale experimental farm (116° E 40° N, Beijing, China) from 7 July 2017 to 7 September 2017. The Hy-line brown laying hens (48-57 weeks) were reared in stacked cages, and the stock was more than 1 000 birds. Due to space limitations, a top-cage area [1.2 m (L) ×1.2 m (W) × 0.6 m (H)] including eighteen hens was selected for device installation (Fig. 1) . The tested cage was modified to an open roof cage covered by Perspex sheets to meet the image collection requirement. There was ad libitum access to food and water during the experiment. The light schedule was from 4: 00 a. m. to 8: 00 p. m.), and the poultry flock was fed twice a day, once between 7: 30 and 8: 30 a. m. and again between 4: 00 and 6: 00 p. m. Temperature and relative humidity parameters were recorded every five minutes. Additionally, production performance data were recorded, such as laying rate, number of dead and culled chickens, feed consumption and water consumption.

Fig. 1　On-site test platform.

2.2　Data collection

Successive 24 h images were recorded over 49 days. A top-view Kinect camera for Windows V1 (Microsoft Corp. , Washington, USA), installed at a height of 1.8 m from the bottom of the cage, was used to monitor the flock (Fig. 2). Depth images (the resolution is 640×480 pixels) were recorded in . txt format at one frame per three seconds . The Kinect was connected to a mini industrial personal computer (IPC) that stored the depth images for subsequent analysis. NI LabVIEW 2015 (American National Instrument Corp. , Texas, USA) was used for preprocessing and analysing depth images based on a VDM (vision development module).

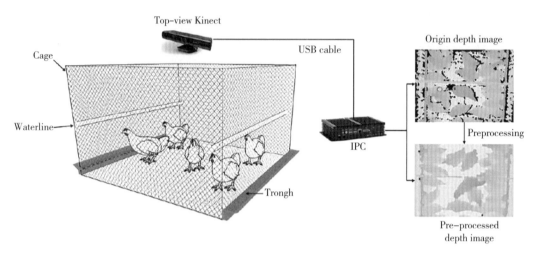

Fig. 2   The schematic of the experiment platform.

2.3 Depth image analysis

Physical activity as defined in this study, quantified by an activity index, is the animal group movement, and it was expressed using the neighbour frame difference in pixel intensity. Original depth images were preprocessed and analysed in LabVIEW software (Fig. 3). The activity index can be used to measure animal movement, and the principle of this technique is to calculate the change in pixel intensity I between two adjacent frames [14]. If chickens are inactive, the change is smaller than that when they are active. The preprocessing steps are as following:

Image cropping: To discard irrelevant edges, it is necessary to crop the original image to remove useless pixel points. For example, the original image size is 640×480 pixels, and some areas, such as chicken feeding trough areas including useless pixel points in one image, were removed (after cropping 510×480 pixels).

Morphological processing: Close, dilate and open operations and a 7×7 median filter were applied to depth image processing to obtain a filtered image.

Binarization processing: Depth images were converted to binary images to extract significant features, including chicken pixels and pixel intensity changes (threshold range: 1 200-1 700 mm).

The change in pixels was calculated as[15]:

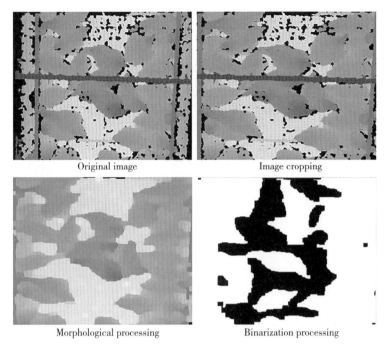

Original image　　　　　　　Image cropping

Morphological processing　　　　　Binarization processing

FIg. 3　Flowchart of depth image processing.

$$I_{mov}(x, y, t) = I(x, y, t) - I(x, y, t-1) \qquad (1)$$

Where:

I_{mov} = neighbour frame difference in pixel intensity;

(x, y) = coordinate of images;

t = time index (s);

I = pixel intensity values.

The image activity index was calculated as the ratio between the total change in pixels and the total number of pixels per image[14]:

$$activity(t) = \frac{\sum\limits_{x, y \in p} I_{mov}(x, y, t)}{\sum\limits_{x, y \in p} I} \times 100\% \qquad (2)$$

Where:

$activity$ = animal activity index (%);

t = time index (s);

I_{mov} = pixel change of two frames;

(x, y) = image coordinates.

The image occupation index was calculated as the ratio between the pixel values of the area with chickens and the total number of pixels in one image to obtain the animal occupation index [14]:

$$occupation(t) = \frac{\sum\limits_{x, y \in p} I_{animal}(x, y, t)}{\sum\limits_{x, y \in p} I} \times 100\% \qquad (3)$$

Where:

$occupation$ = animal occupation index (%);

t = time index (s);

I_{animal} = the total pixel values of areas with chickens;

(x, y) = image coordinate.

Frame difference data were analysed in ways that avoidedthe influence of increasing the body size of the birds (all birds were of similar size over the experimental period because they had grown to a stable body weight between 48 and 57 weeks old).

2.4 Statistical analysis

Statistically, the experimental datawere calculated to obtain the mean value and standard deviation of the activity index as well as the occupation index in different weeks. The mean prediction model (MPM) based on the normal mean value (no human intervention, immune operation and facility breakdown) of both the activity and occupation index of the last week was modelled in the training set (38 days with full 24 h data per day) to detect abnormal activity or occupation events (including human intervention, immune operation and facility breakdown) [16]. When measured values deviated from the mean value, a warning signal occurred. More than 25% of negative or positive deviation from measured values raised the alarm when it lasted for 30 min or longer [5]. An event logbook was recorded by one trained farmer at the same time and was regarded as the ground truth to evaluate the MPM accuracy in abnormal event warnings in the testing set (11 days with full 24 h of data per day) [16]. The performance parameters were calculated as [9]:

$$P_{accuracy} = 1 - P_{FPR} - P_{FNR} \qquad (4)$$

Where:

$P_{accuracy}$ = the accuracy of abnormal event warnings;

P_{FPR} = the rate of misclassifying normal events as abnormal events (i.e., false positives);

P_{FNR} = the rate of misclassifying abnormal events as normal events (i.e., false negatives).

3 Results and discussion

Fig. 4 and Fig. 5 depict the MPM of the activity and occupation quantified index in the training set (no abnormal activities or events included). As shown in these figures, the fluctuation in the amplitude of the activity index was lower than that of the occupation index. A higher activity index and a higher occupation index appeared during the daytime. This can be explained by chickens being active in the daytime along with a high activity index, and they expressed their natural instincts, such as stretching, preening and pacing [17,18]. However, chickens tend to be silent at night and gather together for resting and inactivity, which might result in a lower occupation index [19]. Additionally, the flock expressed a regular change pattern of activity and occupation index per day. Each week also showed a similar change pattern. Chickens tended to be more active during the light period from 4: 00 a. m. to 8: 00 p. m. than during the night period. There were great differences in day and night according to bird activity and the occupation index, which quantitatively confirms diurnal patterns of bird activity. The result of this method of perceiving animal activity patterns is similar to that of a preliminary experiment [16].

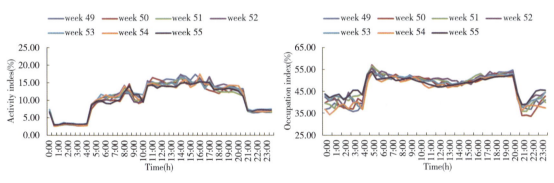

Fig. 4　Week MPM of the activity index in the training set.　　Fig. 5　Week MPM of the occupation index in the training set.

Then, the MPM method was applied to detect abnormal activity events in the testing set to evaluate its performance (no abnormal occupation index occurred). Abnormal events that occurred in the testing set included human intervention, immune operation and feeding restriction, which could impose a negative impact on chicken activity [15,16].

As shown in Fig. 6, abnormal activity events were easily detected by the MPM method. In Fig. 6A, an abnormal activity warning occurred between 4: 30 p. m. and 5: 00 p. m. and it was caused by artificial feeding restriction. It has been proven that if chickens cannot obtain access to food, they might express a strong demand for food, feel frustrated, and produce more food calls [17]. Fig. 6B, 6C, 6D and 6F display similar warnings during feeding time, which might be caused by feeding restrictions. In Fig. 6B and 6G, warning events occurred between 2: 30 p. m. and 4: 30 p. m. , which might be related to flock stress caused by artificial immune operations. Fig. 6B (from 6: 30 p. m. to 7: 00 p. m.), Fig. 6D (from 7: 30 p. m. to 8: 00 p. m.) and Fig. 6G (from 5: 30 p. m. to 6: 00 p. m.) present warning events; these were false positive warning events that were not recorded in the farmer logbook. In Fig. 6C, another warning event occurred between 11: 00 p. m. and 11: 30 p. m. This might be caused by environmental changes that were related to the ventilation control system. The experimental cage was near the air intake, and a change in night ventilation control strategy might influence flock activity [16]. In Fig. 6D, an abnormal activity warning occurred between 4: 30 a. m. and 5: 30 a. m. This might be related to flock stress caused by the manual operation of eliminating weak chickens, as it broke the poultry's circadian clock [20,21]. In Fig. 6E, another abnormal activity warning occurred between 9: 00 a. m. and 10: 00 a. m. This might be related to flock stress caused by noise from pipe maintenance. Compared with the farmer logbook, this MPM method can detect some abnormal activities and might help farmers trace possible reasons for activity event warnings. For example, as shown in Table 1, 5 days into the testing set with abnormal activity displayed that the laying rate was lower than that in the neighbouring two days. The lowest laying rate in the testing period was 83. 7% (24th July), which might be caused by higher average relative humidity (Table 1) . Additionally, abnormal water consumption might exist along with abnormal activity warnings.

Table 1   Hen's production perfermance and environmental parameters in the testing set.

Date set	Date	Laying rate (%)	Dead and culled hen number	Feed-gain ratio (g)	Water consumption (mL)	Environmental parameters	
						Average temperture (℃)	Average relative humidity (%)
Testing set	07-22	91. 9	1	122. 7	28. 5	23. 0	78. 1
	07-24#	83. 7#	0	134. 7	23. 2	23. 0	96. 0
	07-25	92. 5	0	121. 5	34. 5	23. 5	81. 0
	07-26	90. 5#	0	125. 5	30. 4	19. 5	83. 1
	08-15	89. 4#	0	129. 5	38. 3	22. 5	80. 2
	08-20	90. 5#	0	127. 1	63. 4	22. 0	66. 3
	08-23#	89. 9#	1	129. 6	27. 2	23. 0	79. 2

Note: # denotes that the value is lower than that in the neighbouring two days. # denotes an immune operation day.

Whatever type of physical or mathematical model we use, experience shows that the hard work starts when the model is implemented in real livestock houses [8]. One of the main challenges of the MPM method is the lack of enough events in the logbook, especially events recorded at night. Although seven weeks (49-55) of data detected some abnormal activity events through depth image analysis, more abnormally recorded events at daytime and nighttime might be more helpful for the performance assessment of the MPM method and

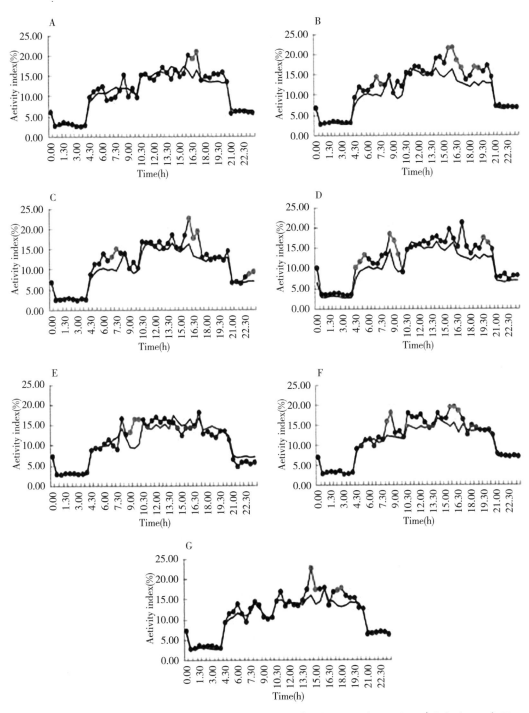

Fig. 6　Results of the MPM in the testing set of A 22nd July in week 49，B 24th July in week 50，C 25th July in week 50，D 26th July in week 50，E 15th August in week 52，F 20th August in week 53，and G 23rd August in week 54. A black line denotes a week MPM. A black-dotted line denotes the normal activity index. Red-dotted lines indicate the abnormal activity index.

feature description of poultry movement. Additionally, the MPM method could be used to predict animal activity conditions of the next week, which is simple, fast and implementable for real-time application. It was found that 32 out of 38 abnormal activity events were correctly detected (the false positive rate was 15.8%, and the false negative rate was 0.0%) (Table 2). The accuracy was higher than 73.9%, which resulted in a preliminary test [16]. The reference methods show accuracies of 95.2% [5] and 61.9% [22]. Compared with the optical flow method, the MPM method could detect animal movement in full 24 h without considering the influence of the change in light levels, and it can also help farmers obtain more details, such as the animal activity index and occupation zones, both day and night. Furthermore, more data are needed to verify that this MPM method would be suitable for monitoring a whole animal production period.

Table 2　MPM accuracy of abnormal event warnings.

Testing set	Abnormol activites number	Number and P_{FER}	Number and P_{FNR}	$P_{accuracy}$
week 49	2	0 (0.0%)	0 (0.0%)	100.0%
week 50	24	4 (16.7%)	0 (0.0%)	83.3%
week 52	3	0 (0.0%)	0 (0.0%)	100.0%
week 53	5	0 (0.0%)	0 (0.0%)	100.0%
week 54	4	2 (50.0%)	0 (0.0%)	50.0%
Total	38	6 (15.8%)	0 (0.0%)	84.2%

This study attempts to monitor animal activity quantitatively to provide early warning of abnormal events that affect production performance. An in-depth exploration into the occurrence rule of abnormal events has not yet been conducted, which is also the next step in this study. Moreover, different image frames can, to some extent, influence the analysis results, such as one frame per second or one frame per five seconds. However, in this study, deep research has not yet been carried out. The key problem may be the smaller number of chickens in the experiment compared with a similar broiler warning system [5], which is due to the difficulty of device installation caused by stacked-cage design and the viewing angle of a depth camera. If the problem can be solved in future research, more detailed and associated information may be explored, and the relationship between animal activity and production performance may be well understood.

4　Conclusions

An MPM method was proposed to continuously and noninvasively monitor hen activities and quantified occupation indices in real time. It can report abnormal activity events caused by human intervention, immune operation and feeding restriction in daily stacked-cage management. This method can help farmers monitor their animals and rapidly detect abnormal events in a 24 h day. In the testing stage, MPM was able to detect abnormal activity events with an average accuracy of 84.2%. By using depth image processing, it is possible to eliminate the interference of environmental light and external shielding, such as cages and cables. Compared to colour images, depth images are more suitable for animal behaviour detection and analysis at low illuminance. Additionally, the flock showed a diurnal pattern in the change in activity and occupation index, and each week presented a similar change pattern. Further research will consider the application of this method in other systems, such as enriched cages and no-cage systems.

References

[1] Kashiha M, Bahr C, Haredasht S A, et al. The automatic monitoring of pigs water use by cameras [J]. Computers and Electronics in Agriculture, 2013a, 90: 164-169.

[2] Norton T, Berckmans D. Developing precision livestock farming tools for precision dairy farming [J]. Animal Frontiers, 2017, 7: 18.

[3] EFSA. Scientific Opinion on the use of animal-based measures to assess welfare of broilers [J] . EFSA Journal, 2012, 10: 2774.

[4] Lao F D, Du X D, Teng G H. Automatic recognition method of laying hen behaviors based on depth image processing [J] . Transactions of the Chinese Society for Agricultural Machinery, 2017, 48 (1): 155-162 (in Chinese) .

[5] Kashiha M, Pluk A, Bahr C, et al. Development of an early warning system for a broiler house using computer vision [J] . Biosystems Engineering, 2013b, 116: 36-45.

[6] Porto S M C, Arcidiacono C, Anguzza U, et al. A computer vision-based system for the automatic detection of lying behaviour of dairy cows in free-stall barns [J] . Biosystems Engineering, 2013, 115: 184-194.

[7] Ni J, Liu S, Radcliffe J S, et al. Evaluation and characterisation of Passive Infrared Detectors to monitor pig activities in an environmental research building [J] . Biosystems Engineering, 2017, 158: 86-94.

[8] Dawkins M S, Cain R, Roberts S J. Optical flow, flock behaviour and chicken welfare [J] . Animal Behaviour, 2012, 84: 219-223.

[9] Lao F D, Brown-Brandl T, Stinn J P, et al. Automatic recognition of lactating sow behaviors through depth image processing [J] . Computers and Electronics in Agriculture, 2016, 125: 56-62.

[10] Gregersen T, Jensen T, Andersen M, et al. Consumer grade range cameras for monitoring pig feeding behaviour [C] . In: 6th European Conference on Precision Livestock Farming, 2013: 360-369.

[11] Van Hertem T, Maltz E, Antler A, et al. Automatic lameness detection based on 3D-video recordings [C] . In: European Conference on Precision Livestock Farming, 2013.

[12] Viazzi S, Van Hertem T, Romanini C E B, et al. Automatic back posture evaluation in dairy cows using a 3D camera [C] . In: European Conference on Precision Livestock Farming, 2013: 83-91.

[13] Springer S, Seligmann G Y. Validity of the Kinect for Gait Assessment: A Focused Review [J]. Sensors, 2016, 16.

[14] Bloemen H, Aerts J M, Berckmans D, et al. Image analysis to measure activity index of animals [J]. Equine Veterinary Journal, 1997: 16-19.

[15] Costa A, Borgonovo F, Leroy T, et al. Dust concentration variation in relation to animal activity in a pig barn [J]. Biosystems Engineering, 2009, 104: 118-124.

[16] Du X D, Cao Y F, Teng G H. A method based on image and sound processing for monitoring abnormal events in a breeder house [J]. Journal of China Agricultural University, 2018a, 12: 114-121 (in Chinese).

[17] Kuhne F, Sauerbrey A F C, Adler S. The discrimination-learning task determines the kind of frustration-related behaviours in laying hens (Gallus gallus domesticus) [J]. Applied Animal Behaviour Science, 2013, 148: 192-200.

[18] Pereira D F, Miyamoto B C B, Maia G D N, et al. Machine vision to identify broiler breeder behavior [J]. Computers and Electronics in Agriculture, 2013, 99: 194-199.

[19] Du X D, Lao F D, Teng G H. A Sound source localisation analytical method for monitoring the abnormal night vocalisations of poultry [J]. Sensors, 2018b, 18: 2906.

[20] Hy-Line. Parent stock management Guides [Z/OL]. Available: http://www.hyline.com/aspx/general/dynamicpage.aspx? id=255, 2016.

[21] Shimmura T, Yoshimura T. Circadian clock determines the timing of rooster crowing [J]. Current Biology, 2013, 23, R231-233.

[22] Pluk A, Cangar O, Bahr C, et al. Impact of process related problems on water intake pattern of broiler chicken [C]. International Conference of Agricultural Engineering. Cemagref. Proceedings, 2010.

Oestrus Analysis of Sows Based on Bionic Boars and Machine Vision Technology[*]
（基于仿生公猪和机器视觉技术的母猪发情分析研究）

Kaidong Lei   Chao Zong   Xiaodong Du   Guanghui Teng   and Feiqi Feng

Abstract：This study proposes a method and device for the intelligent mobile monitoring of oestrus on a sow farm，applied in the field of sow production. A bionic boar model that imitates the sounds，smells，and touch of real boars was built to detect the oestrus of sows after weaning. Machine vision technology was used to identify the interactive behaviour between empty sows and bionic boars and to establish deep belief network (DBN)，sparse autoencoder (SAE)，and support vector machine (SVM) models，and the resulting recognition accuracy rates were 96.12%，98.25%，and 90.00%，respectively. The interaction times and frequencies between the sow and the bionic boar and the static behaviours of both ears during heat were further analysed. The results show that there is a strong correlation between the duration of contact between the oestrus sow and the bionic boar and the static behaviours of both ears. The average contact duration between the sows in oestrus and the bionic boars was 29.7 s/3 min，and the average duration in which the ears of the oestrus sows remained static was 41.3 s/3 min. The interactions between the sow and the bionic boar were used as the basis for judging the sow's oestrus states. In contrast with the methods of other studies，the proposed innovative design for recyclable bionic boars can be used to check emotions，and machine vision technology can be used to quickly identify oestrus behaviours. This approach can more accurately obtain the oestrus duration of a sow and provide a scientific reference for a sow's conception time.

Keywords：machine vision；bionic boar；sow；oestrus detection；video analysis；welfare

1 Introduction

At present，precision livestock farming (PLF) research has attracted the attention of manyresearchers[1]. The breeding of pigs in a noncontact，stress-free，and healthfulmanner

---

＊ 该文章发表于 *Animals*，2021，11（6）：1485。

has always been a research field that scholars worldwide have focused on [2-4]. During the process of pig breeding, sows play an important role. By scientifically understanding sows' oestrus state, one can determine the best mating times of sows and increase the overall embryo implantation rate. The oestrus period is the period in which a sow can accept a boar and achieve ovulation and conception. The oestrus cycle of a sow is composed of four periods: pre-oestrus, oestrus, late oestrus, and dioestrus periods. The average duration of oestrus is 21 days, but there are large individual differences between the phases. The oestrus behaviours of sows are cyclical and transient. When a sow exhibits oestrus behaviour, it needs to be bred in time; otherwise, the breeder will miss the optimal breeding time and must wait for another oestrus cycle. Empty sows lead to increased breeding costs and reduced production efficiency [5,6].

During sow production, oestrus detection is mostly conducted with the manual observation method and boar test method. During the oestrus period, the sow's feed intake is reduced, its sensitivity to environmental changes increases, and the amount of activity in the pigpen increases [7]. The current general observation method for checking sow emotions is to observe the reaction of the sow to a back pressure test (BPT) and any induced changes in the vulva. A back pressure reaction is one in which sow pigs in oestrus appear to stand still and their ears stand still when their backs are manually pressed on, when they are crawled over by boars, when their backs are hunched, and so on. At the same time, the sow's vulva becomes red and swollen, and more mucus is excreted from it [8]. The boar test method is used to observe whether a sow has a static reaction through direct contact between a boar and the sow or through the enclosure. This method is simple and convenient to perform but is time-consuming and labour-intensive.

At present, many scholars' research has shown they have carried out related research on the automatic identification of the process of sow oestrus [9]. Scolari et al. [10] exploredthe changes in the temperatures of the vulvas and buttocks of sows before and after oestrus; the results showed that the temperature of each sow's vulva increased significantly at the beginning of oestrus and decreased significantly before ovulation, while the temperature of the buttock surface did not change significantly. Sykes et al[11]. found that the maximum and average temperatures of the vulva during the oestrus period of sows were higher than those during the oestrus cycle period, but there was no difference between the minimum values. Simões et al[12]. found that the temperatures of the sow vulvas first increased and then decreased; the temperatures of the buttock surfaces were not significantly different. Altmann used a small accelerometer to detect sow activity and found that activity during the oestrus period was twice that of a sow that is not in heat [13].

Bressers [14] used accelerometers and set activity thresholds to detect oestrus in sows. The results showed that the acceleration change range of sows during the oestrus period was significantly higher than that of sows during the non-oestrus period. When the threshold was set to 10 m/s^2, the amount of exercise detected by the sensor was 10 times higher than that during the non-oestrus period [14]. Freson used infrared sensors to detect the oestrus of sows; the accuracy rate of the proposed method was 86%, the sensitivity of oestrus detection was 79%, and the specificity was 68% [15]. Houwers et al. used sensors to automatically record the frequency of sows visiting boars, and the results showed that when the sows had never been in heat and entered the oestrus period, the frequency of boar visits gradually increased [16]. Korthals counted the average length of sow visits. The statistical results showed that the length of sow visits obeyed the 92 s/d Poisson distribution. The sensitivity of heat detection was 76.4%, and the specificity was 80.3% [17]. Therefore, it is feasible to predict the oestrus time of a sow through the frequency and duration of the sow's visits to a boar.

With the development of neural networks, more researchers hope to use the autonomous learning capabilities of neural network models to automatically determine the characteristics of recognition behaviour. The advantage of automatically extracting features lies in the robustness and strong adaptability of this approach. In addition, during the learning process, features with higher degrees of discrimination can be automatically distinguished. However, when the resultant model is established, it is necessary not only to pursue accuracy but also to take the running speed of the model into account so that it can be effectively applied in practice.

There has been some development in the recognition of pig behaviour images. Several scholars have studied sow behaviour image recognition, video analysis, and sow behaviour tracking [18-21]. However, there are few reports on the detection of sow oestrus based on the 'contact windows' of machine-vision based bionic boars, and the current research has not been applied to sow oestrus detection and pig production.

Thus, the analysis of the behaviours of sows after weaning based on a 'contact window' model of a bionic boar using machine vision technology is herein proposed. This research method is intelligent, offers a non-stressful environment, guarantees sow welfare, and ensures biosafety during the detection process. The main contributions of this article are as follows:

(1) An information model was researched, developed, and built based on the 'contact window' of a bionic boar.

(2) The oestrus behaviours of sows were analysed based on machine vision recognition.

2　Materials and methods

To detect the conditions of sows based on machine vision, in this study, a bionic boar detection model was built，which mainly includes an intelligent mobile platform，a bionic model device，an image acquisition device，and a PC processing terminal. Through simulation of the sound and smell of boars，and analysing the response behaviour of sows through machine-vision based imaging technology，a sow oestrus recognition model was established，and the oestrus times of sows can be accurately obtained.

2.1　Animals and housing

To identify the oestrus behaviours of empty-breasted sows，this study was carried out at the pig farms of the Shandong and Chongqing Academy of Animal Husbandry (Rongchang，Chongqing，China) pig farms from July 2019 to December 2020. The test subjects were Yorkshire sows that already been weaned. The sows had 2-3 L. The feeding times were 10：00 a. m. and 4：00 p. m. The test tracked a total of 76 large white sows. The data collection platform included a computer，an intelligent mobile platform，an intelligent detection camera［LRCP106＿1080P（Zhi fei，China）］，and a bionic model device.

In this study，the experimental method mentioned that a total of 76 sows were observed. In other words，68 sows were observed in the preliminary experiment. At the same time，68 sows underwent the same experimental observation，the bionic boar，including sound and smell devices that simulate pigs. The actual verification results are based on conducting an inspection with the bionic boar and performing a simultaneous manual back pressing test（BPT）at the same time. A diagram of the test platform is shown in Fig. 1.

The intelligent mobile platform used in this article was the FBL80E3500＿X linear actuator (FUYU，China)，which can be used to set the speed，position，and time through a program compiled in the C programming language. The FBL80E3500＿X is a synchronous belt-based linear module. It is a medium-sized closed linear module for synchronous belt transmission. The horizontal speed under a full load is 300-1 700 mm/s，the positioning accuracy is 0. 1 mm，the horizontal maximum load is 30 kg，and the maximum thrust is 55 N.

For the image acquisition device，the acquisition frequency was set to 30 fps，the resolution was 1 280×720，and the collected sow images were transmitted to the PC processing terminal through a data line. A camera was arranged on the upper backside of the circle，and the PC processing terminal was an Intel Core i7 9th generation processor. The bionic model device，a built-in ultrasonic atomization device，was a YuWell（YW，China）NM211C miniature piezoelectric atomizer and it was used to simulate the release of odours from boars；the sound module，the external amplifier of a voice recording playback device，was a Sony ICD-TX650，with a noise reduction function and the advantages of a stereo microphone layout. To increase the authenticity of the bionic model，the front end of the device was designed with a silicone pig nose，and the ultrasonic atomization outlet was a silicone pig

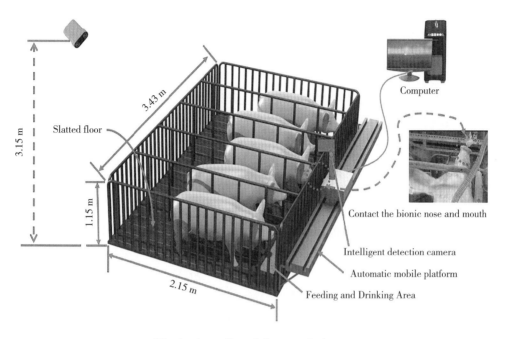

Fig. 1   A rendering of the test platform.

nose, which could effectively capture the interactions between a sow and a boar. MATLAB (MathWorks Inc of America.), Origin8. 0 (OriginLab Microcal of America), LabVIEW (the Laboratory Virtual Instrument Engineering Workbench of America), and other software were used for image analysis.

2.2 Bionic boar for verifying the principle

Studies have shown that the boar test method is often used during production to observe whether the sow's nose and theboar's nose will have a standing reaction after touching it. and then to determine whether the sow is in heat. Most chosen boars are middle-aged and old, with strong salivation abilities and docile temperaments. The boars used to provoke the sows should be alternated to avoid having the sows lose interest in a single boar. When arranging the situation, the routes of the boars are planned in advance, the staff are arranged to drive the boars, and fences/gates are set up to block the boars from escaping [22].At the same time, pressure is put on the back of the examined sow to observe whether the sow has a standing reaction; the knees of the boar are pushed against the sow's udder, abdomen, and crotch to observe whether the sow is close to exhibiting a standing reaction. It is observed whether changes in the shape and colour of the sow's vulva occur and whether the sow has frequent urination, irritability, arching, ear erections, body tremors, an inability to eat, and so forth. According to the above procedure for the identification of the sow's oestrus, it is comprehensively judged whether the sow is in oestrus and the status is marked accordingly [23,24]. Although this process has a high accuracy rate, it is time-consuming and labour-intensive, and there is a certain degree of danger.

It is necessary for boars to be on-site when making love. The boars on pig farms can generally be divided into two categories according to their use: love-making boars and semen-providing boars. Both are essential parts of daily pig farm production. Lovemaking boars refer to a type of boar used for sex checking, sex induction, and assisted artificial insemination. These boars mainly rely on the smell of saliva to stimulate a sow's physiological and psychological responses. These 'odour molecules' at work are boar pheromones [25].

A boar can stimulate a sow with sight, touch, smell, and sound, the most important of which is smell. Pigs have a large number of olfactory receptors [26]. When a boar appears in front of a sow in heat, the pheromones in the boar's saliva stimulate the sow's sexual behaviours, such as standing still, erecting its ears, arching its back, scraping the stall, and so on, within a short time period [27]. After the sow in heat comes into contact with the boar, the pheromone molecules produced by the boar can be sensed through breathing and chewing. This explains why the boar needs to be in contact with the sow's snout and nose for the processes of sex checking, sex induction, and traditional artificial insemination. Building upon the above theoretical basis, we propose a technical route for research on bionic boar-based sow checking, in Fig. 2.

Firstly, according to the technical roadmap and the pheromone of the boar to detect the heat of the sow and at the same time to observe the degree of reaction of the sow. Then, by simulating sound and smell of boar and simulating the principle of contact window of boar, a bionic boar was constructed in the experiment. Finally, SOB, SOS, SOC, SO-SE, SOW and other behaviors were analyzed and recognized. Behavior detection can be used as the evaluation of the effect of bionic boars on sows in the estrus cycle. The bionic boar device developed in this study was mainly an intelligent detection device (consisting of a sound-releasing device, an odour-releasing device, an image acquisition device, and a bionic silicone pig nose). A scent device and sound player were installed on the intelligent detection device to check the oestrus of sows. The saliva, urine, semen, and so forth of a boar were contained in the scent release device, and these items were atomized and released through a miniature atomizer. The experimental results show that the combination of the saliva and voices of middle-aged boars leads to long-duration and high-frequency contact with sows. During the investigation, the weaned sow actively touched the detection device, contacted it frequently, had a long contact duration, stood still with both ears motionless, and so forth. These interactions were captured by the camera above the detection device and transmitted to the PC through the end of the accompanying data cable.

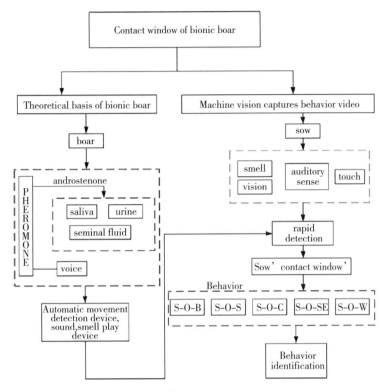

Fig. 2　Technich roadmap.

Note: S-O-B denotes that a sow in the oestours cycle bites the rod; S-O-S denotes that the sow stands still during the cestrus cycle; S-O-C denotes that the sow touches the bionic boar snout during the oestrus cycle; S-O-SE denotes that the sow's ears stand still during the oestrus cycle; S-O-W denotes that the sow moves her head over a wide range during the oestrus cycle.

During the sow's bionic boar check-up, the response degree of the sow and the bionic device was observed and recorded. In this study, a bionic boar was used to check up on sows for a period of three minutes per sow. Through the artificial BPT on the sow, observations were made regarding whether both ears of the sow were standing still, whether it had a hunched back, and whether the sow's pubic area was red and swollen. At the same time, the sow's response behaviour and frequency were recorded through image technology. The response relationship between the sow and the bionic boar was recorded and studied through video analysis.

2.3 Machine vision recognition of oestrus sow behaviour in response to a bionic boar

2.3.1　Pre-processing of video image of bionic boar　To recognize the response behaviour of a sow in a complex environment, the collectedvideo/image set needed to be pre-processed to reduce background interference and improve the contrast between the target sow and the background. In the home environments of pigs, sows cause imaging problems such as occlusion, complex lighting, and difficulty in capturing target images, and the collection of high-quality images requires computer processing hardware. In MATLAB, the avi2img. m program converted each video into a series of single pictures frame by frame. The grey _ frame = rgb2grey (frame) statement was used to convert a colour image into a greyscale image. The contrast between the target sow and the background was improved by transforming the grey level. Establishing different action feature extractions through MATLAB can provide classification data labels for subsequent model classifications.

2. 3. 2   A deep belief network (DBN) **model for oestrus behaviour identification in sows**   A model is proposed based on bionic boar checking and sow response behaviour identification. Responsive sow behaviours include chewing on railings, standing still, touching thesnout of the bionic boar, standing with both ears still, and swinging the head widely. Standing behaviour and contact with the bionic boar snout are the main characteristics that reflect the oestrus behaviour of a sow [3]. By analysing the response behaviour characteristics of sows, a comprehensive evaluation of standing behaviour and combined contact with the snout and nose of the bionic boar is proposed.

(1) Establishment of an oestrus behaviour recognition model for sows   A deep belief network (DBN) was one of the first non-convolutional models to besuccessfully applied to deep architecture training. The basic unit of this network is the restricted Boltzmann machine (RBM) [28]. Hinton et al. [29] first used unsupervised pretraining to initialize the network parameters and then adjusted the network parameters through the backpropagation (BP) algorithm. DBNs are widely used in handwriting recognition, image recognition, speech recognition, and other fields [29]. Compared with the traditional multilayer neural network, which is difficult to train, a DBN abstracts highlevel features from the observed low-level features by applying a layer-by-layer greedy learning method to realize network training [30-32].

DBNs are often trained using the contrastive divergence (CD) algorithm. Assuming that there are d explicit layer neurons and q hidden layer neurons in the network model, let v and h denote the state vectors of the explicit layer and the hidden layer, respectively. In the same layer, each unit satisfies an independent and identical distribution, and the following formulas can be obtained:

$$P(v \mid h) = \prod_{i=1}^{d} P(v_i \mid h) \qquad (1)$$

$$P(h \mid v) = \prod_{j=1}^{d} P(h_j \mid v) \qquad (2)$$

In the CD algorithm, the training samples are first assigned to the explicit layer v, and the score of the hidden layer neuron state is calculated according to Equation (2). The probability distribution is determined, and then sampling is performed according to this distribution to obtain h. Then, v' is calculated from h' according to Equation (1), and v' and h' are subsequently obtained. The update formula for the connection weight is as follows, where ΔW is the update weight and h is the learning rate.

$$\Delta W = \eta \ (vh^{\mathrm{T}} - v'h'^{\mathrm{T}})　　　　　　(3)$$

In image processing, pixels are used as visible layer units to deactivate hidden layer units with a certain probability. By running the divergence algorithm, the network model is trained layer by layer to obtain the weight matrix and offset of the visible unit and the hidden unit of each layer. Finally, according to the weight matrix and offset of each layer, it is judged whether the hidden unit is activated. The distribution of the hidden unit is the abstracted high-level feature.

(2) A sparse autoencoder (SAE) model for female oestrous behaviour recognition

Sparse autoencoder (SAE) neural network technology is an efficient unsupervised feature learning and deep learning classification method. Since it was first proposed, the SAE technology has been widely used in various classification and pattern recognition problems [33]. An SAE neural network classifier includes multiple training layers, and the training output result of the previous training layer is the training input value for the next training layer. The detection process of the SAE neural network classifier described in this section first involves initializing the parameters; then, feedforward conduction is used to train the neural network; finally, a reverse conduction fine-tuning process is used to determine the minimal cost function with the globally optimal parameters, and a collapsed-lane deviation detection classifier is obtained. The SAE neural network integrates all the advantages of a deep neural network and has a strong expression ability.

A stacked autoencoder neural network is composed of multiple layers of sparse autoencoders, and its training process adopts a layer-by-layer greedy training method. The layer-by-layer greedy training method uses the output of the previous layer of an autoencoder as the input of the next layer of the autoencoder to train the network layer by layer in order from front to back. During each step, the first k-11 layers that have been trained are fixed, and then layer k is added; that is, the output of layer k-1 (which has been trained) is the input of layer k to be trained. The feature input set of the stacked autoencoder neural network is called the input layer, the classification result set is called the output layer, and all the layers between the input layer and the output layer are called hidden layers.

Unlike other neural networks, a single-layer SAE is an unsupervised learning algorithm that does not require the training samples to be calibrated. The overall network performance of the single-layer SAE can be obtained via Equation (4):

$$J(W, b) = \frac{1}{m}\sum_{i=1}^{m} J(W, b; x^{(i)}, y^{(i)}) + \frac{\lambda}{2}\sum_{l=1}^{n-1}\sum_{i=l}^{s_l}\sum_{j=1}^{s_l+1}(\omega_{ij}^{(l)})^2 + \beta\sum_{j=1}^{s_l}(\rho\log\frac{\rho}{\hat{\rho}_j} +$$

$$(1-\rho)\log\frac{1-\rho}{1-\hat{\rho}_j}) \tag{4}$$

Among the parameters, W and b are used to fit the data of the original input x (i), y (i), and b is the bias node, also called the intercept. m is the number of feature inputs, and l is the weight attenuation coefficient, which is used to control the relative importance of the first mean squared error term and the second weight attenuation term to prevent overfitting. $w_{ij}^{(l)}$ is the data weight value from the j-th unit of the l-th layer to the i-th unit of layer $l + 1$. s represents the number of neurons in each hidden layer, β is the weight of the sparsity penalty term, ρ is the introduced sparsity parameter, and $\hat{\rho}_j$ is the average value of the m input activations of hidden neuron j. $\beta\sum_{j=1}^{s_l}(\rho\log\frac{\rho}{\hat{\rho}_j}) + (1-\rho)\log\frac{1-\rho}{1-\hat{\rho}_j})$ is the relative entropy between a Bernoulli random variable with a mean of ρ and a Bernoulli random variable with a mean of $\hat{\rho}_j$ (obtained by measuring the difference between the two distributions). $\frac{\lambda}{2}\sum_{l=1}^{n-1}\sum_{i=l}^{s_l}\sum_{j=1}^{s_l+1}(\omega_{ij}^{(l)})^2$ is a regularization term, and $\frac{1}{m}\sum_{i=1}^{m}J$ (W, b; $x^{(i)}$, $y^{(i)}$) is the cost function of model.

(3) Support vector machine (SVM) model for sow oestrus identification　The support vector machine (SVM) method is a machine learning classification algorithm based on statistical learning. It isbased on the principle of achieving the minimum structural risk, and it was first proposed by Vapnik et al. [34]. Its basic model is a linear classifier with the largest interval in a given feature space. With its solid theoretical foundation and accuracy, the SVM is widely used in text classification and face recognition tasks. In the field of text classification, compared with traditional methods, the SVM approach not only has better robustness but also achieves good results when dealing with high dimensionality [35].

The learning strategy of an SVM is to maximize the category interval, that is, to find the hyperplane with the largest interval in a given feature sample space. This hyperplane separates the two compared types of data and maximizes the geometric interval; this can be approximately understood as a convex quadratic programming problem. The purpose of an SVM is to determine the separation hyperplane that can divide the dataset correctly and obtain the largest set interval. The equation for dividing the hyperplane is shown in Equation (5):

$$W^T x + b = 0 \tag{5}$$

The separation hyperplane is determined by the normal vector W = (W1; W2; …; Wd) and the intercept b, which needs to be obtained through training with the sample data. Assuming that the linearly separable training set $T = \{ (x_i, y_i) \}$, $i = 1, 2, \cdots, n$, x_i represents a feature vector in the training set, $y_i \in \{+1, -1\}$, y_i is used to input the classification of x_i (6):

$$\begin{cases} W^T x + b \geqslant 1, & y_i = 1 \\ W^T x + b \leqslant -1, & y_i = -1 \end{cases} \tag{6}$$

The distance between a sample point in the space vector and the hyperplane is denoted as D, which can be expressed as follows (7):

$$D = \frac{|W^T + b|}{||w||} \tag{7}$$

Among the variables, $||w|| = \sum_{i=1}^{n} w_i^2$. The objective function and constraint conditions of an SVM are as follows (8):

$$\begin{cases} \min \dfrac{1}{2} ||w||^2 \\ s, t, y_i(W^T x + b) \geqslant 1 (i = 1, 2, \cdots, n) \end{cases} \tag{8}$$

SVMs are widely used. In addition to text classification, they have also been applied to speech recognition and face recognition tasks, and they have achieved satisfactory results. However, SVMs and other algorithms also have some shortcomings. In cases with small data volumes, the SVM is based on the strategy of minimizing structural risk, and thus, its generalization ability is better. The time required for learning and training in an SVM is longer than for other algorithms.

3 Results and discussions

Regarding weaned sows, mastering detection of the oestrus statuses of sows is a key production process. By checking oestrus with a bionic boar, one can use the degree of response exhibited by a sow as a reference for sow oestrus analysis.

3.1 Model evaluation for detecting the oestrus of sows based on machine vision

To evaluate the actual effect of recognizing oestrus behaviours in sows based on machine vision detection technology, this article preliminarily verifies the response behaviour of a sow when in oestrus with the bionic boar. We utilized a DBN by applying a layer-bylayer greedy learning method to obtain high-level features from the underlying features and to realize the classification of sow behaviours. The accuracy rate of the proposed method was 96. 12%. Fig. 3 shows the changes in recognition accuracy after the DBN was applied; the recognition accuracy changed after 55 iterations and 100 iterations.

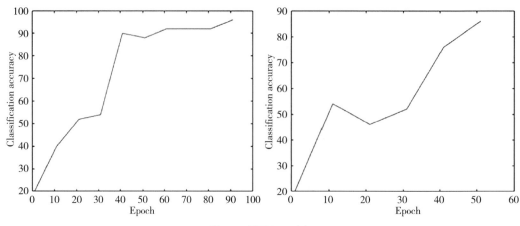

Fig. 3   DBN model.

Note：The accuracy rate of the method reached 96. 12% after 100 iterations and the average error was 3. 88%；it was 84. 21% after 55 iterations and the average error was 15. 79%.

The SAE neural network technology is an efficient and unsupervised feature learningand deep learning classification method. The accuracy rate of sow behaviour recognition using this approach reached 98. 25%. Fig. 4 shows the classification results of the SAE model. After 100 iterations，the accuracy rate reached 98. 25%，while after 55 iterations，the accuracy reached 94. 82%. Compared with the DBN model，the SAE model has a higher accuracy rate.

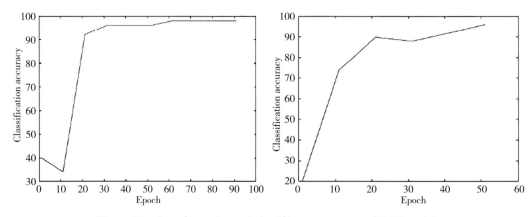

Fig. 4   Number of iteration and classification accuracy of SAE model.

Note：The accuracy after 100 iterations was 98. 25% and the average error was 1. 75%；after 55 iterations the accuracy rate was 94. 82% and the average error was 5. 18%.

In terms of sow behaviour recognition，after 45 iterations the accuracy of the SVM model was 88%，and the average error was 22%，after 40 iterations，the accuracy rate was 90. 00% and the average error was 10%. Compared with the other models，the recognition effect of the SVM model is poor (Fig. 5) .

Based on the DBN，SAE，and SVM models，we identified the response behaviours of weaned sows，including gnawing on railings，standing still，touching the snout of a bionic

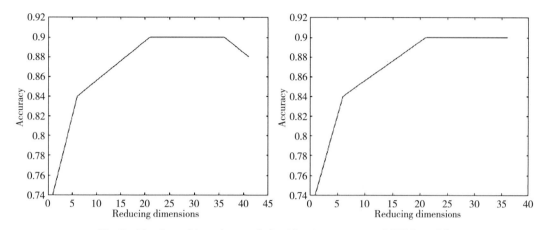

Fig. 5　Number of iterations and classification accuracy of SVM model.

Note：The accuracy of dimensionality reduction after 45 iterations was 88% and the average error was 22%；
after 40 iterations，the accuracy rate was 90. 00% and the avarage error was 10%.

boar，standing with both ears motionless，and swinging their heads wildly. The analysis results are shown in Table 1. Note：S-O-B denotes that a sow in the oestrous cycle bit the rod；S-O-S denotes that the sow stood still during the oestrus cycle；S-O-C denotes that the sow touched the bionic boar snout during the oestrus cycle；S-O-SE denotes that the sow had her ears standing still during the oestrus cycle；S-O-W denotes that the sow moved her head over a wide range during the oestrus cycle.

Table 1　Recognition accuracy rate of three models for five behavious. The average error was 5. 17%.

Model Behaviour	DBN	SAE Accuracy （%）	SVM
S-O-B	90. 00	80. 00	90. 00
S-O-S	100. 00	99. 94	90. 00
S-O-C	60. 86	80. 21	80. 00
S-O-SE	100. 00	100. 00	100. 00
S-O-W	80. 00	80. 00	80. 00

3.2　Response of the sows and bionic boars

In this study，the experimental method mentioned that a total of 76 sows were observed. In the preliminary experiment，68 sows were observed by bionic boars. At the same time，68 sows underwent the same experimental observation，including a bionic boar sound and smell check）. The results show that the bionic boars can achieve the effect of real boars. Finally，8 sows are randomly selected for analysis and verification. The following focuses on the analysis of 8 sows. The statistical analysis of the test data is divided into two parts. In the first part, the sow data was analysed in a pre-test (68 sows), and the latter part was verified by 8 sows, which can better illustrate the reliability of the method.

3. 2. 1   Duration of sow response   The behaviour of a sow is very complicated. Based on video analysis of the response duration of a sow, the physiological condition of the sow, her menstrual cycle, and the influence of the external environment on the sow can be obtained. Some scholars have found that sows usually enter the oestrus cycle within 1-5 days after weaning and are subsequently bred[3]. This article analyses the response durations of sows to bionic boars 1-7 days after weaning. The results show (Fig. 6) that there are regular increases and decreases in the duration of the sow's visit to the bionic boar within those 1-7 days. This is because the average oestrus cycle of a sow is 21 days. During the oestrus cycle, there are regular changes in the hormone levels of sows [luteinizing hormone (LH), folliclestimulating hormone (FSH), oestrogen, and progesterone levels)][36]. Prolonged contact between a sow and a boar can cause the sow to lose interest[37]. As a result, the data obtained are not accurate. The analysis in this discussion is on the response of the sow to the bionic boar within 3 min.

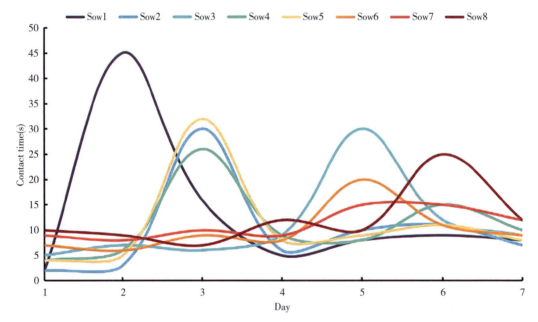

Fig. 6   The contact duration between the sow and the bionic boar 1-7 days after weaning
(a single sow check lates three minuters).

As shown in Fig. 6, Sow 1 frequently contacted the bionic boar device on the second day after weaning, reaching a peak level of contact of 45 s/3 min during the oestrus cycle. The results show that the oestrus of sows is detectable mainly on the third and fifth days. When the bionic boar checked for oestrus and the sow's oestrus behaviour was detected by video analysis, the average duration of the sow's contact with the bionic boar was 29. 7 s/3 min, and the unit detection time was three minutes or less. Thus，there is a correlation between a sow's oestrus status and the duration of her response to the bionic boar. To verify the actual effect of the bionic boar，combined with the daily inspection of the sows，the back of each sow was manually pressed. The indicator of oestrus was the sow's response to back pressure（i. e.，the BPT）[38]. Within 1-7 days after weaning，when the sow was not in heat，the average duration of contact with the bionic boar was 8. 44 s/3 min. On the second day after weaning Sow 1，the sow's response time reached 45 s/3 min. At the same time，the manual BPT was performed，and the test results are consistent；all the sows were in oestrus.

The response durations of weaned sows to the bionic boar were analysed through video analysis. The results show that when a sow was not in heat，the sow was curious about the bionic boar and interested in the 'pheromone'（saliva，urine，semen，etc. ）released by the bionic boar and the recorded sound of the boar. During this period，the average response duration of weaned sows to the bionic boar was 8. 44 s/3 min. However，an excessively long response time was recorded on a certain day during this cycle. When the weaned sow had a higher contact duration on one day than on other days，At the same time，the results of artificial back pressure detection of estrus on sows are often in estrus.

3. 2. 2　Duration of response（frequency）　The target of experimental data analysis，from the first day of weaning to the sow's artificial back pressure detection，was the cut-off of oestrus. The number of responses（frequency）of the weaned sow to the bionic boar was also an indicator of the physiological condition of the sow. The bionic boar model proposed in this paper has real boar smell，sound，and touch devices. The video analysis results show that the number of times that the sow is in contact with the bionic boar after weaning changes regularly with the number of days of post-weaning.

When the sow was detected as being in oestrus under artificial back pressure，the weaned sow frequently came into contact with the bionic boar device. In the unit's 3-min investigation time，the average number of interactions with the bionic boar was 8 times/3 min. When not in heat，the average number of contacts was 3. 8 times/3 min. The data show that when a sow was detected as exhibiting oestrus under artificial back pressure，a crest curve appeared during the oestrus cycle. When the weaned sows were in heat，the average number of exposures to the bionic boars was 8. 3 times/3 min. In Fig. 7，Sow 7 and Sow 8 are represented by the dotted lines. Through video analysis，although Sow 7 exhibited curve fluctuations，there was no oestrus behaviour detected upon the application of manual back pressure. The curve in the figure corresponding to Sow 6 shows a regular rise and fall. The result of manual detection is that the oestrus state was not obvious. Finally，mating was carried out，and the tracking results show that Sow 6 returns to oestrus after mating，but the mating was unsuccessful.

According to the verification data obtained from the pig farm, the number of sow responses to the bionic boar was used to determine whether the sow was in oestrus, and the accuracy rate was 87.5%. Research has shown that in the field, radiofrequency identification (RFID) and other sensors are used to monitor the frequency of sow visits to boars to detect oestrus [39-41]. The result was compared with that of the artificial BPT, and the preliminary statistics in the field test exceeded 60%. It is concluded that compared with manual observation, subjective observation can be quantified into image analysis, which is different from previous studies. The subjective observation of previous studies is time-consuming and laborious. However, real boars are used for checking in this situation. In contrast, this study used a bionic boar device for checking, which offers time savings, labour savings, and automated operation. The most important aspect is that the bionic boar device is a recyclable and sustainable device.

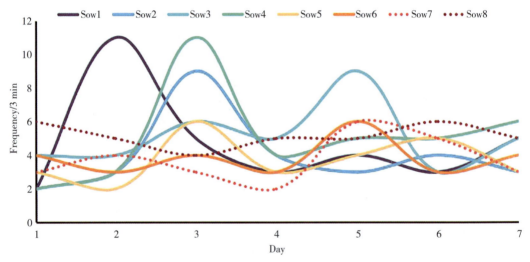

Fig. 7   The number of times the sow was in contact with the bionic boar 1-7 days after weaning
(within three minutes of a single love check).

3.2.3 Duration of the static response of both sow ears   The experimental object in this study was a large white sow that had good reproductive performance and a high number of surviving piglets. During oestrus, large white sows often exhibit some signs that they are actually in oestrus. With a decrease in feed intake and an increase in activity, a sow's vulva appears red and swollen, the sow's limbs often appear to stand still, and the ears are upright.

In a pig house, the staff is required to have experience and theoretical knowledge to be able to judge whether a sow is exhibiting oestrus behaviour based on physical signs. We used time quantification to determine the duration of a sow's body movements. As shown in

Fig. 8, the response of the ears of a weaned sow (erect) to the bionic boar was more obvious during the oestrus period than in other periods. In Fig. 8, the peak of the binaural standing occurred during oestrus, and the duration of binaural standing was higher when the sow was in oestrus than when it was in the non-oestrus state. On the second day after weaning, Sow 1 had both ears upright for 88 s/3 min when the bionic boar performed an oestrus check, with each check carried out once every 3 min. Combined with the artificial BPT, the results all show that the sows were in oestrus. According to the data analysis in Fig. 8, the average standing time of both ears in the sows in heat was 41. 3 s within a single 3-min check. When a sow is not in oestrus, its ears will not stand still for a long time. According to the response behaviour of the sows and the bionic boars during oestrus in the early stage, a preliminary statistical analysis was carried out on the 8 sows in the field verification test. The later tracking results showed that Sow 6 returned to oestrus after unsuccessful breeding. The rest of the verified sow samples all entered the gestation stage. A threshold regarding the length of time that the ears of a sow are erect can be used as the basis for the judgement of oestrus states.

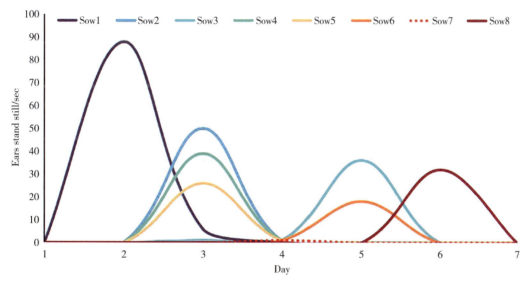

Fig. 8　The time until a sow's ears stand still 1-7 days after weaning
(within three minutes for a single oestrus check).

3. 3　Evaluation rule of the response of a yorkshire sow to a bionic boar from weaning to oestrus

During the period from weaning to oestrus, the behaviour changes of a sow can be determined through video analysis technology. The SAE model can be used to classify and recognize sow behaviour, enabling one to achieve mastery of a sow's oestrus status. For the research in this article, an analysis and a comparison of multiple neural networkbased classification and recognition models with respect to sow behaviours during the oestrus cycle were performed. The maximum accuracy rate reached 98. 25% (minimum 90. 00%), and the detection effects of these models are superior to that of the method of Freson [15] (86%), which uses infrared sensors to detect sow oestrus. At the same time, the use of bionic boars combined with machine vision technology for oestrus detection shows intelligence and mindfulness of sow welfare. This is an obvious advantage over detection with traditional sensors [16]. Through binaural erection analysis of the response degree of a weaned sow to the bionic boar, including the duration of the response to the bionic boar device, the number of visits, and the actual response of the sow, a comprehensive evaluation of the oestrus status of the sow can be achieved with a high accuracy rate.

As shown in Fig. 9, when the bionic boar was checking the oestrus status of a sow, the frequency of the sow's response (standing still with both ears up) and the frequency of contact with the bionic boar's silicone snout device were strong correlation factors related to the sow's oestrus behaviour. However, the duration of the exposure of the sow to the biomimetic boar snout device on its own was a relatively weak factor. The possible cause for this is that the bionic boar did not induce enough sexual attraction in the sow.

Fig. 9 shows the actual verification results of conducting an inspection with the bionic boar and a simultaneous manual BPT at the same time. For the sow and the bionic boar device, if the duration of contact and the time that both ears stand still were significantly higher than the same metrics in the previous test, the sow was judged to be in oestrus (quantify the results of manual observation through the method of data analysis and display). The manual pressure back test was performed simultaneously, and the test results are found to be consistent. Through the continuous tracking of breeding, it was found that Sow 6 returned to oestrus and that breeding was unsuccessful. Sow 7 was determined to have not been in heat. The reason for this is that the body condition of a pig is poor after weaning, and the skeleton of the sow's back can be clearly seen. Poor body conditions or high obesity can directly affect a sow's oestrus status [42]. This method is aimed at a single breed, Yorkshire sows, and three-parity sows with regular oestrus periods were selected. In the future, the method can be extended to the oestrus monitoring of gilts.

4 Conclusions

In this study, a bionic boar model was built to check the conditions of weaned sows in response to the saliva, urine, and semen of a boar. Furthermore, based on video analysis technology, a neural network model was used to classify the behaviours of large white sows

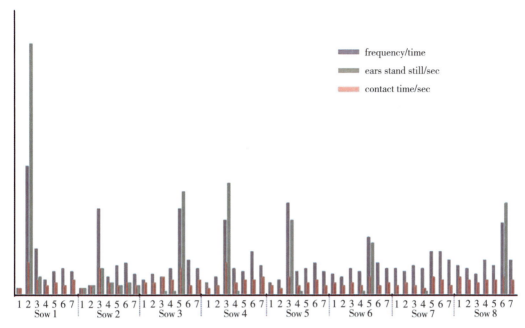

Fig. 9　Date columns of different colours on the Y axis represent different untis, so they are placed under the same Y axis in order to visually display the date distribution. The response of a sow to the bionic boar（frequency, duration of ears standing still, duration of body standing still）. Frequency/time is the number of instances of contact, that is, the frequcy of contact per unit time; ears stand still/sec is the static time of both ears of the sow in reponse to the bionic boar in seconds; contnact time/sec is the duration of contact between the sow and the bionic boar in seconds.

during the oestrus cycle. The results show that the recognition accuracy of the SAE model was 98.25％. It can effectively identify the oestrus behaviours of sows by analysing the response degree of each sow to the bionic boar. The results show that there was a strong correlation between the contact duration of the oestrus sow with the bionic boar and the static behaviours of both ears during the oestrus cycle. The average duration of contact between the sows in oestrus and the bionic boar was 29.7 s/3 min, and the average length of time that the ears of the oestrus sows were upright was 41.3 s/3 min. The interaction between a sow and the bionic boar can be used as the basis for judging the sow' s oestrus state.

References

［1］ Berckmans D. Precision livestock farming（PLF）［J］. Computers and Electronics in Agriculture, 2008, 62: 1.

［2］ Turner S P, Camerlink I , Baxter E M , et al. Breeding for pig welfare. In Advances in Pig Welfare ［J/OL］. Elsevier: Amsterdam, The Netherlands, 2018: 399-414.

［3］ Ostersen T , Cornou C , Kristensen A R. Detecting oestrus by monitoring sows' visits to a boar ［J］.

Computers and Electronics in Agriculture，2010，74：51-58.

［4］ Pietrosemoli S ， Tang C. Animal Welfare and Production Challenges Associated with Pasture Pig Systems：A Review ［J］. Agriculture，2020，10：223.

［5］ Lehrer A，Lewis G，Aizinbud E. Oestrus detection in cattle：Recent developments ［J］. Animal Reproduction Science，1992，28：355-362.

［6］ Kilgour R，Choquenot D. The estrous cycle and induction of estrus in the australian feral sow（Sus scrofa）［J］. Theriogenology，1994，41：1181-1192.

［7］ Oczak M，Maschat K，Berckmans D ，et al. Can an automated labelling method based on accelerometer datareplace a human labeller? —Postural profile of farrowing sows ［J］. Computers and Electronics in Agriculture，2016，127：168-175.

［8］ Thompson R，Matheson S M，Plötz T ，et al. Porcine lie detectors：Automatic quantification of posture state and transitions in sows using inertial sensors ［J］. Computers and Electronics in Agriculture，2016，127：521-530.

［9］ Cornou C. Automated oestrus detection methods in group housed sows：Review of the current methods and perspectives for development ［J］. Livestock Science，2006，105：1-11.

［10］ Scolari S C，Clark S G，Knox R V，et al. Vulvar skin temperature changes significantly during estrus in swine as determined by digital infrared thermography ［J］. Journal of Swine Health and Production，2011，19：151-155.

［11］ Sykes D，Couvillion J ，Cromiak A，et al. The use of digital infrared thermal imaging to detect estrus in gilts ［J］. Theriogenology，2012，78：147-152.

［12］ Simões V G，Lyazrhi F ，Picard-Hagen N，et al. Variations in the vulvar temperature of sows during proestrus and estrus as determined by infrared thermography and its relation to ovulation ［J］. Theriogenology，2014，82：1080-1085.

［13］ Altmann M . Interrelations of the sex cycle and the behavior of the sow ［J］. Journal of Comparative Psychology，1941：31，481-498.

［14］ Bressers H P M . Monitoring Individual Sows in Group-Housing：Possibilities For Automation ［D］. Ph. D. Thesis，Wageningen University & Research，Wageningen，The Netherlands，1993：135.

［15］ Freson L，Godrie S ，Bos N，et al. Validation of an infra-red sensor for oestrus detection of individually housed sows ［J］. Computers and Electronics in Agriculture，1998，20：21-29.

［16］ Houwers H. Locality registration as a way of oestrus detection in an integrated group-housing for sows ［C］. In Proceedings of the International Congress on Applied Ethology in Farm Animals，Skara，Sweden，1988：（17-19）：44-50.

［17］ Korthals R L. The effectiveness of using electronic identification for the identification of estrus in swine ［C］. In Proceedings of the ASAE/CSAE-SCGR Annual International Meeting，Toronto，ON，Canada，1999，7（8-21）：10.

［18］ Kashiha M ，Bahr C，Ott S，et al. Automatic identification of marked pigs in a pen using image pattern recognition ［J］. Computers and Electronics in Agriculture，2013，93：111-120.

［19］ Zhang L ，Gray H ，Ye X，et al. Automatic individual pig detection and tracking in pig farms ［J］. Sensors，2019，19：1188.

［20］ Chen C，Zhu W，Ma C，et al. Image motion feature extraction for recognition of aggressive behaviors among group-housed pigs ［J］. Computers and Electronics in Agriculture，2017，142：380-387.

［21］ Kim J，Chung Y，Choi Y，et al. Depth-Based Detection of Standing-Pigs in Moving Noise Environments ［J］. Sensors，2017，17：2757.

［22］ Daqin W，Haiyan H. A Research on identification and predication of sows' oestrus behavior based on hopfield neural network ［C］. In Proceedings of the 11th International Conference on Measuring Technology and Mechatronics Automation (ICMTMA)，Qiqihar，China，2019，4（28-29）：399-401.

［23］ Gerritsen R，Langendijk P，Soede N，et al. Effects of artificial boar stimuli on the expression of oestrus in sows ［J］. Applied Animal Behaviour Science，2005，92：37-43.

［24］ Hemsworth P H，Barnett J L. Behavioural responses affecting gilt and sow reproduction ［J］. Journal of Reproduction and Infertility，1990，40：343-354.

［25］ McGlone J J，Garcia A，Rakhshandeh，A. Multi-Farm Analyses Indicate a Novel Boar Pheromone Improves Sow Reproductive Performance ［J］. Animal，2019，9：37.

［26］ Guiraudie-Capraz G，Slomianny M C，Pageat P，et al. Biochemical and Chemical Supports for a Transnatal Olfactory Continuity through Sow Maternal Fluids ［J］. Chemical Senses，2005，30：241-251.

［27］ Goudet G，Prunier A，Nadal-Desbarats L，et al. Steroidome and metabolome analysis in gilt saliva to identify potential biomarkers of boar effect receptivity ［J］. Animal，2021，15：100095.

［28］ Hinton G E，Osindero S，Teh Y W. A Fast Learning Algorithm for Deep Belief Nets ［J］. Neural Computation，2006，18：1527-1554.

［29］ Rumelhart D E，Hinton G E，Williams R J. Learning representations by back-propagating errors ［J］. Nature，1986，323：533-536.

［30］ Cumani S，LaFace P. Analysis of Large-Scale SVM Training Algorithms for Language and Speaker Recognition ［J］. IEEE Trans. Audio Speech Lang. Process，2012，20：1585-1596.

［31］ Lei Y，Hansen J H L. Dialect Classification via Text-Independent Training and Testing for Arabic，Spanish，and Chinese ［J］. IEEE Transactions on Audio Speech and Language Processing，2010，19：85-96.

［32］ Hinton G，Salakhutdinov R. Supporting Online Material for Reducing the Dimensionality of Data with Neural Networks ［J］. Science，2006：313，504.

［33］ Gu F，Flórez-Revuelta F，Monekosso D，et al. Marginalised Stacked Denoising Autoencoders for Robust Representation of Real-Time Multi-View Action Recognition［J］. Sensors，2015，15：17209-17231.

［34］ Vapnik V The Nature of Statistical Learning Theory ［M］. Springer：New York，NY，USA，1995.

［35］ Perros H G. Support Vector Machines. In An Introduction to IoT Analytics ［M］. CRC Press：Boca Raton，FL，USA，2021：279-330.

［36］ Downey B R. Regulation of the Estrous Cycle in Domestic Animals—A Review［J］. Canadian Journal of Veterinary Research-Revue Canadienne De Recherche Veterinaire，1980，21：301-306.

［37］ Teele T. Effects of Sexual Preparations on Reproductivity of Boars and AI Sows ［M］. Lambert Academic Publishing：Saarbrücken，Germany，2017.

［38］ Turner A，Hemsworth P，Tilbrook A. The sexual motivation of boars housed adjacent to ovariectomised gilts did not affect the efficiency of detecting hormonally induced oestrus using the back pressure test [J] . Applied Animal Behaviour Science, 1996, 49: 343-351.

[39] Bressers H, Brake J T, Engel B, et al. Feeding order of sows at an individual electronic feed station in a dynamic group-housing system [J] . Applied Animal Behaviour Science, 1993, 36: 123-134.

[40] Bressers H P M, Brake J, Noordhuizen J. Estrus Detection in Group-Housed Sows by Analysis of Data on Visits to The Boar [J] . Applied Animal Behaviour Science, 1991, 31: 183-193.

[41] Bure R G, Houwers H W J. Automation in group housing of sows [J] . Nutrition Environment Agriculture, 1990, 41: 384-386.

[42] Johnston L J, Fogwell R L, Weldon W C, et al. Relationship between Body Fat and Postweaning Interval to Estrus in Primiparous Sows [J] . Journal of Animal Science, 1989, 67: 943-950.

猪场紫外线消毒技术研究进展

赵　铖　樊士冉　杜晓冬　邓启伟　刘　聪

摘要：在养猪业健康发展进程中，生物安全问题一直是人们关注的焦点。紫外线（UV）消毒技术以其消毒效率高、使用成本低及安全性高等特点，在猪场生物安全领域开始受到关注，并取得一些进展。但是紫外线消毒技术仍存在病原体不完全灭活等问题，灭活效果是紫外线消毒作业中最被关心的问题。本文就影响紫外线消毒效果的因素、灭活常见的猪病原体所需紫外线剂量以及紫外线消毒在猪场的应用分别展开论述。同时，针对研究发展中的不足之处进行了总结和思考。研究表明，紫外线消毒技术消毒效果受紫外线灯工作参数、外界条件、被照物材料和病原体种类等多因素影响，因此在使用紫外线消毒时，需要综合考虑多方面的因素；部分常见猪病原体紫外线灭活所需剂量仍需研究；此外，需要进一步改进紫外线消毒技术与应用，解决病原体光复活的问题。

关键词：紫外线消毒；猪病原体；紫外线剂量；应用

生物安全是现代化畜牧业中的重要环节[1,2]。目前，化学消毒方式因其操作方便、杀菌（毒）效果好等特点而被猪场普遍应用[3]。但该消毒方式存在残留，易引发二次污染等问题[4]；且猪场使用的消毒剂种类众多，不同消毒对象所用的消毒剂种类和用法也不同[5]；消毒剂对于手机、药品、电脑等易造成损害。综上所述，化学消毒方式无法满足猪场生物安全的所有场景。

紫外线消毒是一种物理消毒方式，其作用机制是利用254 nm波长的紫外光照射，对微生物的遗传物质（RNA和DNA）产生光化学伤害，进而阻碍遗传物质的复制、转录以及蛋白质的合成，最终致使微生物死亡，达成消毒的目的[6]。相比化学消毒方式，紫外线消毒在消毒效果、使用成本及安全性方面，具有消毒时间短、效率高、广谱杀菌、结构简单占地面积小、运行维护简便、无二次污染等特点[7]。化学消毒剂灭活微生物需要较长的时间，而紫外线消毒仅需几秒钟即可达到同样的灭活效果。并且紫外线消毒的操作简便，其基建投资及运行费用也低于其他几种化学消毒方法。紫外线消毒目前已被广泛应用于许多领域，其中包括空气消毒[8,9]、净水和废水处理[10,11]、食品和饮料的保存及医疗应用[12-15]等。近几年，紫外线消毒技术也逐步应用于猪场生物安全[16,17]。影响紫外线杀菌效果的因素有很多，不同猪病原体灭活所需的剂量也不同，以及紫外线消毒在猪场应用的最佳工作对象和场景尚需探究。将各因素进行综合分析，提高紫外线应用于猪场消毒的效果和效率，使其合理化应用，是当前猪场紫外线消毒研究工作的热点和重点。

1　影响微生物灭活效果的因素

紫外线灭活微生物受多种因素的影响，可归结为三大方面：紫外线灯工作参数、外界条件和被照射物的情况。对于紫外线设备而言，紫外线灯的光照度、照射时间、照射距离、照射角度等因素决定了到达微生物所在位置的紫外线剂量，进而影响对微生物的灭活效果；外界条件如温湿度、是否有遮挡物影响紫外线传播；微生物依附的照射物材料也会影响消毒效果。对紫外线灯灭活微生物的影响因素进行探究和总结有助于提高紫外线消毒在猪场的应用效率。

1.1　紫外线灯的工作参数

微生物所接收到的紫外线照射剂量决定了其灭活的程度，微生物通过紫外线消毒器时接收到的紫外线剂量定义[18]为：

$$Dose = \int_0^t I dt \tag{1}$$

式中：$Dose$＝紫外剂量，mW·S/cm^2 或 mJ/ cm^2；

　　　I＝微生物在其运动轨迹上某一点接收到的紫外线照射光照度，mW/cm^2；

　　　t＝曝光时间或滞留时间，s。

公式（1）表明，紫外线灯的照射时间和光照度决定紫外线照射剂量大小。Runda 等[19]针对非洲猪瘟病毒进行了紫外线消毒试验研究，重点集中于照射时间和照射强度等因素对灭活效果的影响，结果表明紫外线照射时间越长，非洲猪瘟病毒的核酸损伤越严重。对比紫外线强度的灭活效果，当紫外线强度为 110～120 μW/cm^2 时，完全灭活非洲猪瘟所需的辐照时间为 30 min，而当光照度为 3 600 μW/cm^2 时，仅需 3 s 即可完全灭活。Conner-Kerr 等[20]检测了 UVC 照射灭活耐抗生素的金黄色葡萄球菌的有效性，在紫外线灯光照度为 15.54 mW/cm^2、灯与病毒的距离为 25.4 mm、金黄色葡萄球菌（MRSA）达到灭活率为 99.9％时，需照射 5 s，灭活率为 100％时需照射 90 s。由此得出紫外线光照度越大，消毒时间越长，杀菌效率越高。

而不同照射距离处的光照度不一样，光照度与照射距离的关系[21]遵循公式（2）：

$$\frac{I_1}{I_2} = \frac{D_1^2}{D_2^2} \tag{2}$$

式中：I_1＝位置 1 处的光照度，mW/cm^2；

　　　I_2＝位置 2 处的光照度，mW/cm^2；

　　　D_1＝位置 1 至光源的距离，cm；

　　　D_2＝位置 2 至光源的距离，cm。

由公式（2）可知，离紫外线灯越近的物体，光照度越大。Nerandzic 等[22]探讨了不同距离对紫外线灭活金黄色葡萄球菌（MRSA）效果的影响，将承载菌落总数为 5 个对数滴度的载玻片分别放置在距离紫外线灯不同的辐射范围内，照射相同时长，结果表明在0.15 m、1.22 m和3.05 m距离下，金黄色葡萄球菌对数滴度减少分别为3.3、1.8和0.7，增加与紫外线灯的距离，大大降低了紫外线照射的灭活效果。与此同时，距离被照物越近，紫外线灯可照射的范围也越有限，为确保紫外线灭活效果，需平衡好紫外线强度大小与紫外线照射范围，因此在紫外线光源和目标物体之间选定适当的距离至关重要（图1）[23]。

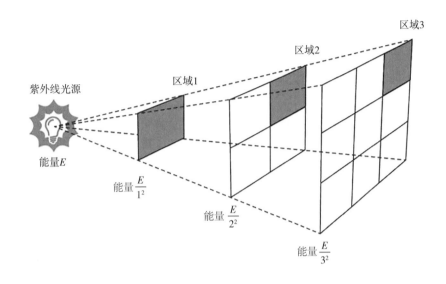

图 1　紫外线传播遵循平方反比定律

根据朗伯余弦定律（Lambert's Cosine Law）[24]可知，随紫外线与被照物表面所在平面的夹角增大，被照物表面上的紫外线辐射功率余弦变化，在紫外线垂直于被照射物表面时，紫外线强度最高。因此，照射角度的变化也影响着病毒灭活效果[25]。Katelyn等[26]和孙巍等[27]对比了大肠杆菌和铜绿假单胞菌在紫外线灯不同角度位置处的灭活效果，发现大肠杆菌和铜绿假单胞菌正对紫外线灯时灭活效果最好，而与紫外线光平行表面灭活效果最差（图2）。

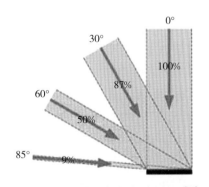

图 2　光照度随照射角度变化情况[25]

1.2　外界条件

温度和湿度是影响紫外线照射强度的重要因素。韩雪玲等[28]观察了温湿度对紫外线光照度的影响，研究表明当温度由 18℃增加到 30℃时，紫外线光照度值则由 103 μW/cm^2 增加到 122 μW/cm^2；叶圣勤[29]报道了在 5～20℃范围，紫外线光照度在 81.5～109.3 μW/cm^2 范围内上升，两者呈正相关（r=0.991）。紫外线光照度值随温度增加而上升，而紫外线光照度又影响着微生物的灭活率。Gayán 等[30]研究表明，紫外线与热处理灭菌产生协同作用，紫外线工作环境温度低于 45℃时灭菌效果保持不变，超过 45℃后随着温度增加灭菌效果呈指数增长。协同作用百分比在 55℃下最高，为 40.3%。

相对湿度对紫外线杀菌力的影响较复杂，Timothy[31]探究了不同湿度范围对猪繁殖与呼吸综合征病毒（PRRS）的灭活效果，在相对湿度为 25%～79%时病毒灭活率最高，相对湿度＜24%时病毒灭活率其次，相对湿度＞80%时病毒灭活率最低。Tseng 等[32]对比不同湿度下被测病毒灭活所需紫外线剂量，相对湿度 85%下的病毒灭活所需的紫外线剂量显著高于相对湿度 55%下所需的紫外线剂量。Peccia[33]对此现象进行了分析，表示高湿度条件下，附着在病毒表面的水导致紫外线光照度的衰减。

紫外线消毒也有一定的局限性。紫外线在直接照射物体表面的情况下消毒效果最佳，而在有遮挡的情况下消毒效果较差。Aaron 等[34]对比了在紫外线灯直接照射和间接照射（培养皿盖遮挡）处理下，猪流行性腹泻病毒（PEDV）的减少量。照射前 PEDV 半数组织细胞感染量（TCID$_{50}$）为 1 万左右，紫外线照射 1min 后，直接照射的 PEDV 半数组织细胞感染量（TCID$_{50}$）下降了 4 个对数滴度，而间接照射的 PEDV 仅下降了 0.68 个对数滴度。导致该现象的原因是由于紫外线的穿透力差，遮挡物使得紫外线光照度变弱，与无遮挡的处理相比，病原体接收到的紫外线剂量大幅度下降，灭活效果随之变差。易附着在紫外线灯管上的灰尘也会降低紫外线光照度，因此须定期擦拭紫外线灯管上的灰尘，防止因灰尘的遮挡而降低紫外线灯的消毒效果[35]。对于遮挡物，除石英玻璃以外，紫外线不能透过普通玻璃和其他非透明材料。因此，石英也被用来制造紫外线灯外层[36]。

1.3　被消毒物品材料

紫外线消毒方式可以快速有效地灭活猪场常见病毒[34,35]。猪场需要消毒的常见物品较多，如橡胶鞋、一次性乳胶或布手套、塑料袋、混凝土地板、装疫苗的玻璃瓶、装载物品的纸箱等，因此需要消毒的材料种类也较多。材料种类及其表面粗糙情况不同，紫外线消毒效果也不同。Chelsea 等[37]以塞内卡病毒（SVA）为试验病毒，对比了该病毒在猪场常见材料上（纸板、塑料和布）的紫外线灭活效果，研究显示紫外线消毒能有效降低塑料处理组的 SVA 病毒对数滴度，其病毒对数滴度下降超过 7，而在纸板和布表面的最大下降对数滴度为 2，没有有效降低 SVA 病毒对数滴度。推测原因，纸板和布表面粗糙，紫外线不易照射到分布在材料表面较凹处的病毒，而塑料表面光滑度较高，紫外线照射情况优于粗糙表面的纸板等材料。

2　灭活常见猪病原体的紫外线剂量研究进展

微生物的灭活程度与所使用的紫外线剂量直接相关，合适的紫外线剂量对大多数致病微生物具有较佳的灭活效果，但在紫外线剂量不足的情况下，有很多失活却未被彻底杀灭的微生物可凭借光的协助作用修复自身被破坏的结构，从而复活[38]；且各种微生物对紫外线的吸收剂量和对紫外线的抵抗力不尽相同[30]，细菌比病毒更容易感光，相同紫外线剂量下，细菌会更快地被完全灭活。

为有效地抑制微生物的修复能力，不同猪病原体（完全灭活）所需的紫外线剂量需要合理设计。微生物的灭活效果用对数尺度进行分级，具体为照射前后微生物浓度比值的对数值 $\lg (N/N_0)$ 来表示灭活率[39]，其中 N_0 为初始微生物浓度，N 为紫外线照射后微生物存活的浓度。表 1 和表 2 分别为文献中获取到的猪场常见细菌减少量（lg）所需的紫外线灯照射剂量和常见病毒减少量所需的紫外线灯照射剂量。表 1 和表 2 中，lg-1 对应灭活率为 90%，lg-2 对应灭活率为 99%，lg-3 对应灭活率为 99.9%，lg-4 对应灭活率为 99.99%[40]。表 1 和表 2 中列出了病原体每一级减少所需要的高频短波紫外线（UVC）照射剂量，以及数据来源的参考文献。

表 1　猪场常见致病性细菌对数滴度下降与 UVC 照射剂量对照表

细菌	对数滴度下降量					参考文献
	lg-1	lg-2	lg-3	lg-4	lg-5	
大肠杆菌	3.0	6.6	9.0	—	—	[41, 42]
鼠伤寒沙门菌	8.0	15.2	—	—	—	[43]
金黄色葡萄球菌	2.6	6.6	—	—	—	[43]
副猪嗜血杆菌	—	—	—	—	—	
猪肺炎支原体	—	—	—	—	—	

注："—"代表无数据。

表 2　猪场常见致病性病毒对数滴度下降与 UVC 照射剂量对照表

病毒	对数滴度下降量					参考文献
	lg-1	lg-2	lg-3	lg-4	lg-5	
猪星状病毒		10.0～12.0	—	—	—	[44]
猪繁殖与呼吸综合征病毒	3.9	4.5	>4.9			[32]
猪流行性腹泻病毒	—	0.7	2.7	2.9	—	[32]
口蹄疫病毒	24.0	48.0	72.0	96.0	120.0	[45]
水泡性口炎病毒	—	—	—	19.0	<75.0	[46]
狂犬病病毒				5.0		[47]
猪细小病毒	—	—	—	—	86.0	[48]
非洲猪瘟病毒						

（续）

病毒	对数滴度下降量					参考文献
	lg-1	lg-2	lg-3	lg-4	lg-5	
猪传染性胃肠炎病毒	—	—	—	—	—	—
经典猪瘟病毒	—	—	—	—	—	—
伪狂犬病病毒	—	—	—	—	—	—
猪圆环病毒Ⅱ型、Ⅲ型	—	—	—	—	—	—

注："—"代表无数据。

　　由表1和表2数据可知，对于已试验的猪场常见细菌和病毒，其灭活所需紫外线剂量都小于 150 mJ/cm²，基本在紫外线剂量较小时，就已被完全灭活。同时，在所了解的文献中，大部分猪病原体，如猪肺炎支原体、猪链球菌、伪狂犬病病毒、猪瘟病毒等常见病原体紫外线灭活所需剂量未见公布。猪病原体灭活紫外线剂量的获取是促进猪场高效利用紫外线消毒的重要环节，虽然相关研究人员在不同猪病原体微生物灭活所需的紫外线剂量方面做了大量工作，但目前该部分研究仍较为缺乏。因此，亟须测试猪场常见病原体的灭活紫外线剂量，为猪场紫外线消毒设备的设计和应用提供参考，以达到更好的消毒效果。

3　紫外线消毒在猪场的应用

　　作为综合生物安全计划的一部分，猪场紫外线消毒设备的应用在过去几年有所增加。紫外线消毒箱是猪场常见的消毒设备（图3），通常它被设计为传递窗形式，放置于猪场入口处或隔离点，具体操作是将被消毒的物品从传递窗一侧（相对脏区）放入，紫外线照射处理后，从传递窗另一侧（相对净区）取出。

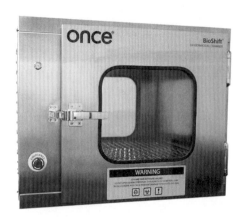

图3　紫外线消毒箱的传递窗

　　使用紫外消毒箱的物品主要为手机、电脑、小工具和药物等小到中等尺寸的物品[35]。化学消毒可能会对这些物品造成腐蚀，而紫外线消毒可避免该情况的发生。这些物品表面通常无脏污，表面可完全被紫外线照射，消毒效果可以得到保证。此外，对于塑料和纸板材质的物品，紫外线消毒方法并不适用，重复暴露在紫外线下的塑料会发生变化，如颜色变浅或产生气味。紫外线对纸板消毒效果有限，其原因是纸板表面粗糙多孔，孔内存在紫外线无法照射到的地方，导致该处病原体无法被有效灭活。

紫外线也可以用来净化猪场污水，李同等[16]探讨了不同透射率、紫外线杀菌装置内水深和水力停留时间对沼液絮凝上清液的杀菌效果。结果表明，当紫外线光照度为 395 $\mu W/cm^2$、沼液絮凝上清液的透射率为 0.69%、水力停留时间为 15 min 和紫外线杀菌装置内水深 2 cm 时，沼液絮凝上清液中大肠菌群的数量从 3.9×10^6 个/L 下降至检出限（3 个/L）以下。John 等[49]对比了氯和紫外线消毒方式对猪场废水的消毒效果，紫外线可灭活氯无法杀死的耐氯菌，因此紫外线灭活废水的细菌数量高于氯消毒后的灭活细菌数量，但因猪场污水未做絮凝处理，达到灭活效果所消耗的能量较高，相对于氯处理，其经济性较差。

紫外线也可净化空气中的病原菌。现代规模化猪场使用空气过滤器拦截病菌[50]，但研究表明，空气过滤系统并未使猪舍空气中的细菌数量显著减少[51]。基于此种情况，Lisa 等[52]研制了安装紫外线灯的空气过滤器，以金黄色葡萄球菌、胸膜肺炎放线杆菌、猪细小病毒和猪繁殖和呼吸综合征病毒等易在空气中传播的病菌为试验病菌，探究该设备的杀菌情况，并与未安装紫外线灯的空气过滤器杀菌效果作对比。数据分析得出，单独使用空气过滤器的猪病原体数量平均减少 67.51%，而紫外线照射与空气过滤器联合使用时，病原体减少 99% 以上。

4 小结和展望

紫外线消毒技术具有独特优势，是猪场生物安全不可或缺的一部分。但是，目前关于猪场紫外线消毒应用的研究是有限的，相关科研人员仍需要在已有研究成果的基础上，进一步探究该技术在猪场生物安全领域的应用方向。其未来研究方向可概括为以下三个方面。

4.1 紫外线工作参数及环境参数对消毒效果的影响

目前，紫外线杀菌仍存在不能彻底灭活等问题，为了保证紫外线在物体表面上有良好的杀菌效果，应针对不同消毒对象，探究紫外线最优的作业参数以及组合形态，包括光照度、光照时长、照射距离，以及灯管数量、位置、间距等，以提高病原体灭活效率。

4.2 进一步开展猪只病原体紫外线灭活所需剂量

充足的紫外线剂量才可将微生物彻底灭活，而不同的病原体对紫外线的敏感度不同，所需剂量也不一样。目前，很多常见猪只病原体所需的紫外线灭活剂量仍未公布。未来需要加大对这些猪只病原体灭活所需紫外线剂量的研究，为猪场紫外线消毒应用设备研发提供指导。

4.3 探究紫外线消毒技术的改进和应用场景

现有的紫外线消毒技术仍面临光复活所带来的挑战，因此，改进紫外线消毒技术在实际中的应用是非常必要的。紫外线联合消毒技术是未来的发展方向，相关研究表明，紫外线与高温、紫外线与臭氧具有协同灭菌的效果，且可以弥补紫外线无法灭活物体内部病菌的不足[30,53]。应用方面，较为成熟的是利用紫外线消毒箱对手机等小物品进行消毒，而紫外线对猪场其他场景和消毒对象的消毒效果未做深入研究，紫外线消毒的优点未被最大化利用。因此，紫外线消毒技术在猪场相关场景的应用有待进一步探索。

参考文献

[1] 张训海，王旋，赵磊，等. 畜禽健康养殖及其生物安全体系的构建 [J]. 中国家禽，2013，35（010）：55-56.

[2] Nauwynck H J，Glorieux S，Favoreel H，et al. Cell biological and molecular characteristics of pseudorabies virus infections in cell cultures and in pigs with emphasis on the respiratory tract [J]. Veterinary Research，2007，38（2）：229-241.

[3] 王爱玲. 四种消毒剂对猪场常见病原微生物的杀灭效果研究 [D]. 西北农林科技大学，2016.

[4] Ghernaout D，Ghernaout B. From chemical disinfection to electro-disinfection：The obligatory itinerary [J]. Desalination and Water Treatment，2010，16（3）：156-175.

[5] 张启祥，卢少达. 猪场常用化学消毒剂的使用方法 [J]. 养殖技术顾问，2013，08（8）：199-199.

[6] Kuluncsics Z，Perdiz D，Brulay E，et al. Wavelength dependence of ultraviolet-induced DNA damage distribution：involvement of direct or indirect mechanisms and possible artefacts [J]. Photochem Photobiol B：Biology，1999，49（1）：71-80.

[7] Curtis J D. Dose Improving surface cleaning and disinfection reduce health care-associated infections [J]. American Journal of Infection Control，2013，41（5）：S12-S19.

[8] Brickner P W，Vincent R L，First M，et al. The application of ultraviolet germicidal irradiation to control transmission of airborne disease：bioterrorism countermeasure [J]. Public Health Reports，2003，118（2）：99-114.

[9] Noakes C J，Beggs C B，Sleigh，et al. Modeling the transmission of airborne infections in enclosed spaces [J]. Epidemiology and Infection，2006，134（5）：1082-1091.

[10] Hiljnen W A，Beerendonk E F，Medema G J. Inactivation credit of UV radiation for viruses，bacteria and protozoan（oo）cysts in water [J]. Water Research，2006，40（1）：3-22.

[11] Lin C H，Yu R F，Cheng W P，et al. Monitoring and control of UV and UV- TiO2 disinfections for municipal wastewater reclamation using artificial neural networks [J]. Journal of Hazardous Materials，2012，209-210：348-354.

[12] Keyser Maricel，Ilze A，et al. Ultraviolet radiation as a non-thermal treatment for the inactivation of microorganisms in fruit juice [J]. Innovative Food Science & Emerging Technologies 9，2008，3：348-354.

[13] 洪雅敏，朱庆庆，刘 清，等. 紫外线在果蔬保鲜方面的研究进展 [J]. 中国野生植物资源，2017，

036 （004）：50-52，59.

[14] Thai T，Keast D，Campbell K，et al. Effect of ultraviolet light C on bacterial colonization in chronic wounds ［J］. Ostomy Wound Manage，2005，51 （10）：32-45.

[15] Marie L，Eva T，Claes L，et al. Ultraviolet-C decontamination of a hospital room：Amount of UV light needed ［J］. Burns，2020，46 （4）：842-849.

[16] 李同，董红敏，陶秀萍. 猪场沼液絮凝上清液的紫外线杀菌效果 ［J］. 农业工程学报，2014，30 （06）：165-171.

[17] Blázquez E，Rodríguez C，Ródenas N，et al. UV-C irradiation is able to inactivate pathogens found in commercially collected porcine plasma as demonstrated by swine bioassay ［J］. Veterinary Microbiology，2019，239：108450.

[18] 韩兴. 污水紫外消毒装置设计及工艺参数优化研究 ［D］. 吉林农业大学，2006.

[19] Runda X，Lang G，Heng W，et al. Disinfection Effect of Short-wave Ultraviolet Radiation （UV-C） on ASFV in Water ［J］. The Journal of Infection，2020，80 （6）：671-693.

[20] Conner-Kerr T A，Sullivan P K，Gaillard J，et al. The effects of ultraviolet radiation on antibiotic-resistant bacteria in vitro ［J］. Ostomy Wound Management 44，1998，10：50-56.

[21] 凯尔·柯克兰德. 光与光学 ［M］. 上海：上海科学技术文献出版社，2008.

[22] Nerandzic M M，Thota P，Sankar C T，et al. Evaluation of a pulsed xenon ultraviolet disinfection system for reduction of healthcare-associated pathogens in hospital rooms ［J］. Infect Control Hosp Epidemiol，2015，36 （2）：192-197.

[23] Simmons S，Dale C，Holt J，et al. Role of Ultraviolet Disinfection in the Prevention of Surgical Site Infections ［M］. Ultraviolet Light in Human Health，Diseases and Environment，2017.

[24] 金伟其. 辐射度光度与色度及其测量 ［M］. 北京：北京理工大学出版社，2006.

[25] Miko. Understanding UV light measurement Part1 The irradiance measurement ［EB/OL］. https：// lightquality. blog/2020/04/23/understanding-uv-light-measurement-part1-the-irradiance-measurement. April 23，2020.

[26] Katelyn R，My Y，Montserrat T. Evaluation of Efficacy of Ultraviolet Germicidal Chambers in Swine Farms ［R］. Swine Disease Eradication Center，2019，8 （4）.

[27] 孙巍，谈智，唐晨晨，等. 某高效能紫外线消毒器对物体表面消毒效果及影响因素 ［J］. 江苏预防医学，2020，31 （04）：378-380.

[28] 韩雪玲，李梅玲，刘芳娥，等. 温湿度对紫外线辐照强度影响的观察 ［J］. 中国消毒学杂志，2006，23 （4）：288-288.

[29] 叶圣勤，贺金方，李克伟. 温度对紫外线灯照射强度的影响 ［J］. 中国消毒学杂志，1993，01：59.

[30] Gayán E，Mañas P，Álvarez I，et al. Mechanism of the Synergistic Inactivation of Escherichia coli by UV-C Light at Mild Temperatures ［J］. Applied and Environmental Microbiology Jun，2013，79 （14）：4465-4473.

[31] Timothy D C，Chong W，Steven J H，et al. Effect of temperature and relative humidity on ultraviolet （UV 254） inactivation of airborne porcine respiratory and reproductive syndrome virus ［J］. Veterinary Microbiology，2012，159 （1-2）：47-52.

[32] Tseng CC，Li CS. Inactivation of virus-containing aerosols by ultraviolet germicidal irradiation ［J］.

Aerosol Science and Technology, 2005, 1; 39 (12): 1136-1142.

[33] Peccia J, Werth H M, Miller S, et al. Effects of Relative Humidity on the Ultraviolet Induced Inactivation of Airborne Bacteria [J]. Aerosol Science and Technology, 2001, 35: 3, 728-740.

[34] Aaron B S. Reduction in Swine Pathogen Numbers by a UVC Germicidal Chamber [J]. https: // www. nationalhogfarmer. com/animal-health/reduction-swine-pathogen-numbers-uvc-germicidal-chamber. Oct 31, 2017.

[35] Derald H, Clayton J, Jacek A K, et al. Ultraviolet C (UVC) Standards and Best Practices for the Swine Industry Project [R]. Swine Health Information Center. Project #19-237, 2020.

[36] Cutler T, Wang C, Qin Q, et al. Kinetics of UV (254) inactivation of selected viral pathogens in a static system [J]. Applied Microbiology, 2011, 111 (2): 389-395.

[37] Chelsea, Zhang J Q, Jenna S, et al. Efficacy of Ultraviolet C disinfection for inactivating Seneca virus A on contaminated surfaces commonly found on swine farms [J]. Veterinary Microbiology, 2019, 7: 18-114.

[38] Blatchley E R, Dumoutier N, Halaby T N, et al. Bacterial responses to ultraviolet irradiation [J]. Water Science and Technology, 2001, 43 (10): 179-186.

[39] Iranpour R, Magallanes A, Loge F, et al. Ultraviolet disinfection of secondary wastewater effluents: Prediction of performance and design [J]. Water Environment Research, 1997, 69: 123.

[40] 郭美婷, 胡洪营, 李莉. 污水紫外线消毒工艺的影响因素研究 [J]. 中国环境科学, 2007, 4: 104-108.

[41] Clordisys. Ultraviolet Light Disinfection Data Sheet [EB/OL]. https://www. clordi- sys. com/appnote-s. php. 2020.

[42] Peschl. Surface disinfectiondose [EB/OL]. http: //peschl-ultraviolet. com/english _ n/all-about-uv/uv-disinfection/uv-dose-lethal-dose/uv-dose-lethal-dose. html

[43] UV Light Technology Limited [EB/OL]. http: //UV-Light. co. UK.

[44] Lytle C D, Sagripanti J L. Predicted inactivation of viruses of relevance to biodefense by solar radiation [J]. Journal of Virology, 2005, 79 (22): 14244-14252.

[45] Nicholson W L, Galeano B. UV resistance of Bacillus anthracis spores revisited: validation of Bacillus subtilis spores as UV surrogates for spores of B. anthracis Sterne [J]. Applied and environmental microbiology, 2003, 69 (2): 1327-1330.

[46] Kariwa H, Fujii N, Takashima I. Inactivation of SARS coronavirus by means of povidone-iodine, physical conditions and chemical reagents [J]. Dermatology, 2006, 212 (Suppl. 1): 119-123.

[47] Weiss M, Horzinek M C. Resistance of Berne virus to physical and chemical treatment [J]. Veterinary microbiology, 1986, 11 (1-2): 41-49.

[48] Chin S, Jin R, Wang X L, et al. Virucidal treatment of blood protein products with UVC radiation [J]. Photochemistry and Photobiology. 1997, 65 (3): 432-435.

[49] John J M, Qiang Z, Adams C D, et al. Disinfection of swine wastewater using chlorine, ultraviolet light and ozone [J]. Water Research, 2006, 40 (10): 2017-2026.

[50] 吴志君, 伍少钦, 肖有恩. 空气过滤技术在养猪生产中的应用 [J]. 猪业科学, 2011, 28 (05): 44-47.

［51］Wenke C，Pospiech J，Reutter T，et al. Impact of different supply air and recirculating air filtration systems on stable climate，animal health，and performance of fattening pigs in a commercial pig farm ［J］. PLoS One，2018，13（3）：e0194641.

［52］Lisa E，Tobias R，Matthias H，et al. Impact of UVC-sustained recirculating air filtration on airborne bacteria and dust in a pig facility ［J］. PloS One，2019，14（11）：e0225047.

［53］陈晓義，刘斌，陈晴，等. 臭氧和紫外联合杀菌在人工干制加工柿饼中的应用 ［A］. 中国食品科学技术学会. 中国食品科学技术学会第十七届年会摘要集 ［C］. 中国食品科学技术学会，2020：2.

基于可编程序控制器（PLC）的人员洗澡智能管控系统研究

张志勇　樊士冉　刘　聪　米凯臣　邓启伟　张瑞雪

摘要： 非洲猪瘟疫情对养猪行业造成了巨大的损失，而提高猪场的生物安全水平对于猪场的非洲猪瘟防控至关重要，其中对于人员跨区域流动必须执行严格的洗澡程序。但是由于洗澡行为的隐私性，目前猪场对于员工的洗澡行为无法进行有效的监控。一款基于 PLC 的人员洗澡智能管控系统，在保证人员处于正确的洗澡位置、执行正确操作的同时，实现了对人员洗澡过程中用水量及用时的管控，大大提高了洗澡的效果，对于防控非洲猪瘟起到了极佳的作用。

关键词： 猪场；洗澡；智能管控系统；PLC

近年来，由非洲猪瘟病毒引起的猪瘟疫情在世界各地广泛传播，其具有发病率高、影响范围广、持续时间长的特点，使得非洲猪瘟成为养猪业的主要威胁[1-3]。2018 年 8 月，中国暴发了非洲猪瘟，给中国养猪业造成了巨大损失[4]。非洲猪瘟病毒的传播途径有直接接触、间接接触两种，直接接触多是健康猪与携带病毒或确诊的猪密切接触所致[5]；间接传播范围广、较复杂，可通过人流（如访客、内部员工）、物品（饲料、食物工具等）、车辆（场内运输车、拉猪车等）、猪群（引种、出猪、调栏等）、生物流动（鼠类、猫、犬等）进行传播[6]。由于目前对于非洲猪瘟尚无有效的疫苗和药物，提高猪场生物安全措施已成为养猪从业者对于猪场疾病控制的主要策略。需要从新猪场的选址、栋舍布局、人员淋浴通道、物资消毒方式、出猪台等硬件设计建造上加入防止病原传入的生物安全理念[7]。

对于人员流动，"脏净分区、单向流动、防止交叉"是养殖场生物安全的核心原则[8]。生物安全等级高的区域为净区，生物安全等级低的区域为脏区[9]。对于人员的跨区域流动，特别是从脏区进入净区，必须进行严格的洗澡消毒程序[5,10]。

然而在目前猪场管理模式下，员工洗澡基本上靠自觉。当员工反复洗澡造成心理抵触时，洗澡效果极大的降低，对生物安全造成极大的隐患。而员工洗澡属于私密行为，人工无法干预，无法通过人工监管的方式强制实施。当前无法通过技术手段、人工方式对洗澡效果进行快速、准确的判断。与此同时，在规模化猪场，信息技术对智慧养猪至关重要[11]。对于目前猪场人员洗澡存在的管控问题，通过自动化、智能化技术的引入，设计一款满足猪场人员洗澡的智能化管控系统势在必行。

1　总体方案设计

1.1　研究对象分析

目前猪场常见的洗澡通道布局如图1所示。人员从脏区进入净区，需经过淋浴间进行淋浴。员工在脱衣间脱下在脏区内穿着的衣物然后进入淋浴间进行洗澡，洗澡完成后在穿衣间更换为净区的衣物。由于在淋浴过程中，淋浴间为隐私密闭环境，员工的洗澡过程无法进行有效监管。需在此通道内设计人员洗澡智能管控系统，保证洗澡效果。

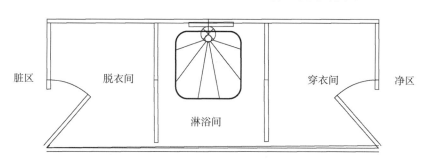

图1　猪场洗澡通道布局

1.2　工作原理

人员洗澡智能管控系统通过装有磁力锁的物理门，确保人员在洗澡通道不可随意流动；借助控制及自动化技术，对人员的在洗澡通道的动作进行监测与管控；通过对洗浴时间和水流量等因素的控制，确保人员在不被直接监督的情况下完成有效洗澡过程。

人员洗澡智能管控系统设计总体方案如图2所示。

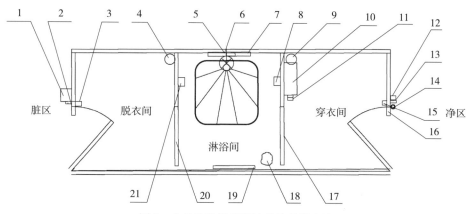

图2　人员洗澡智能管控系统总体方案

1. 开关＋指示灯A　2. 磁力锁A　3. 机械开关　4. 人体感应器A　5. 流量计　6. 电磁阀
7. 水平光栅A　8. 竖直光栅Ⅰ　9. 人体感应器B　10. 控制电箱　11. 电位器　12. 指纹机A
13. 开关＋指示灯B　14. 强制开门　15. 指纹机B　16. 磁力锁B　17. 一号门（带磁力锁和开关）　18. 多路语音播报器　19. 水平光栅B　20. 二号门（带磁力锁和开关）21. 竖直光栅Ⅱ

1.3 关键模块介绍

人员洗澡智能管控系统主要由主控箱模块、门禁模块、传感感应模块、播报与警告模块组成。

1.3.1 主控箱模块 以三菱 FX3U PLC 控制器为核心部件，通过控制器高速扫描刷新机制，确保人员按照指定流程完成整个洗澡过程。指纹机、开关按钮、传感器、流量计、电位器与控制器连接，构成输入通路；控制器与门锁、电磁阀、警报器、播报器连接，构成输出通路。控制器与外围设备连接的示意如图 3 所示。PLC 的输入、输出点的分配如表 1 所示。

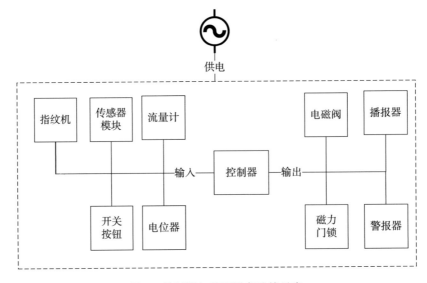

图 3 控制器与外围设备连接示意

表 1 PLC 输入、输出点地址分配

地址编号	元件	地址编号	元件
X000	人体感应器 A	Y000	电磁阀
X001	人体感应器 B	Y001	播报器
X002	竖直光栅 I	Y003	警报器
X003	竖直光栅 II	Y004	一号磁力门锁
X004	水平光栅 A	Y005	二号磁力门锁
X005	水平光栅 B		
X006	开关按钮 A		
X007	开关按钮 B		
X010	电位器		

（续）

地址编号	元件	地址编号	元件
X011	流量计		
X013	指纹机 A		
X014	指纹机 B		

1.3.2　门禁模块　由指示灯、磁力锁、按钮开关、指纹机等组成，系统根据洗澡过程的逻辑决定磁力锁的通断、指示灯的亮灭等。此模块是限定人员轨迹的关键模块。其中两台指纹机布置在靠近净区的进出口处，通过前期录入的指纹信息，保证只有授权的人员可以进入净区。

1.3.3　传感感应模块　人体传感器是以微波多普勒原理为基础，可以判断员工洗澡过程中所处的房间位置。电位计用于设置单次洗澡所需要的用水量，通常设置为 50L。流量计用来记录洗澡过程中的用水量，与电位计相比较，判定洗澡过程中的用水量是否符合要求。计时器用来记录洗澡过程中的用时情况，通常设置为 20min，通过控制洗澡过程中充足的时长来保证洗澡效果。

通过光栅确定人员洗澡所处位置为正确位置，保证花洒喷淋的有效性。光栅设置两组红外线光栅对，分别为竖直光栅对和水平光栅对。两个光栅对两两对应地设置在房间中环绕淋浴喷头的四个水平空间面上。竖直光栅对的两根光栅竖直对称设置在淋浴喷头两侧，竖直光栅的感应范围为淋浴喷头下方；水平光栅设置在与竖直光栅相邻的两个面上，水平光栅和地面的间距为 800～1 200 mm。两组光栅对的联合使用可以用于判断人员是否在洗澡位置。光栅实物如图 4 所示。

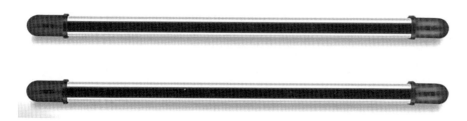

图 4　光栅实物

只有人员在正确洗澡位置时，计时器开始计时，流量计开始记录流量，保证人员洗澡足够的时间与用水量。

1.3.4　播报与警告模块　多路语音播报器根据流程中各节点进行相应播报，可提示流程中的正确操作与违规事项。多路语音播报器如图 5 所示。

警告模块的作用是洗澡人员因紧急情况按下紧急按钮或者系统判定人员洗澡非正常超时，声光报警器启动，同时门禁断电，方便紧急状况处理。人员洗澡非正常超时是指未检测到人（人静止，考虑昏厥等突发情况）的持续时间超过 10 min。声光报警器如图 6 所示。

图 5　多路语音播报器

图 6　声光报警器

2　系统操作流程

2.1　脏区进入净区流程

图 7 为人员从脏区进入净区洗澡智能管控流程，闭门前提下，淋浴时间和用水量须满足洗澡要求，方可开门进入净区，其间若有折返脏区的行为，统计都将复位。具体流程如下：

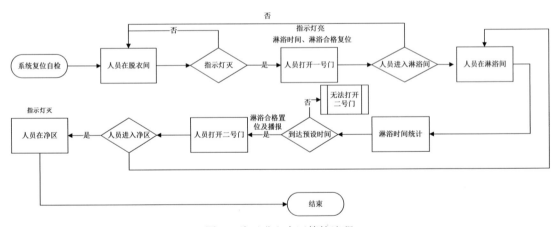

图 7　脏区进入净区管控流程

（1）进出两个门都关好的情况下，若指示灯亮，说明里面有人，此时打不开门；若灯未亮，则说明里面没人，按下按钮，开门进入。

（2）进入脱衣间，关好门（门若未关好，播报器会提示"请关好门"），脱衣间内的人体感应器感知内部有人，播报器提醒人员洗浴注意事项。

（3）从脱衣间到淋浴间，开始洗浴，时间和用水量达到要求后，播报器会播报合格（若未合格，从淋浴间返回脱衣间，播报器会提醒"勿返回"，同时之前的淋浴条件会复位；若未合格，从淋浴间到穿衣间，播报器会提示"您已违规，请返回继续淋浴"，同时记录违规信息到指纹机，此时也无法打开门到达净区）。

（4）合格后，从淋浴间到穿衣间，播报器提醒人员已完成淋浴。被穿衣间的人体传感器感知，则判断人员已到达穿衣间，可刷指纹开门进入净区，同时关好门，否则会播报提醒"请关好门"。系统检测内部无人，整个系统自检复位，一个淋浴过程完成。

2.2　净区进入脏区流程

图 8 为人员从净区返回脏区的洗澡智能管控流程，人员返回脏区时系统对于淋浴时间和用水量不做强制管控。具体流程如下：

（1）进出两个门都关好的情况下，若指示灯亮，说明里面有人，此时打不开门；灯若未亮，则说明里面没人，按指纹，开门进入；

（2）进入穿衣间，关好门。

（3）依次进入淋浴间、脱衣间（不强制洗澡）；此时若想要返回净区，须按规定完成淋浴，否则会语音提醒且无法开门进入净区。

（4）按下按钮，开门，进入脏区并关好门。

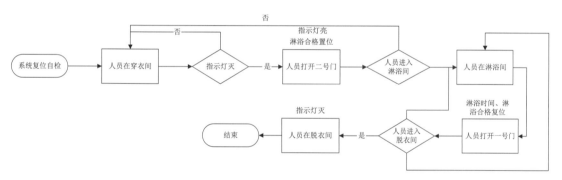

图 8　净区返回脏区管控流程

3　试验验证

人员智能洗澡管控系统验证试验于 2020 年 5 月在河北省唐山市唐山新好农牧有限公司的入场关口进行。关口已有相应的洗澡通道，具有热水器、花洒等基础的洗澡功能。对已有洗澡通道进行改造，改造完成之后的效果如图 9 所示。

现场安装完成后的部分细节图如图 10 所示，安装完成后对管控系统的功能进行试验验证。经过试验，人员洗澡智能管控系统实现了对洗澡过程的智能管控。人员从脏区进入洗澡间，若不按规定要求完成淋浴过程，则无法打开净区门进入内部，实现了管控的智能化、无人化，大大提升了人员跨区域流动的生物安全管理水平。

图 9　人员洗澡智能管控系统实施效果

图 10　人员洗澡智能管控系统完成现场安装

4 结论

本文研究了一款基于 PLC 的人员洗澡智能管控系统，通过控制人员的洗澡位置、洗澡时长、用水量等参数，对人员跨区域流动的洗澡效果起到了非常好的保障作用，解决了目前行业内洗澡无法监管的难题。对于猪场非洲猪瘟等流行性疾病的防控起到非常重要的作用，有极强的应用前景与行业推广价值。

参考文献

［1］Yuan Q Q，Liu W Y . African Swine Fever and Its Prevention and Control Measures［J］. Science and Technology of Food Industry，2019.

［2］Andraud M，Hammami P，Hayes B H，et al. Modelling African swine fever virus spread in pigs using time-respective network data：scientific support for decision-makers［J］. 2021.

［3］Costard S，Zagmutt F J，Porphyre T，et al. Small-scale pig farmers' behavior，silent release of African swine fever virus and consequences for disease spread［J］. Scientific Reports，2015，5：17074.

［4］Ma M，Wang H H，Hua Y，et al. African swine fever in China：Impacts，responses，and policy implications［J］. Food Policy，2021，3：102065.

［5］路鹏云，关翔，李结. 非洲猪瘟疫情下规模化猪场人员与物资入场流程的建立［J］. 中国猪业，2021，16（4）：2.

［6］张巍巍，王新. 规模化猪场非洲猪瘟防控要点［J］. 猪业科学，2021，38（06）：100-103.

［7］刘纪玉. 规模猪场防控非洲猪瘟的生物安全措施［J］. 猪业科学，2021，38（10）：3.

［8］吴海民. 非洲猪瘟形势下猪场的生物安全管理措施［J］. 甘肃畜牧兽医，2021，51（10）：72-73.

［9］贾义平，沈俊俊，李翠玲，等. 非洲猪瘟防控下猪场生物安全体系的建立［J］. 中国畜牧业，2021，14：2.

［10］靳登伟. 非洲猪瘟防控下规模化猪场的生物安全措施［J］. 国外畜牧学（猪与禽），2021，41（02）：56-58.

［11］周菊. 智慧养猪的研究及其应用［J］. 农村科学实验，2021，15：4.

做好猪场产床局部生物安全工作的重要性和改进措施

邓启伟

摘要： 产房生物安全是猪场内部生物安全的重要组成部分，产房内部生物安全的核心在于重视并做好产床局部生物安全。降低人员和物资进出产床的频率，可有效提高产床的生物安全水平，减少病原微生物进入产床以及在产床扩散的情况。严格控制和减少物资与人员进出产床的次数，可有效控制和降低传染病在产房的暴发。

关键词： 生物安全；产房；产床；进出频率

非洲猪瘟在中国发生后，给中国的养猪业造成了巨大的经济损失，同时也给人民的生活造成了很大影响。当前中国养猪业的首要工作是把猪养好，提高出栏商品猪数量，满足人民对猪肉及其相关产品的需求。由于非洲猪瘟尚无有效商品疫苗，所以防控方法以综合防控为主。要在这种情况下将猪养好，做好生物安全工作是关键。为了加强大环境的生物安全工作，各养殖小区养殖场除了在各进出口设立消毒隔离设施外，还在猪场周围筑起了围栏或实体围墙屏障。除了谢绝外人参观，还需严格控制本场人员和物资的出入，因此猪场就成了当地乡镇的"禁区"。但这只是猪场生物安全工作的一部分，做好外围生物安全的同时，还应做好猪场内部的生物安全工作，即使外围生物安全防控失守，内部疫病发生，也能将疫病控制在最小范围内，将损失降到最低。在猪场内部也有一些"禁区"，而产房的产床就是其中之一。产床是仔猪的出生地，仔猪会在这里度过其生命中最脆弱的阶段，产床对于仔猪而言好比人类婴儿的摇篮。做好产床的生物安全工作，可有效控制猪流行性腹泻等传染病在栏舍间的传播。产床生物安全工作做得好与坏，直接影响仔猪的断奶存活率，也就间接影响着商品猪的出栏数量。

1 需进出产床的工作

在现行的产房养殖管理方式下，产床在清洗消毒冲洗干燥后等待临产母猪进入，母猪在产前1周左右进入产床。进入产床前的母猪，一般会有一个清洗和体表消毒的过程，大部分猪场会在妊娠舍限位栏（也有大栏）或专门修建的清洗间内进行这项工作。如果是进入产房再清洗，工作人员就会进入产床操作。也有部分猪场实行母猪转栏时不清洗，进入产房或临产前对母猪进行乳房和阴部的局部清洗消毒，在这种情况下工作人员也会进入产床。母猪进入产床后，工作人员会对母猪进行日常的饲喂和清粪工作。如料槽积有过多的吃剩饲料和水，工作人员会对其进行清理。通常情况下清理料槽工作不需要进入产床，因为大部分猪场产房的料槽是可以拆卸和翻转的，但对于一些料槽固定或栏舍结构不合理的猪场，就需要工作人员进入产床进行清槽工作。对于产床清粪工作，一般也不需要人员进入，但刮粪或扫粪的工具需要进入。

在接产阶段，如果母猪生产顺利，也不需要工作人员进入产床，但如果母猪发生难产，就需要工作人员进入产床进行助产操作。有的猪场会要求接产员对临产母猪进行哺乳前的乳头挤压，以排出乳头内的污物和检测乳头功能是否正常，这就需要进入产床。对于产仔较多的母猪，工作人员会对仔猪进行寄养乳头分配和定位，这时也需要进入产床对仔猪进行调整。如果仔猪在夜间需要保温，在设备较为陈旧的猪场，需要对保温灯（板）进行调节和开关，有时也需要进入产床。仔猪完成初乳的采食后，国内大部分猪场都会对仔猪进行剪牙、断尾的操作，有时会连同仔猪的阉割工作同时进行。这项工作需要抓取小猪并对其单个固定才能顺利完成，因此需要工作人员进入产床对仔猪进行围捕和抓取。除以上情形，母猪的打针、输液，以及仔猪的免疫等工作，在现行的管理模式下都需要进入产床才能完成。

2　频繁进出产床导致的问题

工作人员进入产床会直接或间接带来一系列的问题，首先是生物安全问题。一方面，反复高频率地进出产床，极大地提高了将产床外的病原带到产床内的可能性，由于哺乳期的仔猪抵抗病原的侵袭能力有限，尤其是从出生到吃上第一口初乳的这一特殊时间段，极有可能感染病原而发病[1]。此外，在母猪泌乳不足或母源抗体下降的情况下也容易因营养不良和抗体水平下降而感染发病[2]。另一方面，反复高频地进出产床，加快了病原的传播和扩散[3]，导致疾病在整个圈舍内部快速蔓延，严重的情况可导致整个猪场感染或传播到猪场外。其次是对仔猪和母猪造成应激，由于现有产床空间有限，使猪只所需的安全距离减小，当人员进入产床后，母猪和仔猪时刻处于紧张的状态，任何的细微动作和声响都会对母猪和仔猪造成惊扰。这种人为的惊扰，可引起仔猪四处逃窜，并可诱发仔猪蹄肢扭伤，体表擦伤。对母猪的惊扰将导致母猪突然起身，可能会直接踢伤或踩压仔猪。人为的干扰应激，还可诱发敏感体质的仔猪和母猪的激素水平上升，引起抵抗力或泌乳下降[4]，并进一步加大仔猪感染疾病的风险[5]。除了存在以上问题和隐患，频繁进出和翻越产床，还增加了工作人员受到母猪踩压、啃咬，以及存在摔倒、碰撞、扭伤等安全风险，同时也加大了工作人员的劳动量，对工作人员的体能也提出了要求。

3　如何减少进出产床的频率

在现行的产房管理流程和养殖模式下，母猪上产床前进行彻底的体表清洗和消毒是很有必要的，一方面，这样就不用在产房进行清洗消毒工作，避免了人员进出产床；另一方面，上产床前清洗可清除母猪体表大部分的粪污和死皮，有利于产房的卫生维护，降低了产房有害微生物的环境载量，同时也会清除母猪体表大部分的有害微生物，即使没有完全清除，洁净、干燥的环境也会加速病原微生物的失活[6]。

母猪在产房期间的日常饲养管理过程中，应合理控制饲料的投放，减少饲料残留造成的变质发霉和积水，从而减少人工的清理。有经济条件的猪场，可对产床料槽进行改造，对产床布局进行调整，方便工作人员在产床外就能清理。对于刚上产床的母猪，如果是开放式（或大栏）产床，可不用马上清除粪便，而是将粪便推到方便后续清理的角落，以使母猪习惯定点排粪，待母猪生产前再清除，这样母猪产仔后不用进入产床就能完成清粪工作。

当进行仔猪乳头定位、剪牙、断脐、断尾、阉割、免疫等有关的捕捉活动时，要充分了解哺乳猪群的行为和习惯，利用保温箱隔板等辅助工具，将仔猪赶到产床边缘或角落进行捕捉。也可以自制或购买市售仔猪捕捉器材，借助捕捉工具，就能对仔猪进行栏外捕捉，做到不用进出产床而完成捕捉工作。条件允许的，还可对产床布局和结构进行改造和调整，以达到抓捕仔猪省时省力的目的。

4　总结

做好产床生物安全工作，减少人员进出产床的次数，尤其是在仔猪出生前的1～2d和出生后的1周左右，此时人员的进出极有可能将病原带入产床，引起仔猪出生后的感染和发病。此外，做好产床清洁工具的进出也十分重要，如常规的粪铲和扫把，这类工具最好每个单元使用一套，使用后浸泡消毒并沥干。在发现仔猪有腹泻等症状的产床，禁止清理粪污，可采用干燥粉、锯末、煤灰等材料进行覆盖处理，以免清理过程中形成粉尘颗粒感染邻近产床的仔猪，或通过使用后的工具扩散到同栋舍健康仔猪群。现代化规模猪场的产床，尤其是中间采用无缝隔板的产床，是一个相对独立的单元，如果能控制人员的频繁进出，以及减少生产、医疗、护理等工具的交叉重复使用，或做好器械工具的清洗消毒工作，就能很好地控制高度接触性传染病。对于可通过气溶胶、粉尘、飞沫等传播的疾病，由于减少了传播途径，也能起到很好的控制和延缓作用，为猪场拟定处理方案争取宝贵时间，从而降低疫病对哺乳阶段猪群造成的损失。

参考文献

[1] 胡巧云，喻正军，汪雪，等. 当前规模化猪场产房初生仔猪腹泻病原、病因和防控思考 [J]. 养猪，2014，5：114-118.

[2] 刘春桂. 母源抗体在仔猪疾病防治中的作用 [J]. 畜牧兽医科技信息，2019，2：109.

[3] 包阿东，孙衍立，汪长寿. 北方地区规模化羊场产房生物安全建设 [J]. 四川畜牧兽医，2018，45（10）：33-34.

［4］王国侠．母猪泌乳力影响因素及改进措施［J］．畜牧兽医科学（电子版），2020，1：45-46．

［5］左之才，刘兵，李莉，等．早期断乳应激性腹泻对仔猪肠道形态结构与肠道菌群的影响［J］．中国兽医科学，2012，42（1）：64-68．

［6］殷灿．美国白蛾核型多角体病毒的稳定性研究［D］．泰安：山东农业大学，2014．

Chapter 6

种猪改良篇

基于 GWAS 人工智能算法，降低成本，提高选种时效的猪基因组选择新技术的研究

牛安然

摘要： 种猪的常规育种通常面临选种时间长、表型难测定、选育准确率较低、选育投入高的问题。在种猪常规育种的基础上，引入基于全基因组关联研究（genome-wide association study，GWAS）和人工智能算法的基因组选择，可以有效提高种猪的选育准确率、提升表型测定效率、降低选育成本。本文将从基因组在猪育种中的应用、GWAS 人工智能算法的介绍、家猪经济性状的 GWAS 应用现状三方面，对基于 GWAS 人工智能算法的猪基因组选择的应用价值进行论述。最后，通过比对常规育种与基于 GWAS 人工智能算法基因组选择的实际应用情况，证明该新方法可以在种猪育种过程中降低育种成本，提高选种时效。

关键词： 全基因组关联研究；人工智能算法；基因组选育；猪育种

我国是世界上生猪存栏和出栏数量最多的国家。市场对猪肉旺盛的需求量，推动了我国生猪育种的快速发展。因此，保障准确高效的种猪选育，是满足猪肉市场庞大需求的前提条件。基因组育种相比于常规育种，可以有效缩短测定周期，提高选种的准确性，降低表型测定成本[1-7]。但是，基因组育种高昂的测序费用限制了基因组育种的广泛应用[7-9]。将 GWAS[10] 和人工智能算法引入基因组育种，可针对目标性状及目标群体设计专属基因芯片。基因芯片的应用可以有效降低测序密度、缩减测序成本，突破基因组育种高成本投入的限制[10-12]。总体来说，基于 GWAS 人工智能算法的基因组选择可以在保证种猪选育低投入、高效率、高准确度的基础上，充分适应市场和消费者对猪肉需求的变化。

1 基因组育种在猪育种中的应用

猪只的育种是从遗传角度持续提升群体优良生产性能的重要保障。在过去的几十年中，传统育种方法使猪的生长性状、胴体性状等遗传力较高的经济性状获得了较快的遗传进展。但是，传统育种方法既无法使繁殖性状等低遗传力性状获得快速改良，也不能对繁殖性状进行高效率的选育[1-3]。因此，在传统育种方法的基础上引入群体的基因型数据，不仅可以通过提高选种时的准确率，增大对低遗传力性状的改良速度；同时，还可以对初生猪只进行高准确率的早期选留，有效降低表型测定周期，提升有效测定比例[4-8]。猪的经济性状大多为数量性状，影响性状的基因位点数量较多，因此无法在猪的全基因组中找出影响性状表达的

全部基因位点[13-16]。并且，影响性状的基因位点的遗传效应各不相同，既存在遗传效应较大的主基因，也存在效应较小的多基因。然而，通过在 GWAS 和人工智能算法可以在目标群体全基因组范围内定位与目标经济性状存在强关联的基因位点[10]，将这些强关联的基因位点引入估计育种值的数学模型中，可以在目标群体中进一步提高早期选育的准确性。另外，由于有效地筛选了目标性状的关联基因位点，因此基因芯片的密度也相应降低[11-12]。上述改进可以在保证基因组选择高准确性的同时降低经济及时间成本的投入。

2   GWAS 及人工智能算法

GWAS 是一种在全基因组范围内，以单核苷酸多态性（SNPs）为分子遗传标记，进行全基因组水平上的对照分析或相关性分析，从而确认影响复杂性状或数量性状基因变异的一种策略[16]，具体分析原理见图 1。GWAS 主要利用了分子标记连锁不平衡的生物学原理，来研究目标性状与分子标记间的关系，从而确认与目标性状具有因果关系的 SNP 位点。相比于同样可以定位 SNP 的传统数量性状基因座定位（quantitative trait locus mapping，QTL），GWAS 具备了以下优点：可直接对复杂群体进行分析，无须构建作图群体；可同时对多个等位基因进行分析，将比较范围扩大至全基因组的同时提高统计效率；可对全基因组 SNPs 与表型性状关联进行精细定位，实现高定位分辨率[17-20]。因此，在家畜育种领域，GWAS 已逐渐取代 QTL 用于发掘定位可调节家畜重要经济性状的候选基因和 SNP 标记。

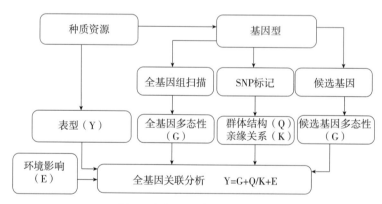

图 1   全基因组关联分析原理

GWAS 一般采用两种分析方法，一种是基于无关个体，另一种是基于家系的关联分析。第一种可以按照用途分为基于随机群体研究数量性状的关联分析和基于病例对照研究设计研究质量性状的关联分析。基于随机群体的关联分析可以采用协方差分析或者线性回归的方法。基于对照的研究对象可以采用卡方检验来比较两者之间 SNP 等位基因频率的差异，并计算 95％的置信区间和相对危险度。第二种是基于家系的关联分析，基于家系的关联分析可以避免遗传背景复杂性，降低对全基因组关联分析结果准确性造成的影响。对具有家系的群体样本可以采用传递不平衡检验来进行研究，分析检测出来的 SNP 位点与关联性状的相关性[11]。

对于不同规模的样本群体及测序要求，GWAS 的试验设计通常也分为两类，分别是单阶段设计和两阶段或多阶段设计。单阶段设计的研究群体相对要足够大，然后对研究群体的 SNP 进行检测和分型，最后对检测出来的 SNP 位点与研究性状进行关联分析。当样本群体较大时，单阶段试验设计的测序成本相对较高，而两阶段或多阶段试验设计测序成本相对较低，因此两阶段或多阶段试验设计被广泛应用。两阶段或多阶段试验设计是在单阶段试验设计的基础上进行的，首先可以在表型差异显著的极端个体中挑选小样本个体进行基因分型，将基因分型的 SNP 位点与性状进行关联分析。对关联的 SNP 位点在更大的样本中进行分型，对分型结果与性状进行关联分析，最终结合两个阶段的结果进行综合分析。两阶段中第一阶段的分型结果既可以采用单个个体单独测序分型，也可以采用混池法，该方法可以极大地降低测序成本。

另外，基于 GWAS 中进行的大量比较或者多次比较的校正过于保守或过于严格，会导致假阳性的发生，因此需要校正假设验证的 P 值来降低多重假设验证造成的假阳性。方法有：Bonferroni 方法，该方法用总的 I 型错误水平除以用以关联分析的 SNP 数量，来确定显著性阈值。如果校正后的 P 值仍低于普遍认为的显著性阈值 0.05，则说明该 SNP 位点与研究性状是关联的。这种方法的缺陷在于忽略了位点之间的相关性，提高了出现假阴性的概率，减少发现与性状相关位点的可能性；控制错误发现率，该方法首先对未进行校正的 P 值按照从小到大进行排序，其中最大的 P 值保持不变，其余的 P 值依次乘以其对应系数，作为该次假设检验的显著性 P 值。如果校正后的 P 值仍小于 0.05，则认为该位点与研究的目标性状存在关联性。该方法相对于 Bonferroni 方法而言，会导致出现更多的假阳性，该方法对 SNP 的筛选相对宽松，允许更多的 SNP 位点参与关联分析。另外还有模拟运算法和递减调整法，这两种方法使用相对少一些，因为模拟运算法和递减调整法相对烦琐，耗时相对较长。总的来说，无论采取以上任何一种方法，均无法从本质上解决多重假设检验造成的假阳性问题，后期还是需要进行功能验证才能保证结果的准确性，再运用到实际生产中才能取得最大的成果。

家猪的重要经济性状均为数量性状，而对于数量性状 GWAS 主要采用一般线性模型（general linear model，GLM）和混合线性模型（mixed linear model，MLM）进行关联性分析。数量性状的一大特点是，性状表型易受多种因素影响。因此，使用在 GLM 上加入作为随机效应的亲缘关系产生的 MLM 进行关联分析，可以更加准确地评估 SNPs 与性状表型间的关联[21]。采用 MLM 进行关联分析的一个假定基础是，SNPs 间的效应均为加性效应。虽然，基因的加性效应是唯一可稳定遗传的影响因素，但是，基因的显性离差和上位效应同样也会影响性状的表型。由于 MLM 仅能对呈线性关系的加性效应进行估算预测，而无法捕捉如显性离差上位效应等非线性关系。因此，应用 LMM 进行 GWAS 分析所得结果的准确性具有一定的局限性。GWAS 分析的另一个特征是样本数量远小于 SNP 的数量，小样本大 SNP 通常会导致过拟合，使 GWAS 所得结果不可用于其他群体。虽然增加样本量是避免过拟合的有效方法，但是受实际生产条件所限，扩大样本群体的可行性较低。到目前为止，GWAS 并没有可以有效避免过拟合的统计模型。

在 GWAS 的基础上引入人工智能可以有效解决 GWAS 目前所面对的挑战。人工智能算法可以通过随机森林算法、梯度提升模型和神经网络等算法[22]，最大化地提升预测结果的准确性。相比于 GWAS 仅考虑了加性遗传组分，人工智能算法不会在关联分析前假定 SNP 的遗传组分。人工智能算法提供了一种更优的方法来学习多 SNP 位点的遗传变异及 SNP 位点间的相互作用，因此人工智能算法不仅可以执行对显著 SNP 的特征选择，同时还可以识别特征 SNP 间的复杂交互作用，从而更加准确地预测 SNP 位点与复杂性状间的关联[23]。同时，对于 GWAS 分析中无法通过增加样本数量来避免的过拟合问题，人工智能算法可以通过有效的模型正则化，在符合数量性状遗传规律的基础上，尽可能减少过拟合的影响，从而提高预测的准确性，使分析结果更加具有普适性[24]。

3　家猪经济性状的 GWAS 应用现状

猪只的关键经济性状主要包括饲料转化率、肉质性状、抗病性状和以窝产活仔数为代表的繁殖性状。目前，QTL 检测和 GWAS 已经从猪的全基因组范围内定位了许多与重要经济性状关联的基因[13,25]。

猪的繁殖性状主要指与母猪繁殖相关的性状。繁殖性状是一类典型的低遗传力性状，性状的表型不仅受到来自公猪、母猪遗传因素的影响，同时也很容易受到环境的影响。因此，相比于肉质性状、生长性状、抗病性状等，猪繁殖性状的 QTL 更难定位。目前研究发现，与繁殖性状相关联的 QTL 仅有 2 129 个，是检测到关联 QTL 最少的性状类别；与产仔数相关联的 QTL 有 1 281 个，其中与窝产活仔数相关联的 QTL 有 238 个[25]。猪的繁殖过程不仅涉及卵子与精子的发育、受精卵的发育，还与母猪子宫发育状态息息相关。在这个过程中，激素的调控作用是主要影响因素之一。因此，定位检测到的 QTL 基因的生物学功能大都与激素表达调控相关，少部分与胚胎发育及子宫发育相关。已有研究证明，通过 GWAS 定位到的位于 6 号染色体的 LEPR 基因、位于 3 号染色体的 NCOA1 基因和位于 8 号染色体的 BMPR1B 等基因的生物学功能可以直接影响母猪产仔数性状的表型[13,25-28]。

猪的生长性状主要指与生长速度相关的性状，平均日增重、活体背膘厚、饲料转化率和校正 115kg 日龄都是衡量生长性状的重要指标。猪的生长性状是一类典型的中等遗传力性状，后代表型受父母遗传因素影响较大，容易通过育种获得较大的遗传改良。因此，生长性状的 QTL 相比于繁殖性状更容易定位。通过 GWAS 分析定位，目前与生长性状相关联的 QTL 已有 2 434 个。其中，与饲料转化率相关联的 QTL 有 454 个[25]。猪的生长过程不单只涉及肌肉的生长发育，同时也涉及骨骼的生长发育和营养代谢水平的调控[29]，另有研究表明，仔猪在母体中的发育状态会直接影响仔猪出生后的生长状态[30,31]。因此，可以对猪只生长性状产生影响的基因的生物学功能较为复杂。已有研究证明，通过 GWAS 定位到的位于 1 号染色体的 MC4R 基因、位于 9 号染色体的 MYOG 基因和位于 14 号染色体的 VRTN 等基因的生物学功能可以直接影响猪的饲料转化率[30,31]。

猪的肉质性状也是生猪的一个重要经济性状，主要包括肌内脂肪、肉色、嫩度、系水力、滴水损失以及熟肉率等指标。肉质性状受到诸多因素影响，但来自父母的遗传因素是主要影响因素。由于肉质性状的指标优劣直接关系到猪只售卖时的经济效益，因此肉质性状也是重点选育的经济性状。肉质性状同生长性状、繁殖性状一样也是数量性状，肉质性状的遗传力范围为 0.15～0.3，为中等遗传力性状[32]。通过 GWAS 分析定位，目前与肉质性状相关联的 QTL 已有 15 087 个，是定位 QTL 最多的性状[25]。影响猪肉品质的因素有很多，包括营养、遗传、环境等。与瘦肉率相关的影响因素包括肌内脂肪含量、不饱和脂肪酸水平，而与色泽、酸肉、质地以及滴水损失等相关的影响因素包括氟烷基因和酸肉基因的表达。已有研究证明，通过 GWAS 定位到的位于 6 号染色体的 FABP3 基因、位于 14 号染色体的 SCD 基因和位于 13 号染色体的 ADIPOQ 等基因的生物学功能可以直接影响猪肉的肉质评分、肌内脂肪含量和瘦肉率等肉质性状的表达[32,33]。

4 应用与展望

在实际生产中，对于饲料转化率、肉质性状及繁殖性状的测定与选育都需要较长的时间和较高的测定成本。因为只有猪只成长到预期售卖体重时，才能得到猪只的饲料转化率数据，而肉质性状只有在猪只屠宰之后才可进行测定。同样，对于繁殖性状，则需要等到母猪第一次产仔后，才能进行测定获得表型数据；且对于种公猪繁殖性能的选育需要其后裔的繁殖表型数据作为参考。由于无法从实际生产过程中及时获得重要经济性状的表型，这就导致了种猪的选育工作无法及时有效开展。然而，通过在种猪的选育过程中引入基因组信息，就可以在仔猪初生阶段及时开展有效的选育工作，缩短选育的时间间隔，降低选育的测定成本。全基因组选择虽然可以有效提高种猪的选育效率，但是猪只的全基因组测序成本较高。这不单是因为猪是多胎动物，每个生产批次需要进行基因测序的个体数较多。还因为全基因测序位点较多，现在广泛采用的是 50k 基因芯片测序。然而 50k 基因芯片所包含的基因位点信息不单有目标经济性状的 SNP 位点，同时也包含了其他非目标性状的 SNP 位点。因此，通过 GWAS 和人工智能算法分析定位提供的关键 SNP 信息，就可以仅仅依据关键 SNP 位点，实现对种猪的基因组选育。基于 GWAS 人工智能算法的基因组选择可以达到全基因组选择的效果，在仔猪出生阶段实现高准确率的早期选留、准确挑选参加饲料转化率测定及用于配种的种猪、缩短表型测定周期时间、加快遗传进展、提高有效测定比率；同时，GWAS 人工智能算法可以大幅缩减测序密度，仅对目标性状的关键 SNP 位点进行测序，有效降低测序成本。基于 GWAS 人工智能算法的精准早期选留可以指导生产进行更加准确的阉割选留和分群，提早淘汰不合格的种猪，为核心群猪只提供最优的饲养密度，保证优良的遗传基因得以表达。

某核心育种场的育种计划是：在大白仔猪初生选留阶段依照其母亲繁殖指数，对公仔猪进行阉割选留，且只保留全群的前 20%。对于母猪则会在断奶阶段和出保育阶段实施两次分群，分群的主要依据是母猪的表型及其母系指数。若在上述选留和分群的过程中引入基于 GWAS 人工智能算法的基因组早期选留，可在优化选育过程的同时提高选育的效率及准确性。

基于 GWAS 人工智能算法的猪基因组选育的成本投入，为每头出生候选仔猪的低密度基因芯片测序费用，相比于现行的全基因组高密度测序费用，可减少 50%～80% 的经济投入[34]。目前，对大白猪的选育重点不仅包括繁殖性状也包括生长性状，因此在大白公猪阉割选留阶段仅将其母亲的繁殖指数作为选留依据，是存在错误阉割优秀公猪的可能性。所以，为了减少错误阉割优秀公猪的情况，可以将重点选育性状的 GWAS 和人工智能算法分析定位提供的关键 SNP 结果用于指导阉割选留，从而实现基因组早期选育。为实现基因组早期选留，计划对繁殖指数排名前 50% 的母猪仔猪进行出生后耳组织采样，进行关键性状 SNP 位点测序，通过测序结果计算公仔猪自身指数，保留每窝排名前 40% 的公仔猪。这样可以在不改动原有生产计划的基础上，对公猪进行较为准确的预选留。2018—2021 年某猪场的数据表明，种猪终测后选留的准确率为 40%～60%，而基因组选择试验在终测前的选留准确性约为 50%。因此，在公仔猪阉割选留时采用基因组早期选留指数作为阉割选留依据，不仅可以实现现阶段终测选留的选种效果，还可以缩短 362 d 的测定周期。在种猪选育初期即实现高效、高准确性的选留。

大白母猪与公猪不同，不涉及阉割的问题，但是在断奶阶段和出保育阶段会依据母猪的表型和母系指数进行两次分群。断奶阶段和出保育阶段的母猪仅为 21 日龄和 55～63 日龄，尚未表现出生长性状，也不具备繁殖能力；因此，处于这两阶段的母猪的母系指数来源于其父母的母系指数。按照现行的生产计划，参考母猪的母系指数进行两次分群，无法达到终测选留时 40%～60% 的准确率，也无法优化即将进入核心群种母猪的饲养管理条件。将基因组早期选留引入母猪分群计划，仅在断奶选留阶段进行一次分群，既可以达到与终测选留时相似的选留准确率，也可以对预备进入核心群的母猪提升饲养管理条件，使其充分发挥优良基因的表现，产生预期的生产性能。将基因组早期选留应用于母猪选留，不但同样可以缩短 362 d 的测定周期，还实现了对生产成本的高效利用，优化了现有生产计划。

总的来说，为了改进现阶段种猪育种遗传进展较慢、有效测定比率较低、育种成本投入较高的现状，采用基于 GWAS 人工智能算法的基因组选择对猪只实行选育，是切实可行的有效方法。

参考文献

[1] Burrow H M, Mrode R, Mwai A O, et al. Challenges and Opportunities in Applying Genomic Selection to Ruminants Owned by Smallholder Farmers [J]. Agriculture, 2021, 11 (11): 1172.

[2] Abdollahi-Arpanahi R, Lourenco D, Misztal I. Detecting effective starting point of genomic selection by divergent trends from BLUP and ssGBLUP in pigs, beef cattle, and broilers [J]. Journal of Animal Science, 2021.

[3] Van Raden P M. Symposium review: How to implement genomic selection [J]. Journal of Dairy Science, 2020, 103 (6): 5291-5301.

[4] Xu Y, Liu X, Fu J, et al. Enhancing genetic gain through genomic selection: from livestock to plants [J]. Plant Communications, 2020, 1 (1): 100005.

［5］Wartha C A，Lorenz A J. Implementation of genomic selection in public-sector plant breeding programs：Current status and opportunities ［J］. Crop Breeding and Applied Biotechnology，2021，21.

［6］Xiang T. Combined purebred and crossbred information for genomic evaluation in pig ［D］. Institut Agronomique，Vétérinaire et Forestier de France；Aarhus Universitet (Danemark)，2017.

［7］Gimelfarb A，R Lande. Marker assisted selection and marker QTL associations in hybrid populations ［J］. Theoretical and Applied Genetics，1995，91 (3)：522-528.

［8］Zhang H，Yin L，Wang M，et al. Factors affecting the accuracy of genomic selection for agricultural economic traits in maize，cattle，and pig populations ［J］. Frontiers in Genetics，2019，10：189.

［9］Salek Ardestani S，Jafarikia M，Sargolzaei M，et al. Genomic Prediction of Average Daily Gain，Back-Fat Thickness，and Loin Muscle Depth Using Different Genomic Tools in Canadian Swine Populations ［J］. Frontiers in Genetics，2021，12：735.

［10］Hayes B，Goddard M. Genome-wide association and genomic selection in animal breeding ［J］. Genome，2010，53 (11)：876-883.

［11］Piles M，Bergsma R，Gianola D，et al. Feature Selection Stability and Accuracy of Prediction Models for Genomic Prediction of Residual Feed Intake in Pigs Using Machine Learning ［J］. Frontiers in Genetics，2021，12：137.

［12］Badke Y M，Bates R O，Ernst C W，et al. Accuracy of estimation of genomic breeding values in pigs using low-density genotypes and imputation ［J］. G3：Genes，Genomes，Genetics，2014，4 (4)：623-631.

［13］胡闪耀. 大白猪 ESR FSHβ RBP4 LEF-1 和 PPARD 基因多态性与重要经济性状的相关性研究 ［D］. 华中农业大学，2016.

［14］张豪，张沅，张勤. 后裔测定青年公牛的标记辅助 BLUP 选择 ［J］. 科学通报，2002，20：1566-1571.

［15］吴常信. 分子数量遗传学与动物育种 ［J］. 遗传，1997，19 (1)：1～3.

［16］Daetwyler H D，Capitan A，Pausch H，et al. Whole-genome sequencing of 234 bullsfacilitates mapping of monogenic and complex traits in cattle ［J］. Nature Genetics，2014，46 (8)：858-865.

［17］Visscher P M，Wray N R，Zhang Q，et al. 10 years of GWAS discovery：biology，function，and translation ［J］. The American Journal of Human Genetics，2017，101 (1)：5-22.

［18］Yano K，Yamamoto E，Aya K，et al. Genome-wide association study using whole-genome sequencing rapidly identifies new genes influencing agronomic traits in rice ［J］. Nature Genetics，2016，48 (8)：927-934.

［19］Li C，Sun B，Li Y，et al. Numerous genetic loci identified for drought tolerance in the maize nested association mapping populations ［J］. BMC Genomics，2016，17 (1)：1-11.

［20］Zhang X，Zhang H，Li L，et al. Characterizing the population structure and genetic diversity of maize breeding germplasm in Southwest China using genome-wide SNP markers ［J］. BMC Genomics，2016，17 (1)：1-16.

［21］Yu J，Pressoir G，Briggs W H，et al. A unified mixed-model method for association mapping that accounts for multiple levels of relatedness ［J］. Nature Genetics，2006，38 (2)：203-208.

［22］Enoma D O，Bishung J，Abiodun T，et al. Machine learning approaches to genome-wide association studies［J］. Journal of King Saud University-Science，2022：101847.

［23］Okser S，Pahikkala T，Airola A，et al. Regularized machine learning in the genetic prediction of complex traits［J］. PLoS genetics，2014，10（11）：e1004754.

［24］Okser S，Pahikkala T，Aittokallio T. Genetic Variants and Theirinteractions in disease risk prediction-machine learning and network perspectives［J］. BioData Mining，2013，6（1）：1-16.

［25］https：//www. animalgenome. org/cgi-bin/QTLdb/SS/index.

［26］周臣，胡群，贺艳娟，等. 猪重要经济性状 QTLs 定位研究进展［J］. 中国畜牧杂志，2019，55（10）：35-41.

［27］Wang Y，Ding X，Tan Z，et al. Genome - wide association study for reproductive traits in a Large White pig population［J］. Animal Genetics，2018，49（2）：127-131.

［28］Bergfelder-Drüing S，Grosse-Brinkhaus C，Lind B，et al. A genome-wide association study in large white and landrace pig populations for number piglets born alive［J］. PloS One，2015，10（3）：e0117468.

［29］Horodyska J，Hamill R M，Varley P F，et al. Genome-wide association analysis and functional annotation of positional candidate genes for feed conversion efficiency and growth rate in pigs［J］. PLoS One，2017，12（6）：e0173482.

［30］Bergfelder-Drüing S，Grosse-Brinkhaus C，Lind B，et al. A genome-wide association study in large white and landrace pig populations for number piglets born alive［J］. PloS One，2015，10（3）：e0117468.

［31］Casas-Carrillo E，Kirkpatrick B W，Prill-Adams A，et al. Relationship of growth hormone and insulin-like growth factor-1 genotypes with growth and carcass traits in swine［J］. Anim Genet，2015，28（2）：88-93.

［32］Kim J M，Park J E，Lee S W，et al. Association of polymorphisms in the 5′ regulatory region of \ r，LEPR \ r，gene with meat quality traits in Berkshire pigs［J］. Anim Genet，2017，48（6）：723-724.

［33］Salas R C D，Mingala C N. Genetic factors affecting pork quality：Halothane and rendement napole genes［J］. Anim Biotechnol，2016，28（2）：148-155.

［34］Tribout T，Larzul C，Phocas F. Economic aspects of implementing genomic evaluationsin a pig sire line breeding scheme［J］. Genetics Selection Evolution，2013，45（1）：1-16.

系谱校正在种猪引种和生产中的研究进展

杨雨婷

摘要： 系谱校正是指通过利用动物基因组 DNA 标记如 STR 和 SNP 等验证亲子对间亲子关系或计算个体间基因组亲缘相关系数，利用这种亲子关系或个体间亲缘关系对群体系谱进行校正。近年来随场间联合育种和外引种猪逐渐增多，加之猪场现场生产管理难度大，猪群系谱缺失或错漏的情况频发，对育种工作造成一定影响，对核心猪群进行系谱校正已成为选育自有纯种猪的基础。本文主要从系谱校正的步骤、系谱校正的方法及发展历程、亲子鉴定原理、影响系谱和系谱校正准确性的主要因素和系谱校正在猪育种中的研究应用，总结系谱校正在猪育种中的优势和应用情况，对系谱校正技术在我国猪育种中的应用提出建议。

关键词： 亲子鉴定；系谱校正；遗传标记；动物育种

系谱是计算育种值、基因型填充、GWAS 的重要信息来源，对育种工作至关重要。种猪错误的系谱会通过"金字塔"式的猪只繁育体系在群体中大量积累，快速放大，导致后代近交快速上升、选育进展缓慢，对育种带来巨大的负面影响[1-3]。真实、完整、准确的系谱是实现种猪自繁育、自更新，摆脱外国种猪公司对国内生猪发展的制约、培育出中国自有种猪的基础之一[4-5]。

生产中误配精液、仔猪未记录调栏、耳牌脱落、耳号识读录入错误等都会造成猪只个体信息及父母信息记录错误，从而导致记录的系谱错误或缺失[6]。在新投产猪场引种建群时需要引入大批同代次猪只，引种猪只通常由供种场提供系谱，这种系谱通常代次有限，引种猪群存在的部分潜在近交无法完全从提供的系谱中体现出来。种猪的培育过程需要定期从外部群体引种来避免因闭群繁育、猪群高度近交带来的后代性能退化和畸形频发，场间联合育种也不可避免地需要进行猪只交流，这两种引种都需要真实准确的系谱来保证育种效果[7]。

本文主要从系谱校正方法、影响系谱和系谱校正准确性的主要因素、系谱校正在猪育种中的优势和研究应用等方面进行综述，总结目前常用的系谱校正方法，为不同场景下对种猪进行系谱校正提供参考。

1 系谱校正步骤

系谱校正的主要步骤主要包括位点选择、群体基因组分型、亲子鉴定及系谱校正 3 个步骤[8]。根据使用的系谱校正方法不同，具体的操作流程也有一定差异，大体可分为特定位点测序和高密度测序两种[9]。特定位点测序首先需要根据群体参考基因组序列及前人研究确定进行基因分型的位点及引物序列，设计并合成引物后用于待进行系谱校正群体的特定位点基因分型。完成群体基因分型后根据记录的系谱逐个进行亲子鉴定，发现亲子对间存在冲突后结合仔猪出生记录、母猪分娩记录和分型结果确定正确的亲子关系，逐个修正亲子对后完成系谱校正。高密度测序不需要事先选择特定位点，而是使用高密度测序方法对猪群进行基因分型[10]，获得全基因组信息，根据全基因组信息进行系谱校正或重构。

2   系谱校正方法

2.1   系谱校正发展历程

系谱记录是系谱校正的依据和根本，从猪只出生即佩戴耳标、记录父母信息，对耳牌脱落或污损的猪及时补打耳牌，种猪配种时单一精液配种、校对首配和复配精液无误是保证系谱准确最基本最有效的方法，但由于猪场生产人员操作不规范、管理有漏洞、记录遗失等问题，完全依靠记录保证系谱完整准确存在一定难度，因此不依靠单纯人力记录，而是借由动物本身所携带的遗传信息进行亲子鉴定的方法逐渐发展起来，用于弥补系谱记录的错漏。

动物的血型和血液蛋白型是最早应用于亲子鉴定的方法，其遗传遵循孟德尔分离和自由组合定律，但血型和血液蛋白型的多样性较少，用于大量群体的亲子鉴定假阳性率很高（即一个子代存在多个可能亲本），目前已几乎不将其作为系谱校正的依据[11]。

随着分子生物学的发展，DNA 分子标记逐渐成为亲子鉴定、系谱校正和系谱重构的主流方法。分子标记（molecular markers）是以个体间遗传物质内核苷酸序列变异为基础的遗传标记，是 DNA 水平遗传多态性的直接的反映[12]。与其他几种遗传标记，如形态学标记、生物化学标记、细胞学标记相比，DNA 分子标记具有以下优点[13]：

（1）大多数分子标记为共显性，对隐性性状的选择十分便利。

（2）基因组变异极其丰富，分子标记的数量几乎是无限的。

（3）在生物发育的不同阶段，不同组织的 DNA 都可用于标记分析。

（4）分子标记揭示来自 DNA 的变异。

（5）表现为中性，不影响目标性状的表达，与不良性状无连锁。

（6）检测手段简单、迅速。

利用分子标记进行系谱校正按发展时间先后大致可分为三代：限制性片段长度多态性、微卫星标记和单核苷酸多态性[14]。

2.2   限制性片段长度多态性在系谱校正中的应用

限制性片段长度多态性（restriction fragment length polymorphisms，RFLP）是最早应用的分子标记技术，它是检测 DNA 在限制性内切酶酶切后形成的特定 DNA 片段的大小，反映 DNA 分子上不同酶切位点的分布情况，因此 DNA 序列上的微小变化，甚至 1 个核苷酸的变化，也能引起限制性内切酶切点的丢失或产生，导致酶切片段长度的变化。随机扩增多态 DNA（random amplified polymorphic DNAs，RAPD）技术通过其独特的检测 DNA 多态性的方式使得 RAPD 技术很快渗透于基因研究的各个领域。扩增片段长度多态性（amplified fragment length polymorphisms，AFLP）是 1992 年由 Zabeau 和 Vos 发明的一种 DNA 分子标记新技术[13]。AFLP 融合了 RFLP 和 RAPD 两种技术的优点，既具有 RFLP 的可靠性，又具有 RAPD 的灵敏性，是分子标记技术重大突破，目前被认为是一种十分理想的、有效的分子标记技术。

在 RAPD 和 RFLP 技术基础上建立了序列特异性扩增区域（sequence characterized amplified regions，SCAR）酶切扩增多态序列（cleaved amplified polymorphic sequence，CAPS）和 DNA 扩增指纹（DNA amplified fingerprints，DAF）等标记技术。这些技术的出现，进一步丰富并完善了 RFLP 技术，增加了人们对 DNA 多态性的研究手段。

RFLP 的优点在于其多态性来自自然变异，试验结果比较稳定可靠；但由于这种方法原理的限制，只能对内切酶识别位点上的变异进行检测，因此能获得的信息量十分有限，多态性低、检测周期长、方法烦琐等不利因素都限制了 RFLP 在系谱校正中的应用，目前已基本不采用这种方法进行系谱校正。

2.3 微卫星标记在系谱校正中的应用

在真核生物基因组中存在许多非编码的重复序列，如重复单位长度在 15～65 个核苷酸左右的小卫星 DNA（minisatellite DNA），重复单位长度在 2～6 个核苷酸左右的微卫星 DNA（microsatellite DNA）[15]。由于重复单位的大小和序列不同以及拷贝数不同，从而构成丰富的长度多态性。因此，在基因组多态性分析中，采用了数目可变串联重复多态性（varible number of tandem repeats，VNTR）标记技术来区别。

短串联重复序列（short tandem repeat，STR）标记技术，又称简单重复序列（simple sequence repeat，SSR）标记技术。STR 也称微卫星 DNA，是一类由几个（多为 1～5 个）碱基组成的基序（motif）串联重复而成的 DNA 序列，其中最常见是双核苷酸重复即（CA）n 和（TG）n。STR 标记的基本原理：根据微卫星重复序列两端的特定短序列设计引物，通过 PCR 反应扩增微卫星片段[16]。由于核心序列重复数目不同，因而扩增出不同长度的 PCR 产物，这是检测 DNA 多态性的一种有效方法[17]。

1994 年 Zietkiewicz 等对 SSR 技术进行了发展，建立了加锚微卫星寡核苷酸技术，用加锚定的微卫星寡核苷酸作引物，在 STR 的 5′端或 3′端加上 2～4 个随机选择的核苷酸，这可引起特定位点退火，从而导致与锚定引物互补的间隔不太大的重复序列间的基因组节段进行 PCR 扩增[18]。这类标记又被称为简单重复序列间多态性（inter simple sequence repeat，ISSR）或扩增片段长度多态性（anchored simple sequence repeats，ASSR）。

微卫星多重 PCR 是指在一个 PCR 反应体系内加入多对微卫星引物进行特异性扩增，通过多个微卫星结果的组合进行亲子鉴定[19]。由于这种方法简单便捷、所需设备较少、检测周期快、特异性高，因此被广泛应用于畜禽系谱校正中。但由于这种方法仅能对假定的亲子关系进行确认或排除，在无法获得可能亲本基因的情况下对同代次动物进行系谱校正非常困难，因此几乎不将其用于畜群的系谱重构。

2.4 单核苷酸多态性在系谱校正中的应用

单核苷酸多态性（single nucleotide polymorphism，SNP）被称为第三代 DNA 分子标记，是指同一位点的不同等位基因之间个别核苷酸的差异，这种差异包括单个碱基的缺失或插入，更常见的是单个核苷酸的替换，且常发生在嘌呤碱基（A 与 G）和嘧啶碱基（C 与 T）之间[20]。SNP 标记可帮助区分两个个体遗传物质的差异，被认为是应用前景最好的遗传[21]。

由于连锁不平衡的存在，SNP 可以推断出个体间的共祖片段，即后代的某段染色体片段是否由某一共同祖先获得，这种共祖片段可以在基因层面有效反映个体间的亲缘关系，由此可以对缺失系谱的群体进行系谱构建[1]。同时，由于不同品种间几乎不存在亲缘关系较近的共同祖先，因此由大量个体间亲缘关系组成的集合可以有效划分品种，当某一个体同时与多个品种的集合存在亲缘关系时说明这一个体的祖先包含多个品种，可以据此进行品种鉴定[7]。

3 亲子鉴定原理

家畜的亲子鉴定常用的三种方法包括排除法、似然法和基因型重构法。

3.1 排除法

排除法原理为孟德尔遗传定律，利用各位点的等位基因频率计算非父排除概率，即子代与亲本为不同的纯合基因型的位点数/位点总数，通过计算等位候选父母与子代间的基因频率的排除概率来确定亲子关系[22]，排除概率越低是真实亲本的可能性越高。排除法简单快速，适用于位点多态性高候选亲本少的情况，无效等位基因的存在和基因型判定误差会导致亲子鉴定错误或不准确，影响系谱校正[23]。

3.2 似然法

似然法是通过建立候选父亲与子代的似然函数，通过计算假设亲本是真实亲本或无关个体似然函数比值的自然对数值（LOD）进行亲子鉴定，将多个独立位点的似然比累加后求自然对数值得到 LOD 值，LOD 值大于 0 则认为是真实亲本，LOD 值小于 0 则认为不是真实亲本[24]。这一方法充分考虑了基因型判别错误造成的影响，仅需少量位点即可进行亲子鉴定，测序成本低[25]。

3.3 基因型重构法

基因型重构法根据亲属之间共享的 IBD0、IBD1、IBD2 片段比例计算亲缘关系系数，推断亲属关系。IBD（identical by descent）即同祖片段，指两个个体的同一段基因区域来自相同的祖先。在这一段基因区域中，如果不考虑各自独特的染色体变异，这两个个体应该有一套完全相同的单倍形片段[26]。所以，在理想的情况下只要找出相同的单倍形片段就能推断出同祖片段[27]。这一方法适用于较大规模的群体，但对测序密度要求较高，低密度测序结果会降低准确性。

4　影响系谱和系谱校正准确性的主要因素

系谱校正的准确率受多种因素影响，原始系谱记录、生产操作和管理水平、亲子鉴定所用位点类型和数量、使用的系谱校正方法等都对系谱校正准确性有影响。

4.1　生产操作和记录

系谱校正需要根据生产记录的系谱辅助判断亲子对是否准确，通过每一对亲子对的鉴定完成对整个系谱的校正。没有出生记录的猪由于难以确定可能的父母信息，父母一方没有采样分型的情况下就很难匹配出正确的父母关系继而确定系谱，难以确定系谱的种猪只能降级或作为二元猪使用，造成巨大损失。因此，详细准确的生产记录是保证系谱准确的基础，规范管理采精、稀释、配种、分娩记录、调栏、打耳牌和耳牌读写记录等生产操作是保证系谱真实准确最简单可行的方法。

4.2　分型位点选择

SNP 位点通常使用芯片获得，每个个体都能获得上千甚至上万个位点，因此单个位点分型错误对整体结果影响较小，芯片测序密度对系谱校正影响较大[28]。对于使用 STR 位点的系谱校正而言，选择的 STR 位点本身在群体内的分布和多态性以及 STR 位点分型准确性是影响系谱校正准确性的两个主要因素。等位基因频率和有效等位基因数量是筛选 STR 位点的主要标准，同时每个 STR 位点的多态信息含量、观察杂合度和期望杂合度也是影响排除概率计算的重要因素[29]。由于通常选择的 STR 位点数量较少，因此由 PCR 扩增反应链滑脱产生"影子带"或电泳电压不稳导致条带模糊等造成分型错误[30]，导致某一位点分型错误或不可用也会影响系谱校正效果。

4.3　系谱校正方法

计算亲子间关系的方法也是影响系谱校正准确性的重要因素。例如，似然法通过计算非父排除概率判断亲子关系，将非父排除概率低于某一阈值的亲子对判定为真实亲子对[31]。由于不同群体平均的次等位基因频率不同，由此计算的平均错误率也不同，因此这一阈值通常是根据错误率的群体经验分布确定的，真实亲子对与随机个体对间错误率的分布是连续的[32]，极少数个体对会被错误判定关系，因此这一阈值的确定会影响系谱校正准确性。

5 系谱校正在猪育种中的研究应用

5.1 系谱校正在猪育种中的优势

通过DNA分子标记校正或重构系谱可以解决生产管理不规范导致的系谱错漏，通过建立分子系谱的形式提高估计遗传力、减少育种值估计误差；通过测序信息计算父母双方实际对子代的遗传贡献取代系谱计算时父母各贡献1/2的假定，提高选种选配准确性相较于过去降低群体近交，提高性状选择准确性以加快遗传进展。

5.2 系谱校正在不同猪种中的应用

谢苏等[19]使用PAGE法对48头长白猪、大白猪和杜洛克猪的55个微卫星位点进行扩增和排父率分析，筛选了14个适合的微卫星并设计了三组多重PCR微卫星组合，分别达到了0.978 50、0.998 40、0.999 99的亲子鉴定准确率；余国春[9]使用毛细管电泳法利用15对通用STR引物对长白猪、大白猪和长大杂交猪进行亲子鉴定得到12个微卫星的组合，可以达到80%以上的置信度；吴旭东等[33]使用毛细管电泳法利用14个微卫星位点对96头安庆六白猪进行亲子鉴定，可以达到99%以上的累积排除概率；Ba等[34]使用19个微卫星标记对9个品种638头猪进行亲子鉴定并确定家系。

张哲等[22]使用Illumina 50k芯片对191头杜洛克猪进行测序和亲子鉴定，发现即使是管理严格规范的核心育种场，其系谱错误率也会达到6%以上；余国春[9]使用Illumina 60k芯片对24头藏猪、杜藏二元猪和杜长大三元猪进行测序，确定猪亲子鉴定所需位点最少为30个，最少需要120个位点以达到99.99%置信度最优效果。

6 建议与展望

系谱校正可通过DNA分子标记的方法确定猪只系谱，在纯繁种猪构建核心群和外引种猪时均能发挥作用，是计算育种值、基因型填充、GWAS的重要基础，对育种工作至关重要[35]。

使用多重PCR组合对选择的STR位点分型，以每个组合3个STR，使用3个组合进行快速亲子鉴定为例，每头猪亲子鉴定所需的试剂耗材成本可以控制在5元以内，相较于SNP芯片100~200元/头的测序成本低数十倍，可以广泛应用于纯种猪生产核心群的系谱校正，通过校正系谱提高估计遗传力和育种值估计准确性，降低群体近交、保证选种选配准确性，进而加快遗传进展。畜群中系谱错误率一般在5%~30%[19,22,30]，系谱错误的种猪配种产生的后代可能出现近交上升、畸形率增加等问题，且系谱错误对种猪生产繁育带来的负面影响是不可控的，不能通过生产管理手段消除。

使用高密度芯片测序虽然单头价格较高，但其可以在没有父母信息的情况下重构同代次群体系谱，同时测序数据可以建立参考群用于后代早期选育和基因组选择。引种建群等情况下通过高密度芯片测序重构系谱，以此鉴别引进种猪的品种，避免将二元猪作为种猪使用导致后代性状分离，也能避免系谱不准确导致的近亲配种，防止后代近交快速上升、退化。

目前在猪中利用 DNA 分子标记进行系谱校正尚不及在人或牛中普及，但通过多重 PCR 对生产核心群 STR 分型和高密度 SNP 芯片对新引种群体测序相结合，可以较低成本实现对纯种猪群的系谱校正。

参考文献

[1] 周磊，杨华威，赵祖凯，等. 基因组选择在我国种猪育种中应用的探讨 [J]. 中国畜牧杂志，2018，54（3）：4-8.

[2] Chao W，WangH，Yu Z，et al. Genome-wide analysis reveals artificial selection on coat colour and reproductive traits in Chinese domestic pigs. [J]. Molecular Ecology Resources，2015，15（2）：414-424.

[3] 王晨，秦珂，薛明，等. 全基因组选择在猪育种中的应用 [J]. 畜牧兽医学报，2016，47（1）：1-9.

[4] Hill W G. Applications of population genetics to animal breeding，from wright，fisher and lush to genomic prediction. Genetics，2014，196（1）：1-16.

[5] Sanders K，Bennewitz J，Kalm E. Wrong and Missing Sire Information Affects Genetic Gain in the Angeln Dairy Cattle Population [J]. Journal of Dairy Science，2006，89（1）：315-321.

[6] Powell G. Impact of Paternity Errors in Cow Identification on Genetic Evaluations and International Comparisons [J]. Journal of Dairy Science，2001.

[7] Nechtelberger D，Kaltwasser C，Stur I，et al. DNA microsatellite analysis for parentage control in austrian pigs [J]. Animal Biotechnology，2001，12（2）：141-144.

[8] 殷彬，岳书俭，俞英，等. 奶牛分子系谱构建或亲权鉴定的微卫星标记筛选 [J]. 畜牧兽医学报，2017，48（4）：595-604.

[9] 余国春. 微卫星与 SNP 标记技术在猪亲子鉴定中的有效性研究 [D]. 四川农业大学，2014.

[10] Gurgul A，Semik E，Pawlina K，et al. The application of genome-wide SNP genotyping methods in studies on livestock genomes [J]. Journal of Applied Genetics，2014，55（2）：197.

[11] Visscher P M，Woolliams J A，Smith D，Williams JL. Estimation of pedigree errors in the UK dairy population usingmicrosatellite markers and the impact on selection [J]. Journal of Dairy Science，2002，85（9）：2368-2375.

[12] 鲁绍雄，吴常信. 动物遗传标记辅助选择研究及其应用 [J]. 遗传，2002（24）.

[13] 宣之兴，张守纯. DNA 分子标记在动物亲子鉴定中的研究进展 [J]. 畜禽业，2009（6）：3.

[14] Weiss K M. How many disease does it take to map a gene with SNPs [J]. Nature Genet，2000，26（2）：151-155.

[15] 李何君，刘炜，吴昊旻，等. 12 个微卫星标记在大约克夏种猪品系鉴定中的应用研究 [J]. 上海畜牧兽医通讯，2014，6：3.

[16] 马洪雨，岳永生，刘源. 微卫星 DNA 分子标记及其在动物遗传育种中的应用 [J]. 畜牧兽医杂志，2004.

[17] 郭晓令，徐宁迎，Looft Christian，等. 猪微卫星标记多重 PCR 扩增组合 [J]. 遗传，2004，26（1）：5.

[18] James L. Weber，Informativeness of human（dC-dA）n • （dG-dT）n polymorphisms [J]. Genomics，

1990，7（4）：524-530.

[19] 谢苏，沈永巧，陈焱森，等．利用微卫星标记进行猪亲子鉴定［J］．中国兽医学报，2019，39（8）：8.

[20] 宋玉朴，孙永峰，冯自强，等．SNP分型检测技术及其在畜禽遗传和育种中的应用研究进展［J］．中国畜牧杂志，2021，57（7）：6.

[21] 王晨，马宁，郭春和，等．基于SNP芯片分析的蓝塘猪遗传群体结构［J］．广东农业科学，2018，45（6）：7.

[22] 张哲，罗元宇，李晴晴，等．一种基于高密度遗传标记的亲子鉴定方法及其应用［J］．遗传，2014，36（8）：7.

[23] 罗元宇，吴鹏，贺金龙，等．畜禽群体中基于SNP标记的亲子鉴定及亲本推断方法［J］．华中农业大学学报，2016，5：7.

[24] Kalinowski S T，Taper M L，Marshall T C．Revising how the computer program CERVUS accommodates genotyping error increases success in paternity assignment（vol 16，pg 1099，2007）［J］．Molecular Ecology，2007，16（5）：1099-1106.

[25] Kumar J，Hekrujam M，Sharma k，et al. SRAP and SSR marker-assisted genetic divrity，popuation structure analysis and sex identification inJojoba（Simmondsia chinensis）［J］．Ind Crop Prod，2019，133：118-132.

[26] Manichaikul A，Mychaleckyj J C，Rich S S，et al．Robust relationship inference in genome-wide association studies. Bioinformatics，2010，26（22）：2867-2873.

[27] 张静静，高会江，吴洋，等．利用SNP标记估计西门塔尔牛亲缘关系系数的准确性［J］．畜牧兽医学报，2016，47（2）：8.

[28] Goodnight K F，Queller D C. Computer software for performing likelihood tests ofpedigree relationship usinggenetic markers［J］．Mol Ecol，1999，8（7）：1231-1234.

[29] Purcell S，Neale B，Todd-Brown K，et al. PLINK：a tool set for whole-genome association and population-based linkage analyses.［J］．The American Journal of Human Genetics，2007，81（3）：559-575.

[30] 胡明月，刘正喜，赵中利，等．绵羊（ATAG）n四碱基微卫星位点的筛选及其在亲子鉴定中的应用［J］．中国畜牧兽医，2021，48（07）：2484-2494.

[31] 初芹，张毅，孙东晓，等．应用微卫星DNA标记分析荷斯坦母牛系谱可靠性及影响因素［J］．畜牧兽医学报，2011，42（2）：163-168.

[32] 郭刚，周磊，刘林，等．利用SNP标记进行北京地区中国荷斯坦牛亲子推断的研究［J］．畜牧兽医学报，2012，43（1）：44-49.

[33] 吴旭东，周忍，章会斌，等．安庆六白猪14个微卫星位点遗传多样性分析［J］．中国畜牧杂志，2021，57（S01）：6.

[34] Ba N V，Le Q N，Do D N，et al. An assessment of genetic diversity and population structures of fifteen Vietnamese indigenous pig breeds for supporting the decision making on conservation strategies［J］．Tropical Animal Health and Production，2020，52（3）：1-9.

[35] 王继英，成建国，林松，等．枣庄黑盖猪亲缘关系分析及分子系谱构建［J］．猪业科学，2019，36（8）：136-138.

Chapter

7

经营管理篇

"现代养猪企业＋家庭农场"商品猪养殖模式的投资收益分析——基于管理会计视角

王晓燕　闫之春　蒋福涛（山东省阳谷县畜牧局）

摘要： 本研究基于管理会计视角，在畜牧业养殖密集的山东、河北及四川 3 个省份，选择与企业合作养殖商品猪的家庭农场为研究对象，选取不同投资水平的 18 户家庭农场为样本，以猪舍投资地板类型为投资水平的基本划分标准，对投资建设数据、生产经营数据以及合作企业的收益数据等进行统计分析，核算"现代养猪企业＋家庭农场"合作模式下的投资回报情况，并探讨出几种最佳的家庭农场经营模式，旨在对我国农牧企业的经营管理实践提供参考。

关键词： 家庭农场；管理会计；经营分析；投资收益

近几年，国家环保政策不断出台，我国畜牧养殖业进入行业洗牌阶段，多家大型农业企业都开始涉足畜牧养殖业。规模化养殖与现代化企业的结合使得养猪行业对精细化管理的要求越来越高，家庭农场作为主要的养殖合作模式，其经营分析也亟须更加专业化、精准化[1]。管理会计能更细致地分析真正影响养殖效果和养殖效益的成本因素，使财务管理对企业的生产经营起到真正的支撑、提升作用。本研究从管理会计角度，在畜牧业养殖密集的山东、河北及四川 3 个省份，选择与企业合作商品猪养殖的家庭农场为研究对象，选取不同投资水平的 18 户家庭农场为样本，以猪舍投资地板类型为投资水平的基本划分标准，对投资建设数据、生产经营数据以及合作企业的收益数据等进行统计分析，核算"现代养猪企业＋家庭农场"合作模式下的投资回报情况，并探讨出几种最佳的家庭农场经营模式，以期对我国农牧企业的经营管理实践提供借鉴。

1　家庭农场养殖基本划分标准及投资收益模型

家庭农场按照投资的地面类型划分为不同的标准，分别代表不同的投入水平和养殖密度（表 1），低投入代表中国传统的投入模式，通常为实体地面，养殖密度较高；中等投入与高投入分别为半漏缝地板和全漏缝地板。地面类型的不同直接决定了后期养殖方式、雇工方式以及环保投入方式等综合管理模式的差异，故以地面类型为划分标准来进行研究具有重要意义。

表1　家庭农场发展阶段及投入水平划分

发展阶段	地面类型	理想养殖密度（m²/头）
低投入	实体地面	1.5
中等投入	半漏缝地板	1.2
高投入	全漏缝地板	1.0

从管理会计角度，"现代农业企业＋家庭农场"合作模式下，家庭农场的投资收益模型框架包括头均固定成本、头均可变成本、代养费收入，并通过变动成本、固定成本来计算家庭农场的毛利。其中，头均固定成本中包括前期投入、建筑投入、设备投入等；头均可变成本中包括人工、水电、运费及取暖消耗等。

2　不同区域家庭农场成本投入分析

本研究共调研18户家庭农场，覆盖实体地面、半漏缝地板、全漏缝地板3类投资类型，其中剔除3户数据不完整的农场，共采用剩余15户家庭农场的数据作为分析依据。

2.1　前期投资、固定成本及猪舍投资

不同区域家庭农场的土建等前期投入差别非常大。通过调研发现，南方土建等前期投入远高于北方。这主要是由于山东省、河北省交通条件好，主要投入只需用于平整猪场地面，投入相对平稳，而四川省交通条件较差，家庭农场多建在山顶，投入主要用于修路。由图1可知，南北方前期投入的差别在几万元到几十万元不等。限于四川省所处地理位置和地理条件的差异，前期的土建投入是沉没成本，因此，交通便利程度是影响养殖成本的重要条件，也是企业在选取养殖区域时需要考虑的重要因素之一。

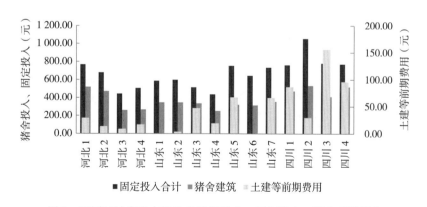

图1　不同区域家庭农场头均前期投入、固定投入、猪舍建筑投入

家庭农场的固定投入中，绝大部分是猪舍投入，猪舍建筑建设比较先进的家庭农场其投入相对较大，如使用全漏缝地板、保温外墙、自动料线、自动通风系统等；而传统的实体地面猪舍采用简单通风等，投入很少。另外，猪舍建筑投入的现代化程度不同，对雇工数量、

环保效果以及猪只的生长情况都会有不同程度的影响。

2.2　运营中的可变成本投入

经调研发现，头均可变成本的投入参差不齐，人工投入与硬件设施先进程度相关性大，其中，山东省实体地面的养殖户头均人工投入最高，部分半漏缝地板的家庭农场和全漏缝地板的家庭农场是采用自动料线，人工投入相对比较低，原因在于在地面清理和喂料等工作相对轻松，尤其规模较大的家庭农场在节约人工方面就更加明显（图 2）；水电投入更是千差万别，通过规范化管理来优化成本的空间非常大。由此可见，高投入尽管在前期来看投入成本高，但从长期来看，通过农户养殖水平以及规范化管理能力的提升，运营成本一定会逐步下降。

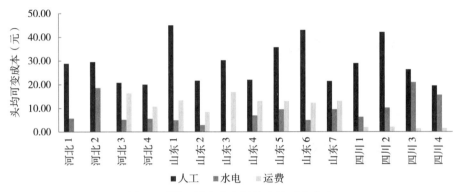

图 2　不同区域家庭农场的头均可变成本

2.3　猪苗、饲料、药品成本投入

在猪只育肥期的成本投入上，猪苗、饲料、药品的投入成本差别非常大，这更加能表明各家庭农场的生产管理水平差别大，养殖水平还处于严重参差不齐阶段。饲料成本在四川省的成本投入中相差尤其明显，普遍在 700 元以上；药品投入各家庭农场之间差别最大，最低在 30 元以下，最高达 80 元以上（图 3）。

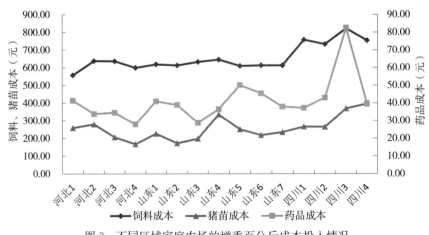

图 3　不同区域家庭农场的增重百公斤成本投入情况

3　家庭农场合作与企业的投资回报分析

3.1　家庭农场收入基本稳定，企业收益波动大

由图4可知，家庭农场尽管前期投入大，且承担着猪只死亡损失的风险，但代养费和毛利基本比较稳定。而企业需要承担来自市场价格波动以及家庭农场养殖成绩不稳定带来的双重风险，其收益波动在头均0～500元，有时甚至会出现负收益。

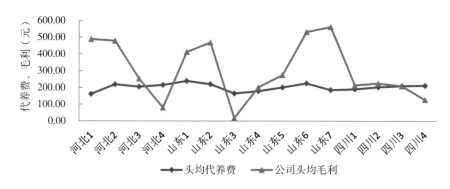

图4　不同区域家庭农场的代养费和公司毛利情况

由此可见，家庭农场与大型养殖企业合作，对农户来说是一种规避风险的模式，能获得稳定收益，尤其在行情低迷时自养农户可能会大量亏损，家庭农场的收益优势会更加凸显。

3.2　家庭农场投资回报分析

按头均代养费和投资回报率从高到低、投资回收期从短到长依次选择排名前40%的家庭农场，分别为河北4、山东1、山东2、山东6和四川4（表2）。另外，山东2的头均代养费尽管在前40%，但是由于投资成本较高，因此投资回报率略低于其他家庭农场。

表2　家庭农场投资收益分析

家庭农场	养殖规模（头）	地面类型	头均代养费（元）	头均固定投入（元）	投资回报率（%）	投资回收期（年）
河北1	1 700	半漏缝	161.62	768.42	25.99	3.80
河北2	3 700	半漏缝	219.42	680.16	47.64	2.10
河北3	1 000	半漏缝	205.00	444.60	62.6	1.60
河北4	1 500	半漏缝	215.46	507.47	65.12	1.54
山东1	400	实体地面	239.58	587.99	57.92	1.73
山东2	1 700	半漏缝	219.96	598.86	56.69	1.76

（续）

家庭农场	养殖规模（头）	地面类型	头均代养费（元）	头均固定投入（元）	投资回报率（%）	投资回收期（年）
山东 3	1 700	半漏缝	164.64	516.03	40.14	2.5
山东 4	2 000	半漏缝	178.80	437.77	56.57	1.8
山东 5	2 100	半漏缝	171.54	755.10	34.57	2.9
山东 6	1 100	半漏缝	225.16	644.59	60.36	1.7
山东 7	1 800	半漏缝	193.84	735.17	35.43	2.8
四川 1	380	实体地面	190.53	763.09	36.33	2.8
四川 2	2 300	半漏缝	203.06	1 055.11	32.87	3.0
四川 3	2 400	全漏缝	120.00	781.84	26.44	3.8
四川 4	3 800	全漏缝	211.67	772.77	70.7	1.41

3.3 企业投资回报分析

按照公司头均毛利从高到低的顺序，选择排名前40%的家庭农场，分别为河北1、河北2、山东1、山东2、山东6和山东7（表3）。综合家庭农场代养费、投资回报率、投资回收期等各项指标以及企业收益指标来看，收益最好的模式分别为山东1、山东2和山东6的家庭农场。

表3 家庭农场合作模式下企业的收益情况

家庭农场	养殖规模（头）	地面类型	头均毛利（元）	头均死淘成本（元）	头均利润（元）
河北 1	1 700	半漏缝	488.80	33.87	454.93
河北 2	3 700	半漏缝	478.42	28.59	449.83
河北 3	1 000	半漏缝	253.63	9.95	243.68
河北 4	1 500	半漏缝	79.65	21.43	58.22
山东 1	400	实体地面	411.95	24.01	387.94
山东 2	1 700	半漏缝	468.19	9.73	458.46
山东 3	1 700	半漏缝	17.22	32.21	−14.99
山东 4	2 000	半漏缝	200.86	60.74	140.12
山东 5	2 100	半漏缝	274.55	83.93	190.62
山东 6	1 100	半漏缝	531.25	64.53	466.72
山东 7	1 800	半漏缝	561.80	18.68	543.12
四川 1	380	实体地面	213.85	33.45	180.40
四川 2	2 300	半漏缝	225.00	22.02	202.98
四川 3	2 400	全漏缝	209.10	58.57	150.53
四川 4	3 800	全漏缝	126.95	20.17	106.78

注：在计算头均利润时，考虑头均死淘成本，未考虑期间费用。

4　结论与分析

4.1　收益较高的模式具有明显的比较优势

经过分析发现，企业和家庭农场双方都获得较高收益的几种模式下，各自都有着比较突出的优势，如表 4 所示。

表 4　企业和家庭农场双方均收益最高的几种模式

家庭农场	养殖规模（头）	地面类型	通风类型	人工投入模式	供料方式	粪污收集
山东 1	400	实体地面	半自动控制	女老板＋雇工 1 人	人工喂料	人工清粪
山东 2	1 700	半漏缝	自动控制	夫妻 2 人＋雇工 1 人	人工喂料	自动刮粪机
山东 6	1 100	半漏缝	自动控制	男老板兼厂长＋雇工 2 人	自动喂料	自动刮粪机

4.1.1　传统的"小而精"模式　分析发现，编号山东 1 的家庭农场，养殖规模小，猪舍建设投入低，采用实体地面，人工投入模式为女老板再雇佣 1 人，女老板勤劳、细心、认真负责，亲自包办很多工作，养殖成绩一直比较好。这种传统的"小而精"模式，规模小，花费精力较少，管理效率更高，只要责任心强、精力投入多，养殖成绩明显比较好。但问题在于，这种养猪模式采用实体地面，人工清粪，在气味处理、环境卫生清理上受限，不符合未来国家环保政策的扶持方向。

4.1.2　中小型规模的"夫妻式"模式　编号山东 2 的家庭农场，规模为中小型，设施设备部分进行了升级，如半漏缝地板、自动通风系统、自动刮粪机，但喂料还是人工操作；人工投入模式上，除夫妻 2 人之外再雇佣 1 人，养殖成绩非常好。这种 2 栋猪舍左右的中小型养殖模式下，夫妻 2 人为主要劳动力，认真负责，时间精力投入多，一些关键点都进行了设施设备升级，提高了管理效率，提升了管理效果，是这种模式下养殖成绩好的重要原因。

4.1.3　中小型规模的"现代化投入"模式　编号山东 6 的家庭农场，设施设备投入水平较高，如半漏缝地板、自动通风系统、自动喂料系统、自动刮粪机等；人工投入模式上是男老板兼厂长，雇工 2 人。这种家庭农场设施设备投入比较现代化，自动化系统能大大节省人员在体力上的投入，把更多的精力花费在猪只养殖、观察等工作上，是这种模式养殖成绩好的重要原因。

此外，编号河北 2 的家庭农场的养殖模式也需要重点关注，这户家庭农场的毛利比较高，经营成绩好，老板不参与生产，雇佣人数多，包括专职厂长、4 个工人、专职财务、厨师、打杂共计 8 人。与老板沟通后了解，其思路在于后勤及辅助工作由专人负责，工人可以有更多的时间、精力用于观察猪只日常生长情况、及时发现猪只异常等。这种"高硬件投入、高人工投入、高经营成绩"的经营思路值得探讨与反思。

4.2　实体地面模式短期收益高，发展生命力不持久

山东 1 和四川 1 是采用实体地面的家庭农场，由于规模小，管理用心，养殖成绩不错。

但实体地面的家庭农场养殖环境、通风条件都比较差，养殖存在着较大的潜在风险，而且粪污处理简单、环境臭味无法处理等问题表现得日趋严重，并不符合越来越严格的环保要求，再加上大企业规模化养殖已成为趋势，小规模、设施落后的模式很明显不符合市场走向，因此进行升级改造也是必然趋势。

4.3   半漏缝地板为主流，全漏缝地板是猪舍建筑迭代升级的必然趋势

目前，半漏缝地板是多数现代迭代升级的家庭农场所采用的主要地板类型，因其既能在很大程度上解决实体地面面临的卫生和环保处理问题，成本投入又相对可以承受。因全漏缝地板的家庭农场多数都在建设中，个别已经建成的家庭农场在管理上也尚不完善，在本研究中四川省的 2 个全漏缝地板家庭农场，其中四川 3 在建第二栋猪舍，感染风险大，导致养殖成绩不理想，四川 4 的老板认为通风、喂料等都是自动化设施，所以减少了雇工数量，自己兼任很多体力工作，管理上较为粗放，而这种人工投入模式必然导致对猪只管理工作无法特别精细，因此养殖成绩明显不理想。

从实际情况来看，全漏缝地板能进一步提高养殖密度，如果生产管理水平也能进一步提高，且疫病防控更加规范，这种方式可以提高出栏数量、更加节约人工，是迎合时代发展需要的有利选择，尤其在人口老龄化趋势、人工工资不断提升的背景下，更高效地利用人工也是未来的必然考虑和选择。

4.4   树立对人工专业化利用的思维将是实现生产高效管理的关键

家庭农场的运营投入成本是可变成本，直接影响家庭农场的收益，其中，人工投入对养殖成绩的影响是直接的，也是显著的。但人工投入数量和管理模式受家庭农场主的观念意识、个人风格及专业化水平等影响很大，有些家庭农场为节省人工大包大揽，有些家庭农场雇佣的人工多数从事喂料、打扫等体力工作。因此，要实现高效管理，关键还是要把人工从"体力活"中解放出来，提高养猪人的专业水平，在猪只的日常管理、观察等方面多花精力，这才是有效提高生产成绩的关键所在。

参考文献

[1] 闫之春 . 现代养猪企业的管理模式及核算——以专业从事大猪生产的规模养猪企业为例［J］. 中国畜牧杂志，2016，52（10）：26-30.

非洲猪瘟常态下，大型母猪场人工成本投入模式优化研究

王晓燕　闫之春

摘要： 文章选取国内大型规模母猪场，以主要车间配怀舍和产房为例，在分析人工数量投入、人工成本投入现状的基础上，从投入产出视角，分析如何进行最优人员分配及人工成本投入，并建立了人员流动模型。在当前非洲猪瘟疫情长期困扰中国养猪业的形势下，行情是企业所不能把控的，而科学管控成本是企业所能做到的。因此，文章探索科学、精确的管理模式，无论从企业管理角度使人工价值最大化，还是从员工发展角度使个人价值最大化，都具有极其重要的意义。

关键词： 人工投入成本；经营分析；人工最优配置模式

自 2018 年 8 月我国首次报道发生非洲猪瘟（ASF）疫情，到 2019 年下半年开始的生猪行情屡创新高，我国生猪养殖的产业格局已经发生了颠覆性的改变。生猪市场行情震荡可能会是未来相当一段时间内的主旋律。养猪企业是经营管理的主体，养猪企业的管理从猪场建设开始，主要的经营责任是投资预算以及管理[1]，面对新形势，科学管控生产成本是大型母猪场获取利润的关键，而人工成本作为重要的成本投入之一，不仅关乎成本，也是提升经营管理效率的核心要素。规模化养殖与现代化企业的结合使得养殖行业对精益化管理的要求越来越高，其经营分析也亟须更加专业化、精准化。[2]

文章选取国内部分大型规模母猪场，以主要车间配怀舍和产房为例，在分析人工成本投入现状的基础上，从投入产出的角度，分析如何进行最优的人工成本投入，并建立了人员流动模型。研究对象选取了新式场和传统猪场两类，新式场是指在 2016 年之后投资建设，并按照标准生产线投产的母猪场。

1　大型母猪场人工数量投入的基本特点

随着这几年大型母猪场的新建、扩建和改建，对养猪人才的需求量剧增，在调研的大型母猪场中，从类别上看，直接参与生产的人员主要构成为大学生和蓝领工人（专科以下学历的社会招聘人员）两大部分；从人员结构比例上看，应届大学生的比例大幅增加，很多母猪场员工中的大学生比例超过了 50%，有的母猪场甚至大学生比例高达 100%。

本研究中的大学生占比是按照标准岗位配置统计，不含储备/实习人员，具体情况如图 1 所示。

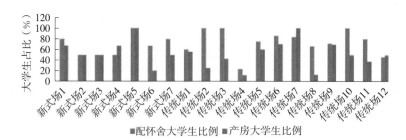

图 1   猪场人员组成中大学生占比情况

2   大学生占比与母猪场主要车间生产效率的相关性分析

2.1   新式场

在配怀舍中，以大学生占比低于 50%（含）的 3 个场为对照组，大学生占比超过 50% 的 4 个场为试验组（图 2）。由图 2 可知，配怀舍的人均养殖效率与大学生占比的相关性不明显，出现略微负相关，这是因为其中有 1 个大学生占比 50% 的新式场，其人工效率远高于其他场所致，不具有代表性，而其他场的人均养殖效率均比较接近。

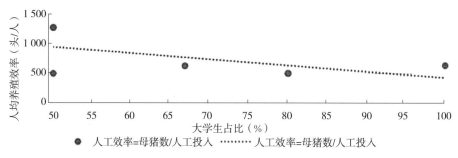

图 2   新式场人均养殖效率与大学生占比的相关性（配怀舍）

在产房中，以大学生占比低于 50%（含）的 4 个场为对照组，大学生占比超过 50% 的 3 个场为试验组（图 3）。由图 3 可知，产房仔猪断奶前成活率与大学生占比无明显相关性。

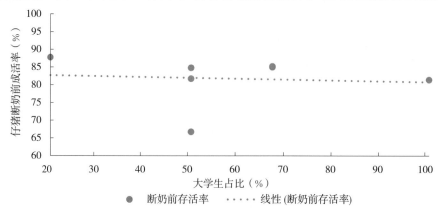

图 3   新式场仔猪断奶前成活率与大学生占比的相关性（产房）

结果表明，在新式场中，配怀舍和产房增加大学生占比，并没有明显增加生产效率和生产成绩。

2.2　传统场

在配怀舍中，以大学生占比低于50%（含）的2个场为对照组，大学生占比超过50%的7个场为试验组（图4）。由图4可知，配怀舍的人均养殖效率与大学生占比的相关性不明显。

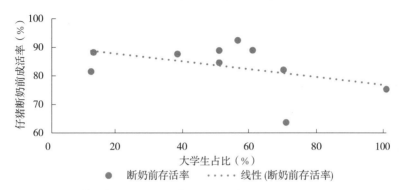

图4　传统场人均养殖效率与大学生占比的相关性（配怀舍）

在产房，以大学生占比低于50%（含）的5个场为对照组，大学生占比超过50%的5个场为试验组（图5）。由图5可知，仔猪断奶前成活率与大学生占比的相关性不大，甚至个别场出现大学生占比高，但仔猪断奶前成活率反而更低的情况。

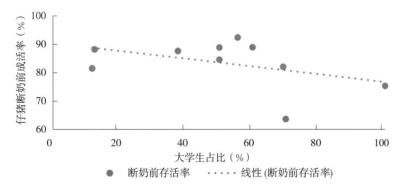

图5　传统场仔猪断奶前成活率与大学生占比的相关性（产房）

因此，研究结果显示，在生产一线岗位上，更多地配置大学生并没有明显有助于提高生产效率。

3　大型母猪场人工成本投入的差异及关键点——以新式母猪场为例

3.1　主要生产车间直接人工配置标准

在大型母猪场的人工配置中，按照3 000头的生产线来考虑，配怀舍和产房岗位设置包括主管、副主管、技术员和饲养员四个级别，配怀舍的技术员和饲养员各2人，产房的技术员和饲养员各 4 人。具体如表 1 所示。

表 1　3 000 头标准生产线主要车间人员配置标准（人）

岗位	配怀舍	产房
主管	1	1
副主管	1	1
技术员	2	4
饲养员	2	4
合计	6	10

3.2　人工成本投入差异的关键点

选择 1 年时间内人员、生产相对稳定的新式母猪场，用生产人员的实际年收入数据进行分析，结果如图 6 所示。

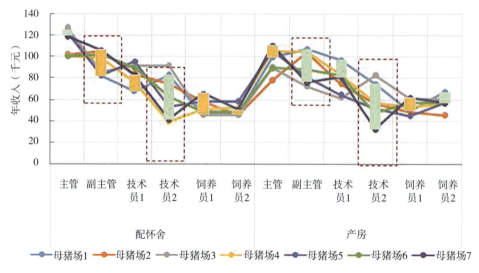

图 6　新式母猪场各岗位人工成本投入差异

从人工成本投入的数据来看，同样的标准化新式母猪场一条生产线，人工成本投入差异大，而分析其根源，在于技术员岗位的人员配置上，大学生和蓝领工人的比例差异大，而大学生工资高于蓝领工人，所以导致总的人工成本差异大。

3.3　结论

通过以上两部分的分析，可以得出以下几点结论：

（1）在生产岗位上配置高比例的大学生并没有与生产成绩呈现正相关性。

（2）目前新式母猪场投入人工成本差异，主要在于技术员岗位投入的大学生数量差异大。

（3）根据实际调研反馈，在生产岗位上配置蓝领工人，在人员稳定性、技术成熟度等方面，都更加符合实际情况。

（4）研究结果显示，考虑大学生的综合素质、知识储备以及自身的职业生涯发展需求，应当将大学生安置在主管及以上的岗位为宜，这类岗位对电脑操作、信息化系统的使用要求较高，需要配备文化水平较高、学习能力较强的人员。

因此，比较科学的人工配置模式应该为，将大学生定位在主管以上的管理岗位，而生产岗位则全部配置蓝领工人，这样既可以满足生产实际需要，又可以实现母猪场的人工成本最优。

4　大型母猪场车间各岗位不同配置模式下的最优成本测算

4.1　三种假设模式下的人工投入成本差异

结合实际情况，假设目前母猪场的配置有三种情况，此处的人工成本数值是在调研数据的基础上，取各岗位的平均值。

4.1.1　配怀舍　主管和副主管岗位全部配置大学生，技术员岗位配置有三种假设，如表2、图7所示：

表2　不同人工配置下的三种模式（配怀舍）

岗位	模式一	模式二	模式三
主管	大学生	大学生	大学生
副主管	大学生	大学生	大学生
技术员1	大学生	大学生	蓝领工人
技术员2	大学生	蓝领工人	蓝领工人
饲养员1	蓝领工人	蓝领工人	蓝领工人
饲养员2	蓝领工人	蓝领工人	蓝领工人

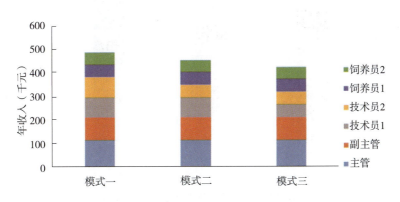

图7　不同人工配置模式下的人工成本对比（配怀舍）

配怀舍的三种配置模式下，技术员岗位全部配置大学生和全部配置蓝领工人，人工投入差异率在14.78%，即如果技术员岗位全部配置蓝领工人的话，成本要比全部配置大学生降低14.78%，超过6.2万元，如表3所示。

表3   三种配置模式下的人工成本差异（配怀舍）

项目	模式一	人工成本	模式二	人工成本	模式三	人工成本
差异值（元）	—	—	—	−32 948	—	−62 476
差异率（%）	—	—	—	−7.28	—	−14.78

注："—"代表无差异。

4.1.2　产房　主管和副主管岗位全部配置大学生，技术员岗位配置有三种假设，见表4；三种模式的成本差异对比如图8、表5所示。

表4   不同人工配置模式下的三种模式（产房）

岗位	模式一	模式二	模式三
主管	大学生	大学生	大学生
副主管	大学生	大学生	大学生
技术员1	大学生	大学生	蓝领工人
技术员2	大学生	大学生	蓝领工人
技术员3	大学生	蓝领工人	蓝领工人
技术员4	大学生	蓝领工人	蓝领工人
饲养员1	蓝领工人	蓝领工人	蓝领工人
饲养员2	蓝领工人	蓝领工人	蓝领工人
饲养员3	蓝领工人	蓝领工人	蓝领工人
饲养员4	蓝领工人	蓝领工人	蓝领工人

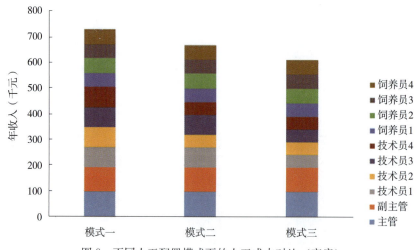

图8   不同人工配置模式下的人工成本对比（产房）

表5   不同人工配置模式下的人工成本对比（产房）

项目	模式一	人工成本	模式二	人工成本	模式三	人工成本
差异值（元）	—	—	—	−59 249	—	−116 767
差异率（%）	—	—	—	−10.62	—	−23.33

注："—"代表无差异。

产房的三种配置模式下，技术员岗位全部配置大学生和全部配置蓝领工人，人工投入差异率在 23.33%，即如果技术员岗位全部配置蓝领工人的话，成本要比全部配置大学生降低 23.33%，超过 11.7 万元。

4.2　最优模式与员工合理配比

经过调研发现，蓝领工人在技术员岗位上的胜任度非常高，稳定性强，从三种模式的配置下可以看到，技术员岗位全部使用蓝领工人的人工成本降低的总额接近 18 万元。因此，模式三是人工成本投入和产出最优的模式，如表 6 所示。

表 6　母猪场车间最优人工配置模式

配怀舍		产房	
岗位	最优配置	岗位	模式三
主管	大学生	主管	大学生
副主管	大学生	副主管	大学生
技术员 1	蓝领工人	技术员 1	蓝领工人
技术员 2	蓝领工人	技术员 2	蓝领工人
饲养员 1	蓝领工人	技术员 3	蓝领工人
饲养员 2	蓝领工人	技术员 4	蓝领工人
—	—	饲养员 1	蓝领工人
—	—	饲养员 2	蓝领工人
—	—	饲养员 3	蓝领工人
—	—	饲养员 4	蓝领工人
节约成本（元）	−62 476	节约成本（元）	−116 767
合计节约成本（元）		−179 243	

注："—"代表无数据。

如果作为大型农牧企业，所属母猪场的生产线数量多，则人工成本节约总量将是惊人的数字。假设全场共设 100 条生产线，按照模式三配置人员节约的人工成本可达到 1 792.4 万元，如表 7 所示。

表 7　大型农牧企业多个母猪场人工成本优化预测（万元）

项目	30 条生产线	50 条生产线	80 条生产线	100 条生产线
年度人工成本节约金额	537.7	896.2	1 433.9	1 792.4

假设大型母猪场采用模式三的最优人工配置模式，即技术员岗位全部配置蓝领工人，那么可以确定大学生的最优配置比例：①在配怀舍中大学生占比为 33%，蓝领工人占比为 67%；②产房的大学生占比为 20%，蓝领工人占比为 80%，如表 8 所示。

表 8　母猪场车间最优人工配置模式

配怀舍			产房		
岗位	最优配置	人数	岗位	最优配置	人数
主管	大学生	1	主管	大学生	1
副主管	大学生	1	副主管	大学生	1
技术员 1	蓝领工人	1	技术员 1	蓝领工人	1
技术员 2	蓝领工人	1	技术员 2	蓝领工人	1
饲养员 1	蓝领工人	1	技术员 3	蓝领工人	1
饲养员 2	蓝领工人	1	技术员 4	蓝领工人	1
—	—	—	饲养员 1	蓝领工人	1
—	—	—	饲养员 2	蓝领工人	1
—	—	—	饲养员 3	蓝领工人	1
—	—	—	饲养员 4	蓝领工人	1
合计		6	合计		10
大学生占比（%）		33	大学生占比（%）		20
蓝领工人占比（%）		67	蓝领工人占比（%）		80

注："—"代表无数据。

5　基于最优配置模式下，大型母猪场之间的人员流动模型

目前大型母猪场所属的大型农牧企业，互相之间输出人才和人员流动比较频繁。结合现有的人员配置最优模式（图 9），可以建立人员流动的模型。

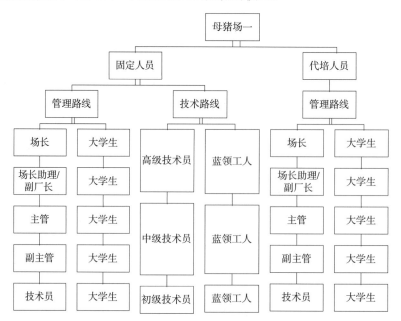

图 9　大型母猪场人工配置组织结构示意

5.1　大学生流动模型

5.1.1　方案一：沿管理岗位向上流动　大学生入职即定位为未来担任管理岗位，从技术员开始学习日常工作，直线向上，按照管理路线流动。

5.1.2　方案二：作为代培人员，输出到其他母猪场　在大型农牧企业中有多个母猪场，大学生入职即定位为代培人员，从技术员开始学习日常工作，之后进入副主管以上的管理岗位，同时根据需要输出到其他母猪场。

5.2　蓝领工人流动模型

将蓝领工人的岗位统一为技术员。划分为初级技术员、中级技术员和高级技术员。

5.2.1　方案一　蓝领工人按照技术路线流动，从初级技术员开始，直线向上，按照技术路线流动，到中级技术员和高级技术员。

5.2.2　方案二　当蓝领工人晋升到中级技术员后，个别有管理能力的蓝领工人，可以转向副主管或主管岗位，并继续按管理线路向上流动。

在当前非洲猪瘟疫情长期困扰中国养猪业的情况下，养猪行情是企业所不能把控的，而科学地管控成本是企业所能做到的，因此，无论从企业管理角度上使人工价值最大化，还是从员工发展角度上使个人价值体现最大化，不断探究科学、精确的管理模式，都具有极其重要的意义。

参考文献

[1] 闫之春. 现代养猪企业的管理模式及核算——以专业从事大猪生产的规模养猪企业为例 [J]. 中国畜牧杂志，2016，52 (10)：26-30.

[2] 王晓燕，闫之春. "现代养猪企业＋家庭农场"商品猪养殖模式的投资收益分析——基于管理会计视角 [J]. 中国畜牧杂志，2018，54 (12)：124-127.

Introduction to an Index Measuring Grow-finishing Efficiency in Swine Production (猪群育肥综合效率指标介绍)

Zhichun Jason Yan   Jing Chen   Xiaowen LI   Zhikai Zeng   Peng LI

Abstract：As we all know，wean-to-finishpigs are the most prominent motivator to keep the farms for the farmers. Farmers always want to cost less and get more meat in order to generate more profits. However，there is not a sole indicator such as ADG，FCR，etc，can reflect the true profitability. For instance，higher ADG is often correlated with higher feed and higher FCR. Therefore，we propose a new and comprehensive indicator called "FSA Index" to solve the dilemma.

Keywords：growing efficiency；profitability；ADG；FCR；survival rate

1 Materials and methods

We define "FSA Index" mentioned before as below：

FSA Index＝SR×ADG / FCR

FSA Index：The index considers Feed conversion ratio，Survival ratio，and Average daily gain.

SR：Survival Ratio of a batch of wean-to-finish pigs.

ADG：Average daily gain（g/d）；i. e.，ADG＝ weight gain per pig（kg）/feeding days× 1 000

FCR：feed conversion ratio；i. e.，FCR＝ total feed amount consumed per pig/ weight gain per pig

The higher the indicator，the greater growing efficiency the batch of pigs has got. That is to say，under a high FSA，ADG and FCR can reach a balance between good weight-gain speed and reasonable feed consumption，therefore reducing feed costs.

We collected cost and production data of 113 batches of finisher barns in Sichuan province，China，from the same company during year 2021，and calculated the FSA Index to demonstrate the effectiveness of the indicator. The quality criteria of analyzed data are defined as follows：

5 kg≤ average initial weight≤10kg，20days≤average weaning age≤40days，101 kg≤

average marketed weight ≤ 150kg，159days ≤ on-feeding days ≤ 232days，and survival ratio ≥ 90%.

The data of FSA and weight-gained feed costs are analyzed as shown in Fig. 1.

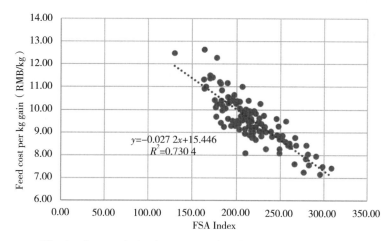

Fig. 1　the correlation between FSA and Feed cost per kg gain.

Note：Feed cost per kg gain (RMB/kg) ＝ total feed cost per marketed pig (RMB) / weight gain from weaning to market (kg).

2　Results

As shown in Fig. 1，there is a significantly negative correlation between FSA and feed cost per kg gain ($P < 0.01$ and $R^2 = 0.730\ 4$) . In other words，FSA had a statistically significant impact on feed cost per kg gain，and was able to interpret 73% variation of feed cost per kg gain.

3　Discussion and conclusion

This indicator mainly illustrates the growing trend of a batch of finishing pigs and is fairly enough to rank the growing efficiency of different grow-finishing batches with various indices (i. e. ，ADG，FCR，and SR) .

FSA is an easy way for us financial analysts to compare performance of various batches and help management to monitor performance. For example，if FSA of a certain batch is substantially low，it most likely indicates that pigs experienced health challenges，including spread of severe disease. Meanwhile，under different ingredients and all else equal，nutritionists may need to adjust their formulations. Moreover，FSA is able to assist us in genetic selection and to discover sub-health condition in grow-finishing herds.

附录　专利情况

专　利

1. 一种可卷扬收放的移动式连廊

　　本发明提供了一种可卷扬收放的移动式连廊，包括底板、竖直设置在底板上的固定架组、用于供猪只进入且一端与底板的上端面铰接连接的升降通道组、一端与升降通道组的一端接触的连接通道组、一端与连接通道组另一端铰接连接的小通道组、用于驱动升降通道组另一端升降的第一驱动机构以及用于驱动小通道组另一端升降的第二驱动机构，小通道组的另一端设置在底板上；第一驱动机构设置在固定架上，第二驱动机构设置在小通道组的上端。本发明提供的可卷扬收放的移动式连廊，使得猪场猪只更加方便地跨越不同区域，转移输送更加便捷，优化了畜牧业养殖场畜禽的转移运送流程。

证书号第10189665号

实用新型专利证书

实用新型名称：一种可卷扬收放的移动式连廊

发　明　人：闫之春;樊士冉;郭景哲;张彦明;邵博;刘聪

专　利　号：ZL 2019 2 1013882.X

专利申请日：2019 年 07 月 02 日

专 利 权 人：西藏新好科技有限公司;夏津新希望六和农牧有限公司
　　　　　　　新希望六和股份有限公司;山东新希望六和集团有限公司

地　　　址：851400 西藏自治区拉萨市经济技术开发区林琼岗支路 2 号
　　　　　　　新希望大厦二层 209 房

授权公告日：2020 年 03 月 31 日　　　授权公告号：CN 210214185 U

　　国家知识产权局依照中华人民共和国专利法经过初步审查，决定授予专利权，颁发实用新型专利证书并在专利登记簿上予以登记。专利权自授权公告之日起生效。专利权期限为十年，自申请日起算。

　　专利证书记载专利权登记时的法律状况。专利权的转移、质押、无效、终止、恢复和专利权人的姓名或名称、国籍、地址变更等事项记载在专利登记簿上。

局长
申长雨

2020 年 03 月 31 日

第 1 页（共 2 页）

其他事项参见续页

2. 一种可伸缩的移动式连廊

本发明提供了一种可伸缩的移动式连廊，包括固定连廊，设置在固定连廊一端内部且可进行升降调节的顶升节，一端与固定连廊另一端连接的伸缩连廊，一端与伸缩连廊另一端连接的驱动连廊，驱动连廊上设置有用于驱动连廊移动的驱动机构，伸缩连廊的一端设置在固定连廊的另一端内部，驱动连廊的一端设置在伸缩连廊的另一端内部；固定连廊内设置有用于顶升节升降的动力机构；固定连廊伸缩连廊和驱动连廊水平设置在地面上。本发明提供的可伸缩的移动式连廊，优化了畜牧业养殖场禽畜等转移运送流程，使得猪场猪只跨越不同区域转移运送成为可能。

3. 一种针对畜牧业生物安全的粪污跨区转运车

本发明提供了一种针对畜牧业生物安全的粪污跨区转运车，包括主车架与主车架固定连接的副车架、设置在副车架上的转架、设置在转架上的液压举升机构以及与液压举升机构连接的罐体，副车架上设置有用于驱动转架旋转的回转驱动机构，罐体设置在液压举升机构的上端部；罐体内设置有用于传输粪污的传输机构。本发明提供的针对畜牧业生物安全的粪污跨区转运车，使得畜牧业传统粪污跨区运输时，保证了高效清洁和生物安全。

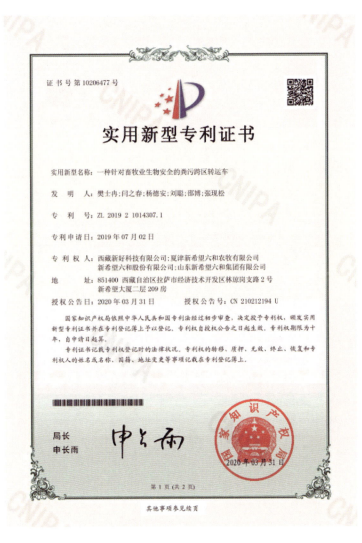

4. 一种用于畜牧业的物资空中运输装置

本发明提供了一种用于畜牧业的物资空中运输装置，包括无人机本体、设置在无人机本体下端的投放装置以及设置在投放装置下端的物资运输箱，无人机本体的下端两侧分别设置有支撑腿，物资运输箱与投放装置活动连接，物资运输箱包括挂载筐和设置在挂载筐内的箱体；投放装置包括与挂载筐活动连接的释放机构和设置在无人机下端部的缓震组件。本发明提供的用于畜牧业的物资空中运输装置，解决了畜牧业精液疫苗等关键物资运输的生物安全问题，节省了人力和时间成本。

5. 一种带有清理机构的圈舍

本发明提供了一种带有清理机构的圈舍，属于养殖技术领域。其技术方案为：包括饲养场和圈舍，圈舍包括设置在饲养场内部地面以下矩形凹坑，矩形凹坑的中间位置设置有矩形隔柱，矩形隔柱上方两侧设置有轨道板，矩形凹坑的上方四周设置有挡板，挡板的上方设置有漏粪板，挡板的上端面高于轨道板的上端面，两个轨道板之间形成轨道且轨道两端均开通，还包括设置在轨道内的带有驱动机构的小车，小车的顶部设置有清扫机构，清扫机构设置在矩形凹坑内，清扫机构的前方且位于矩形凹坑内设置有刮粪板，矩形凹坑的两端且位于矩形隔柱的两侧设置有排粪口。本发明的有益效果为：带有清理机构的圈舍减轻了饲养人员的工作负担，省时省力。

6. 一种畜牧业粪沟清洗机器人

本发明提供了一种畜牧业粪沟清洗机器人，属于畜牧业专用设备领域。其技术方案为：一种畜牧业粪沟清洗机器人，其特征在于包括带有驱动机构的底盘，所述底盘一端设置有推运机构，所述底盘上且位于所述推运机构的上方设置有喷水机构，所述喷水机构与供水装置连接，还包括供电装置。本发明的有益效果为：解决畜牧业牧场中粪沟清洗的问题，减轻人员工作量。

7. 一种可卷扬收放的移动式连廊

本实用新型提供了一种可卷扬收放的移动式连廊，它包括底板、竖直设置在底板上的固定架组、用于供猪只进入且一端与底板的上端面铰接连接的升降通道组、一端与升降通道组的一端接触的连接通道组、一端与连接通道组另一端铰接连接的小通道组、用于驱动升降通道组另一端升降的第一驱动机构以及用于驱动小通道组另一端升降的第二驱动机构，小通道组的另一端设置在底板上；第一驱动机构设置在固定架上，第二驱动机构设置在小通道组的上端。本实用新型提供的可卷扬收放的移动式连廊，使得猪场猪只更加方便的跨越不同区域，转移输送更加便捷，优化了畜牧业养殖场畜禽的转移运送流程。

8. 一种可伸缩的手推移动式连廊

本实用新型提供了一种可伸缩的手推移动式连廊，属于畜牧业专用设备技术领域。其技术方案为：一种可伸缩的手推移动式连廊，其中包括轴向两端通透、依次套接呈伸缩配合的若干节连廊体，设置在若干节连廊体底部的若干个万向轮，设置在若干节连廊体底部的若干个调节支撑件，以及设置在若干节连廊体上的密封组件。本实用新型的有益效果为：方便转移存放及稳定地跨区转运禽畜，便于调整使用尺寸，作业效率高，稳定性好，节约人力成本，结构及使用简单。

9. 一种针对畜牧业生物安全的粪污跨区转运车

本实用新型提供了一种针对畜牧业生物安全的粪污跨区转运车，包括主车架、与主车架固定连接的副车架、设置在副车架上的转架、设置在转架上的液压举升机构以及与液压举升机构连接的罐体，副车架上设置有用于驱动转架旋转的回转驱动机构，罐体设置在液压举升机构的上端部；罐体内设置有用于传输粪污的传输机构。本实用新型提供的针对畜牧业生物安全的粪污跨区转运车，使得畜牧业传统粪污跨区运输时保证高效清洁和生物安全。

10. 一种用于畜牧业的物资空中运输装置

本实用新型提供了一种用于畜牧业的物资空中运输装置，包括无人机本体、设置在无人机本体下端的投放装置以及设置在投放装置下端的物资运输箱，无人机本体的下端两侧分别设置有支撑腿，物资运输箱与投放装置活动连接，物资运输箱包括挂载筐和设置在挂载筐内的箱体；投放装置包括与挂载筐活动连接的释放机构和设置在无人机下端部的缓震组件。本实用新型提供的用于畜牧业的物资空中运输装置，解决了畜牧业精液疫苗等关键物资运输的生物安全问题，节省了人力和时间成本。

11. 一种带有清理机构的圈舍

本实用新型提供了一种带有清理机构的圈舍，属于养殖技术领域。其技术方案为：包括饲养场和圈舍，圈舍包括设置在饲养场内部地面以下矩形凹坑，矩形凹坑的中间位置设置有矩形隔柱，矩形隔柱上方两侧设置有轨道板，矩形凹坑的上方四周设置有挡板，挡板的上方设置有漏粪板，挡板的上端面高于轨道板的上端面，两个轨道板之间形成轨道且轨道两端均开通，还包括设置在轨道内的带有驱动机构的小车，小车的顶部设置有清扫机构，清扫机构设置在矩形凹坑内，清扫机构的前方且位于矩形凹坑内设置有刮粪板，矩形凹坑的两端且位于矩形隔柱的两侧设置有排粪口。本实用新型的有益效果为：带有清理机构的圈舍减轻了饲养人员的工作负担，省时省力。

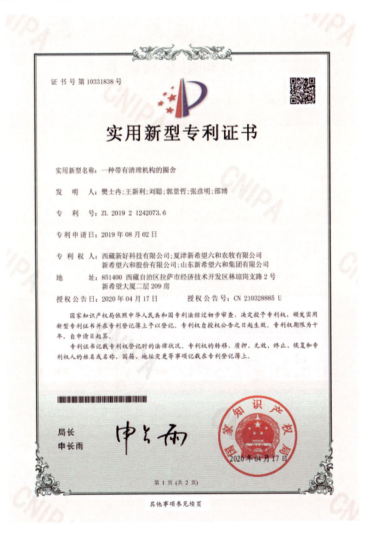

12. 一种智能监测淋浴房

本实用新型提供了一种畜牧业粪沟清洗机器人，属于畜牧业专用设备领域。其技术方案为：一种畜牧业粪沟清洗机器人，其特征在于包括带有驱动机构的底盘，所述底盘一端设置有推运机构，所述底盘上且位于所述推运机构的上方设置有喷水机构，所述喷水机构与供水装置连接，还包括供电装置。本实用新型的有益效果为：解决畜牧业牧场中粪沟清洗的问题，减轻人员工作量。

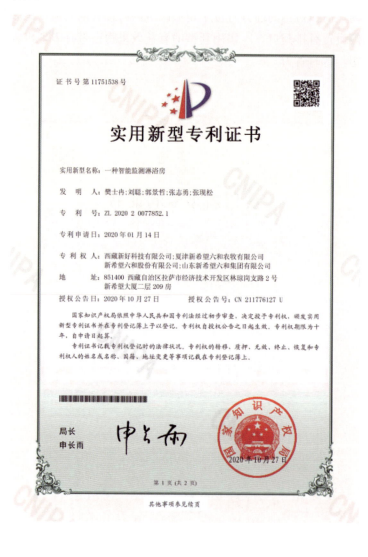

13. 一种人员强制洗消控制系统

本实用新型提供了一种智能监测淋浴房，包括控制器、传感器、开关、电位器、流量计、电磁阀、门锁、播报器、警报器以及与上述各模块（装置）电性相连的用于供电的电源模块。本实用新型提供的一种智能监测淋浴房，可靠性高，抗干性能好，解决了传统洗消用房洗消效果不理想的问题。

14. 一种人员强制洗消控制系统

本实用新型提供了一种人员强制洗消控制系统，包括依次排列构成通道的脱衣间、沐浴间、穿衣间，脱衣间和穿衣间远离沐浴间的墙体上分别设置房门，沐浴间和穿衣间与脱衣间通过隔断墙分隔；房门和墙体上对应设置门锁，墙体上设置有用于控制门锁开闭的开关，房门所在墙体的两侧分别设置有开关；位于脱衣间沐浴间及穿衣间的墙体上分别设置有用于检测人体的传感器。本实用新型提供的一种人员强制洗消控制系统，可靠性高，抗干性能好，解决了传统洗消用房洗消效果不理想的问题。

15. 一种用于物品自动消毒的消毒柜

本实用新型提供了一种用于物品自动消毒的消毒柜，涉及消毒设备领域。其技术方案为：消毒机构包括设置在箱体内部的传送装置，传送装置为链式传送机构，传送装置的传送链路上设置有用于承载待消毒物品的物品箱，物品箱为网孔式镂空箱体，物品箱的运动路径上分布设置若干喷嘴组成的喷嘴组。本实用新型的有益效果为：人员使用消毒柜对物品进行消毒，提高了消毒效率，改善了消毒效果。

16. 一种用于畜牧养殖场的粪污中转运输系统

本实用新型提供了一种用于畜牧养殖场的粪污中转运输系统，属于畜牧养殖技术领域。其技术方案为：一种用于畜牧养殖场的粪污中转运输系统，其中包括集粪池，一端置于集粪池内、另一端露于地面上的传送绞龙，靠近场围墙设置在场内的中转池，以及跨设在场围墙上、一端置于中转池内、另一端位于场外的跨墙绞龙。本实用新型的有益效果为：粪污转运效率高，生物安全问题发生概率低，省时省力，运输成本低。

17. 一种自动单节式移动连廊

本实用新型提供了一种自动单节式移动连廊，属于畜牧业专用设备技术领域。其技术方案为：一种自动单节式移动连廊，其中包括：转猪间，位于猪舍外且一端与猪舍输出端相对应；以及连廊主体，位于转猪间与猪舍输出端之间，一端沿轴向穿设于转猪间内且与转猪间的内壁伸缩配合，另一端与猪舍输出端相配合。本实用新型的有益效果为：实现了猪只暂时存放或跨区转移不堵塞，原有通道使用寿命延长，可以进行分区管理，结构及操作简单。

18. 一种可灵活拖拽的移动式连廊

本实用新型提供了一种可灵活拖拽的移动式连廊，属于畜牧业专用设备技术领域。其技术方案为：一种可灵活拖拽的移动式连廊，其中包括连廊主体，设置在连廊主体轴向一端的搭接组件，设置在连廊主体底部的若干个万向轮，以及设置在连廊主体底部的若干个调节支撑件。本实用新型的有益效果为：可以根据工作与否放置不同位置，不堵塞原有道路，结构简单，可靠性高，适应性好，可维护性强，节约人力成本。

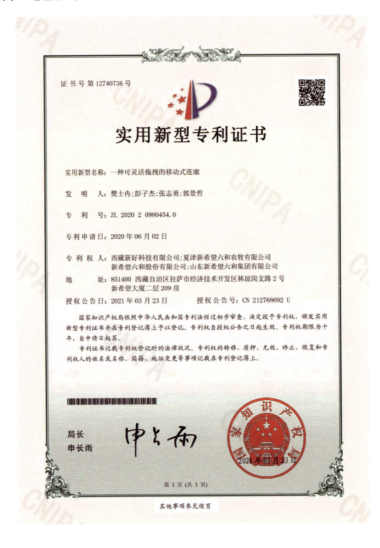

19. 一种可伸缩的手推移动式连廊

本实用新型提供了一种可伸缩的手推移动式连廊，属于畜牧业专用设备技术领域。其技术方案为：一种可伸缩的手推移动式连廊，其中包括轴向两端通透、依次套接呈伸缩配合的若干节连廊体，设置在若干节连廊体底部的若干个万向轮，设置在若干节连廊体底部的若干个调节支撑件，以及设置在若干节连廊体上的密封组件。本实用新型的有益效果为：方便转移存放及稳定地跨区转运禽畜，便于调整使用尺寸，作业效率高，稳定性好，节约人力成本，结构及使用简单。

20. 一种便于精准喂料和清洗的猪用食槽

本实用新型提供了一种便于精准喂料和清洗的猪用食槽，有效地解决了目前精准喂料的料槽容易被猪只拱出浪费污染以及不便清洗等问题。其技术方案为：包括槽体，槽体设置于水平托盘上，托盘上方设置有保护壳，保护壳的上方设置有防护架，槽体连接有清洗组件，槽体底部外侧壁中心处设置有监测机构，槽体的下方设置有排放通道。本实用新型的有益效果为：提供了一种能有效避免猪只拱料浪费踩踏污染，以及可以方便对槽体进行彻底清洗的猪用食槽。

21. 一种便于精准喂料和清洗的猪用食槽

　　本发明提供了一种便于精准喂料和清洗的猪用食槽，有效地解决了目前精准喂料的料槽容易被猪只拱出浪费污染以及不便清洗等问题。其技术方案为：包括槽体，槽体设置于水平托盘上，托盘上方设置有保护壳，保护壳的上方设置有防护架，槽体连接有清洗组件，槽体底部外侧壁中心处设置有监测机构，槽体的下方设置有排放通道。本发明的有益效果为：提供了一种能有效避免猪只拱料浪费踩踏污染，以及可以方便对槽体进行彻底清洗的猪用食槽。

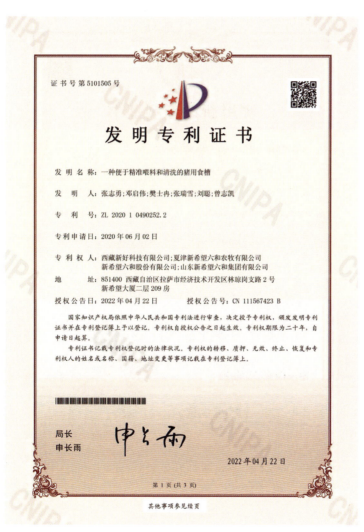

22. 一种针对畜牧业生物安全的自卸式转粪车

本实用新型提供了一种针对畜牧业生物安全的自卸式转粪车，包括主车架、设置在主车架上的液压举升机构以及与液压举升机构连接的防泄漏罐体，罐体顶部设置有冲洗机构以及防溢板，罐体后门设置有液压开门机构，罐体后门设置有密封胶条。本实用新型的有益效果为：设计合理，结构简单，安全可靠，极大程度地解决了使用粪污传统运输方式带来的生物安全问题，且节省人力和时间成本，具有明显优势。

23. 一种多功能沐浴房

本实用新型提供了一种多功能沐浴房，属于养殖加工场辅助设备领域。其技术方案为：包括沐浴房，设置在沐浴房中的沐浴装置，沐浴装置包括旋转站台，旋转站台左右两侧对称设置有弧形护栏，两侧弧形护栏上均设置有弧形滑动架，弧形滑动架上设置有沐浴组件，沐浴组件包括S形沐浴管，S形沐浴管上设置有若干喷淋头，S形沐浴管上设置有进水管，进水管通过多路连接管连接有烘干机高压水源消毒水源。本实用新型的有益效果为：设计了一种对工作人员便捷清洗的沐浴间，大大方便了工作中的使用。

24. 一种人员采样辅助监测装置

本实用新型提供了一种人员采样辅助监测装置，包括壳体，壳体内设置有机体，机体连接有输入模块与输出模块，输入模块包括录像模块与读卡模块，输出模块包括显示模块与播放模块，壳体外部设置有打印纸出口，打印纸出口联通打印机输出端，壳体外设置有出袋口，出袋口连通壳体内部的装袋腔，装袋腔内设置有袋体，壳体外设置有站板，站板位于壳体前侧且对应录像模块，站板水平设置于地面且内部设置有重力传感器。本实用新型的有益效果为：设计合理，结构简单，安全可靠，可节省人力时间，有效提高采样效率。

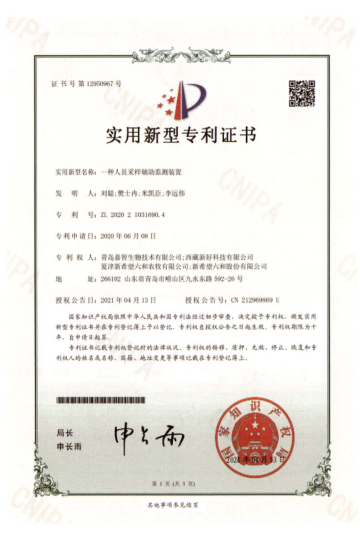

25. 一种双端可自由升降的转运通廊车

本实用新型提供了一种双端可自由升降的转运通廊车，属于物资中转过渡设备技术领域。其技术方案为：一种双端可自由升降的转运通廊车，包括底盘，底盘底部设置底轮；通廊本体，位于底盘正上方；举升系统，设置在底盘上，包括分别位于底盘两端的两对举升机构，每一对举升机构相对应位于通廊本体两侧；以及两个导向部件，相对应设置在通廊本体两端，且分别通过轴承与通廊本体转动配合，每一个导向部件的两端分别与一对举升机构固定连接。本实用新型的有益效果为：实现不同高度车辆之间的物资转运，资源损耗低，中转过渡安全高效，适用性强，节省人力，结构及使用简单。

26. 一种猪场用红外测温及注射多功能小车

本实用新型提供了一种猪场用红外测温及注射多功能小车，属于畜牧业专用设备技术领域。其技术方案为：一种猪场用红外测温及注射多功能小车，包括车体、车架、水平设置在车架中部的置物台、设置在车架顶部的工作台以及设置车架底部的若干个底轮；工控机，设置在工作台顶面上；红外热像仪，通过调节组件设置在工作台上；以及电源箱，内置电源，设置在车架下端部。本实用新型的有益效果为：有效识别猪场内猪只体温，测量无死角，实时记录数据并在线处理，结构简单及操作方便。

27. 一种猪只转运淘汰车

本实用新型提供了一种猪只转运淘汰车，属于畜牧业专用设备技术领域。其技术方案为：一种猪只转运淘汰车，包括一端部通过铰接组件相接的两个车体，以及至少在其中一个车体另一端可拆卸设置一个车门；车体包括底盘，设置在底盘底部的底轮，以及设置在底盘顶部的车架；车架包括沿底盘边缘，相对应设置在底盘两侧的隔栅，在同一侧的隔栅顶部各设置一个横杆，设置在两个横杆相对应一端之间的第一加强杆，以及设置在两个横杆相对应另一端之间的第二加强杆。本实用新型的有益效果为：可以在狭窄的过道内实现转向弯折，适用性强，结构简单，设计巧妙，拆卸和冲洗方便。

28. 一种猪用高产床导流平台

本实用新型提供了一种猪用高产床导流平台，属于家畜养殖辅助设备技术领域。其技术方案为：一种猪用高产床导流平台，包括顶面呈斜坡状底部设有若干个底轮的基台，可折叠置于基台顶面上，一端与基台的坡底端铰接连接的第一折叠体，可折叠置于基台顶面上包括一端与基台的坡顶端铰接连接的第二板体以及设置在第二板体一侧面上的若干个支腿的第二折叠体，以及设置在第二板体顶部另一端，且可折叠置于第二板体上的栏杆。本实用新型的有益效果为：避免妊娠母猪应激，可以在狭窄且弯折的过道内转移，便于在小空间内收纳，结构简单，安全性高，导流过程省时省力，操作便捷且使用灵活。

29. 一种畜用转运连接通廊

本实用新型提供了一种畜用转运连接通廊，属于畜禽转运装置技术领域。其技术方案为：一种畜用转运连接通廊，其中包括纵向两端贯通侧壁可折叠的通廊本体，设置在通廊本体底部的支撑转移单元，可折叠相对应设置在通廊本体两端的搭接单元，以及设置在通廊本体上的通风单元。本实用新型的有益效果为：既不堵塞通道又可实现小空间收纳，转运效率高，适用性强，方便操作。

30. 一种可折叠猪用连接通廊

本实用新型提供了一种可折叠猪用连接通廊,属于家禽转运装置技术领域。其技术方案为:一种可折叠猪用连接通廊,包括纵向两端贯通侧壁可折叠的通廊本体,设置在通廊本体底部的支撑转移单元,可折叠相对应设置在通廊本体两端的搭接单元,以及设置在通廊本体上的通风单元;通廊本体包括廊底两个折叠侧板廊顶,以及相对应设置在廊底与廊顶之间控制折叠侧板展开与折叠,并在折叠侧板展开状态对其进行支撑的两个收放部件。本实用新型的有益效果为:结构简单,操作方便,不占用通道,便于在小空间内进行收纳,稳定性好。

31. 一种推、刮两用畜禽清粪工具

本实用新型提供了一种推刮两用畜禽清粪工具，属于畜牧场清理工具技术领域。其技术方案为：一种推刮两用畜禽清粪工具，包括手把主体，呈S形，包括长杆短杆以及设置在长杆一端与短杆一端之间的连接杆；推刮板，固定连接在长杆另一端；配重手柄，圆杆型，一端固定连接在短杆的另一端；以及防滑保护套，设置在配重手柄上。该推刮两用畜禽清粪工具的构造简单，操作简便，使员工在不弯腰的情况下轻松完成清污工作，并且该工具具备能推能刮的特点，使得清粪工作更灵活，可以满足不同清粪场景的需求。

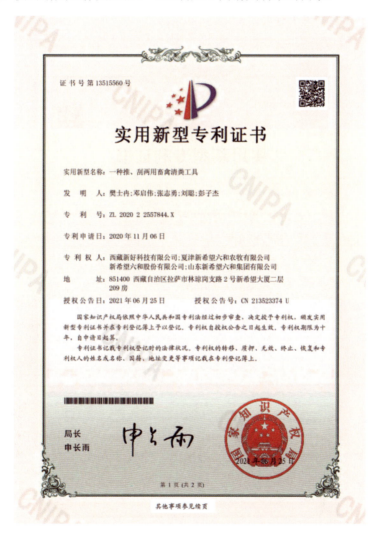

32. 一种育肥舍大栏环境用测温装置

本实用新型提供了一种育肥舍大栏环境用测温装置，涉及农业领域动物表面温度测量设备技术领域。其技术方案为：包括手持的测温杆，测温杆的一端设置有热成像相机模块，测温杆的另一端设置有图像显示模块；热成像相机模块和图像显示模块之间构成数据反馈通路。本实用新型的有益效果为：该装置应用场景之一是规模化猪场中的育肥舍大栏环境，在成本方面有较好的权衡，伸缩式测温杆同价格昂贵的设备对比，在测量偏差不大的情况下，可自行调节测温距离，保证每一个目标的有效测量，操作方便，随着生猪养殖规模的不断扩大，未来该产品成本将会越来越低。

33. 一种便携式可折叠风量测量仪

本实用新型提供了一种便携式可折叠风量测量仪，属于风量测量装置技术领域。其技术方案为：一种便携式可折叠风量测量仪，包括手持杆伸缩杆，设置在手持杆一端部与伸缩杆一端部之间的折叠连接架，沿伸缩杆中心线方向均匀布置通过若干个连接组件与伸缩杆固定相连的若干个风速探头，以及分别与若干个风速探头电连接内置风量计算模块的控制箱；若干个风速探头的进风口方向相互平行。本实用新型的有益效果为：适用范围广，使用方便，测量准确性高且可实时显示，作业效率高，体积小且可折叠。

34. 一种猪用食槽的清洗装置

本实用新型提供了一种猪用食槽的清洗装置，属于家畜养殖技术领域，有效解决了猪用食槽清洗不便、不彻底、易积水，以及清洗水混合粪尿污水增加污水处理量等问题。该技术方案包括食槽上方设置的盖罩，盖罩连接有升降机构，盖罩底部设置有喷头，喷头顶部连接有往复机构，升降机构活动连接有设置在食槽下方的排放机构。本实用新型的有益效果为：在实现对食槽封闭高效清洗的同时，也使清洗水能完全单独排放，有利于减轻粪尿污水处理压力和对残余饲料的回收。

35. 巡检机器人

外观设计产品的名称：巡检机器人。

外观设计产品的用途：本外观设计产品用于巡检，如用于畜牧场巡检。

外观设计产品的要点：在于产品的形状和构造。

36. 一种适用于规模化养猪的加药装置

本实用新型提供了一种适用于规模化养猪的加药装置，属于生猪养殖技术领域，有效解决了目前药物与饲料混合效率低、劳动强度大以及混合不充分、不均匀、易造成药物浪费和环境污染等问题。该技术方案包括平板推车，推车上设置有加药罐，加药罐的输出端通过输药管连接加药计量泵的输入端，输药管上设置有过滤器，计量泵的输出端连接有送药管的输入端，送药管的输出端连接有若干雾化喷头，对应每个雾化喷头在输料管的顶部沿轴向开设有若干个加药孔，输送管道的外侧套设有滑动架，滑动架与加药孔及雾化喷头活动配合。本实用新型的有益效果为：提供了一种方便快捷的可将药物精准有效地均匀混入饲料的加药装置。

37. 一种可便携拆装的畜用微创采样器

本实用新型提供了一种可便携拆装的畜用微创采样器，包括手柄与针头，针头上设置有取样孔，手柄一端设置有弹性夹头，针头尾端位于弹性夹头内部，手柄柄体与弹性夹头之间设置有连接块，连接块外部螺纹连接有连接件，手柄外部还设置有防护套，防护套一端通过弹簧与手柄尾端连接，防护套外壁设置有限位结构。本实用新型的有益效果为：设计合理，结构简单，安全可靠，可便携操作，方便作业人员使用，并可提高工作效率。

38. 一种可快速拆装的畜用微创采样器

本实用新型提供了一种可快速拆装的畜用微创采样器，包括手柄与针头，针头设置有取样孔，尾端设置有固定块，手柄的一端设置有外螺纹，手柄与针头之间通过连接件固定，手柄的尾端设置有限位管，限位管内固定连接有弹簧，弹簧的另一端固定连接有套环，套环转动连接有防护套，防护套包裹与手柄外壁，防护套靠近套环一端设置有限位板，限位管靠近防护套一侧设置有限位壁，限位壁上设置有与限位板配合的通槽。本实用新型的有益效果为：设计合理，结构简单，安全可靠，可便携操作，方便作业人员使用，并可提高工作效率。

39. 一种动物口腔液采样用装置

本实用新型提供了一种动物口腔液采样用装置，包括手柄与取样结构，取样结构包括与手柄固定连接的储液容器，储液容器连接于手柄的一端，储液容器顶部活动连接有取液板，取液板为弧形板，板面上均匀分布有若干滤液孔，取液板通过限位结构与储液容器活动连接，限位结构包括卡扣与卡板，卡板为条状的弧形板，位于储液容器的内壁顶部，卡扣位于取液板底边，且卡扣与卡板配合卡接。本实用新型的有益效果为：设计合理，结构简单，安全可靠，可解决动物口腔采样难度大、时间久、成本高的问题。

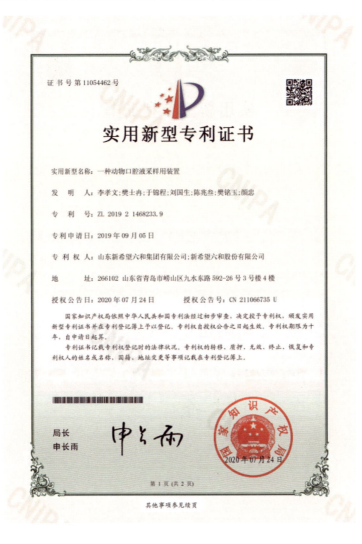

40. 一种鼠洞取样装置

本实用新型提供了一种鼠洞取样装置，包括固定架，固定架顶部连接有固定杆，固定架铰接有旋转架，旋转架底部固定连接有吸水件，固定架内侧设置有两个挤水结构，两个挤水结构的间距小于吸水件的宽度，挤水结构内部中空，且外壁设置有若干滤水孔，旋转架上设置有与固定架铰接的旋转轴，旋转轴的一端连接有电机。本实用新型的有益效果为：设计合理，结构简单，安全可靠，可有效并准确采集老鼠携带的病毒。

41. 一种用于猪喉部液体快速采样装置

本实用新型提供了一种用于猪喉部液体快速采样装置，包括辅助筒，辅助筒的一端为封闭，另一端为开口，辅助筒内固定连接有弹簧，辅助筒的外壁活动连接有若干控制件，每个控制件的顶端均设置有限位块，每个限位块均贯穿辅助筒外壁的通孔延伸至辅助筒内部；本装置还包括取样棒，取样棒的一端设置有吸水棉。本实用新型的有益效果为：设计合理，结构简单，安全可靠，可快速对猪只喉部液体取样，且可有效避免交叉感染。

42. 微创采样器

本发明提供了一种畜用组织微创采样器，属于畜牧业专用器械技术领域。其技术方案为：一种畜用组织微创采样器，其中包括针筒，贯穿针筒两端设置的推杆，设置在针筒尾端、活动连接在推杆相对应端部的封头，设置在针筒头端、套接在推杆相对应端部、且与推杆沿其中心线方向滑动配合的套针。本发明的有益效果为：采样高效，采样时间短，活体组织流血少，结构及操作简单，设计巧妙，适应性好，省时省力，应用价值高。

43. 一种用于猪喉部液体快速采样装置

本发明提供了一种畜用组织医疗采样器，属于畜牧业医疗器械领域。其技术方案为：一种畜用组织医疗采样器，其中包括针筒，穿设于针筒尾端、轴线与针筒的轴线平行的推杆，轴线与针筒的轴线平行，一端设置在针筒头端的套针，以及一端设置在推杆内端、另一端穿设于套针内的顶杆。本发明的有益效果为：结构简单，可靠性高，适应性好，可维护性强，使得对淋巴组织的采样工作更加高效便捷。

证书号第11054934号

实用新型专利证书

实用新型名称：一种用于猪喉部液体快速采样装置

发　明　人：李孝文；樊铭玉；邵振文；颜忠；陈兆叁；樊士冉；刘清源

专　利　号：ZL 2019 2 1975670. X

专利申请日：2019 年 11 月 15 日

专利权人：西藏新好科技有限公司；夏津新希望六和农牧有限公司
　　　　　　新希望六和股份有限公司；山东新希望六和集团有限公司

地　　　址：851400 西藏自治区拉萨市经济技术开发区林琼岗支路 2 号
　　　　　　新希望大厦

授权公告日：2020 年 07 月 24 日　　授权公告号：CN 211066750 U

　　国家知识产权局依照中华人民共和国专利法经过初步审查，决定授予专利权，颁发实用新型专利证书并在专利登记簿上予以登记。专利权自授权公告之日起生效，专利权期限为十年，自申请日起算。

　　专利证书记载专利权登记时的法律状况。专利权的转移、质押、无效、终止、恢复和专利权人的姓名或名称、国籍、地址变更等事项记载在专利登记簿上。

局长　申长雨

2020 年 07 月 24 日

第 1 页（共 2 页）

其他事项参见续页

44. 一种新型物料混合机

本实用新型提供了一种新型物料混合机，属于物料混合技术领域。该新型物料混合机包括一级混合机构、二级混合机构和机架，二级混合机构设置于机架的顶部；一级混合机构设置于二级混合机构的顶部；共分为两级混匀，第一级为初步混匀，物料在一级混合舱内通过搅拌方式混匀，第二级采用双绞龙结构，对物料起到打散和传输的作用，使原料和辅料形成均匀混合比，且混合后没有大块污泥团产生，可完全适用絮凝脱水污泥黏稠的特性，且混合后没有大块污泥团产生，处理效率高，大大地提高了好氧堆肥发酵的效率，且发酵后的物料均匀松散、腐熟度好，设备混合效率高，占地面积小，智能化程度高，操作简单方便，制造成本低，经济效益好。

45. 一种种猪表型测定装置

本实用新型提供了一种种猪表型测定装置，包括通过栅栏组成的行走通道和限位栏，行走通道一端设为入口，另一端收缩变窄后连接限位栏的入口端，限位栏的出口端有闸门，限位栏底部设有称重台；行走通道的外侧设有摄像头组一，摄像头组一能够捕捉猪的行走姿态；限位栏外侧设有摄像头组二，摄像头组二能够捕捉猪的静态的表型特征。本实用新型的有益效果为：本装置能够将种猪表型测定的过程自动化，大大减少人力成本；实现标准化测定，降低人为误差。

46. 一种种猪测定笼称

本实用新型提供了一种种猪测定笼称，包括称重台以及设置在称重台上的测定笼；测定笼包括设置在称重台上方两侧的护栏，两个护栏之间形成测定通道，测定通道一端设有入口门，另一端设有出口门；称重台上且在护栏之间居中设有保定装置，保定装置包括固定在称重台上的剪刀式升降机构以及设置在剪刀式升降机构上的横杆，横杆贯穿测定通道，猪的四肢能够位于横杆的两侧。本实用新型的有益效果为：本实用新型设有保定装置，通过升起的横杆将猪提起来，对种猪进行保定，使测定更加稳定；通过推杆对出口门进行开合，方便高效。

证书号第15628365号

实用新型专利证书

实用新型名称：一种种猪测定笼称

发　明　人：杨季凡；荆晓燕；杜晓冬；刘强；陶强强；杨雨婷；樊上冉；张兴

专　利　号：ZL 2021 2 1894569.9

专利申请日：2021 年 08 月 13 日

专 利 权 人：四川新希望六和猪育种科技有限公司
　　　　　　　江油新希望海波尔种猪育种有限公司

地　　　址：610000 四川省成都市中国(四川)自由贸易试验区天府新区正兴街道蜀州路 2828 号附 OL-05-202012006 号

授权公告日：2022 年 01 月 25 日　　　授权公告号：CN 215639718 U

　　国家知识产权局依照中华人民共和国专利法经过初步审查，决定授予专利权，颁发实用新型专利证书并在专利登记簿上予以登记。专利权自授权公告之日起生效。专利权期限为十年，自申请日起算。

　　专利证书记载专利权登记时的法律状况。专利权的转移、质押、无效、终止、恢复和专利权人的姓名或名称、国籍、地址变更等事项记载在专利登记簿上。

局长　申长雨

2022年01月25日

第 1 页 (共 3 页)

其他事项参见续页

47. 一种物品消毒系统

本发明提供了一种物品消毒系统，涉及消毒设备领域。其技术方案为：包括设置在箱体内的链式传动机构，链式传动机构链条上设置有置物筐，链式传动机构由步进电机驱动；位于链式传动机构的传送路径上，设置有喷嘴，喷嘴通过管路与喷雾剂连通；包括控制器，控制器的输入端和传感器模块启动按钮及时间继电器分别电连接，形成输入通路；控制器的输出端与步进电机，液位指示灯运行指示灯喷雾机分别电连接，形成输出通路；还包括作为供电来源的供电模块。本发明的有益效果为：通过对物品全自动消毒，节约人力时间成本，改善消毒效果，具有可靠性高、抗干扰性强、方便维护等优势。

48. 一种用于物品自动消毒的消毒柜

本发明提供了一种用于物品自动消毒的消毒柜，涉及消毒设备领域。其技术方案为：消毒机构包括设置在箱体内部的传送装置，传送装置为链式传送机构，传送装置的传送链路上设置有用于承载待消毒物品的物品箱，物品箱为网孔式镂空箱体，物品箱的运动路径上分布设置若干喷嘴组成的喷嘴组。本发明的有益效果为：人员使用消毒柜对物品进行消毒，提高了消毒效率，改善了消毒效果。

49. 一种粪污中转畜牧养殖场及粪污中转方法

本发明提供了一种粪污中转畜牧养殖场及粪污中转方法，属于畜牧养殖技术领域。其技术方案为：一种粪污中转畜牧养殖场，其中包括集粪池，一端置于集粪池内、另一端露于地面上的传送绞龙，靠近场围墙设置在场内的中转池，以及跨设在场围墙上、一端置于中转池内、另一端位于场外的跨墙绞龙。本发明的有益效果为：粪污转运效率高，生物安全问题发生概率低，省时省力，运输成本低。

50. 一种畜牧业粪污转运系统

本发明提供了一种畜牧业粪污转运系统，包括装载机构与对接机构，装载机构包括输入端位于粪污中转池、输出端位于承载箱体的螺旋输送机，承载箱体为钩臂车的可卸箱体，对接机构包括升降装置，升降装置位于养殖区内围墙内，升降装置自下至上包括升降组件平移组件与倾斜组件，倾斜组件上设置承载箱体，围墙外为洁净区且设置有转粪车。本发明的有益效果为：设计合理，结构简单，安全可靠，可解决畜牧业传统粪污跨区运输所带来的生物安全及效率问题，且保证粪污跨区转运时清洁高效，同时也可节省人力和时间成本。

51. 一种多功能沐浴房

本发明提供了一种多功能沐浴房，属于养殖加工场辅助设备领域。其技术方案为：包括沐浴房，设置在沐浴房中的沐浴装置，沐浴装置包括旋转站台，旋转站台左右两侧对称设置有弧形护栏，两侧弧形护栏上均设置有弧形滑动架，弧形滑动架上设置有沐浴组件，沐浴组件包括S形沐浴管，S形沐浴管上设置有若干喷淋头，S形沐浴管上设置有进水管，进水管通过多路连接管连接有烘干机高压水源消毒水源。本发明的有益效果为：设计了一种汇集清洗消毒烘干为一体的沐浴间，大大方便了工作中的使用。

52. 一种双端可自由升降的转运通廊车

本发明提供了一种双端可自由升降的转运通廊车，属于物资中转过渡设备技术领域。其技术方案为：一种双端可自由升降的转运通廊车，包括底盘，底盘底部设置底轮；通廊本体，位于底盘正上方；举升系统，设置在底盘上，包括分别位于底盘两端的两对举升机构，每一对举升机构相对应位于通廊本体两侧；以及两个导向部件，相对应设置在通廊本体两端，且分别通过轴承与通廊本体转动配合，每一个导向部件的两端分别与一对举升机构固定连接。本发明的有益效果为：实现不同高度车辆之间的物资转运，资源损耗低，中转过渡安全高效，适用性强节省人力，结构及使用简单。

53. 一种育肥猪个体精准饲喂系统及方法

本发明提供了一种育肥猪个体精准饲喂系统及方法，属于家畜养殖技术领域。其技术方案为：一种育肥猪个体精准饲喂系统，包括饮水区饲喂区和售卖区，设置在饮水区售卖区与饲喂区之间设有第一射频识别器的分栏秤，设置在饮水区与饲喂区之间的单向门，设置在饲喂区相互间隔且相互屏蔽分别设有第二调频识别器的若干个精准饲喂装置，与分栏秤第一射频识别器精准饲喂装置电连接的控制中心，以及佩戴在猪身上的电子标。本发明提供了一种精准饲喂且减少饲料浪费、精准满足个体猪只的最经济的营养需求、出栏体重均匀的育肥猪个体精准饲喂系统，以及提供了一种猪只自动精准饲喂且达到均匀体重、自动出栏节省人工的育肥猪个体精准饲喂方法。

54. 一种猪用高产床导流平台

本发明提供了一种猪用高产床导流平台，属于家畜养殖辅助设备技术领域。其技术方案为：一种猪用高产床导流平台，包括顶面呈斜坡状底部设有若干个底轮的基台，可折叠置于基台顶面上，一端与基台的坡底端铰接连接的第一折叠体，可折叠置于基台顶面上包括一端与基台的坡顶端铰接连接的第二板体以及设置在第二板体一侧面上的若干个支腿的第二折叠体，以及设置在第二板体顶部另一端，且可折叠置于第二板体上的栏杆。本发明的有益效果为：避免妊娠母猪应激，可以在狭窄且弯折的过道内转移，便于在小空间内收纳，结构简单，安全性高，导流过程省时省力，操作便捷且使用灵活。

55. 一种畜用转运连接通廊

本发明提供了一种畜用转运连接通廊，属于畜禽转运装置技术领域。其技术方案为：一种畜用转运连接通廊，其中包括纵向两端贯通侧壁可折叠的通廊本体，设置在通廊本体底部的支撑转移单元，可折叠相对应设置在通廊本体两端的搭接单元，以及设置在通廊本体上的通风单元。本发明的有益效果为：既不堵塞通道又可实现小空间收纳，转运效率高，适用性强，方便操作。

56. 一种畜禽场专用消毒传递窗

本实用新型提供了一种畜禽场专用消毒传递窗，属于畜牧养殖技术领域。其技术方案为：包括一种畜禽场专用消毒传递窗、传递窗本体、分别设置在传递窗本体的传送侧以及接受侧的传递窗门，且沿传送方向相对应设置在传递窗本体内侧壁上部的两根货托滑槽，纵切面呈 U 形开口端分别与货托滑槽配合封闭端水平设置的货托篮，设置在传递窗本体侧壁上的出风口，以及作用于传递窗本体内的消毒结构。该消毒传递窗操作简便，采用多种综合消杀方式，满足了畜牧场生物安全要求，同时采用伸缩式推拉门作为整个传递窗的传递窗门，更为贴合畜牧场环境需求。

57. 一种猪场用超滤水处理设备

本实用新型提供了一种猪场用超滤水处理设备，涉及畜牧业标准化舍内水处理设备技术领域。其技术方案为：包括依次排列的膜体阀体及流量计泵体三部分，膜体阀体及流量计和泵体分别设置在一个单独的框架内，相邻两部分框架连接处及框架内的仪器仪表前后端均采用活动连接的方式，框架间采用长螺栓紧固连接。本实用新型的有益效果为：方便设备分体进入猪场并在舍内迅速组装，解决了传统水处理由于体积较为庞大很难将其安装在舍内工作的难题。

58. 一种猪场生物安全用物资转运箱

本实用新型提供了一种猪场生物安全用物资转运标准箱，属于生物转运标准箱技术领域。其技术方案为：一种猪场生物安全用物资转运标准箱，其中包括开口向上的箱体，可拆卸设置在箱体上用于密封封闭箱体的盖体，可拆卸设置在箱体和/或盖体上内置 RFID 芯片的芯片卡，以及设置在箱体与盖体之间的防误开结构，箱体和/或盖体由透明材质制成。本实用新型的有益效果为：结构简单，保证物资信息的可追溯性，转运全程可视且密封无接触风险，物资转运时效性高，转运管理效率高。

59. 一种猪场生物安全用防开箱密封箱

本实用新型提供了一种猪场生物安全用防开箱密封标准箱，属于生物转运标准箱技术领域。其技术方案为：一种猪场生物安全用防开箱密封标准箱，包括开口向上的箱体，可拆卸设置在箱体上用于密封封闭箱体的盖体，以及设置在箱体与盖体之间的防误开结构；箱体和/或盖体由透明材质制成；盖体包括边缘处内扣且与箱体扣接配合的盖本体，以及沿盖本体内侧面边缘部位连续且首尾相接设置且延伸至盖本体边缘处的密封条；密封条与箱体内侧壁贴紧配合。本实用新型的有益效果为：结构简单且密封效果好，减少或避免物资与外界接触的风险，提高物资到达绿区的时效性，方便搬运。

60. 一种养殖场用消毒房

本实用新型提供了一种养殖场用消毒房，涉及畜牧技术领域。其技术方案为：包括为与一层的工作区和位于二层的生活区，工作区域包括位于主通道两侧的物资区和食材区，主通道一端朝向厂区外，另外一端朝向厂区内，食材区远离主通道的一侧设置人员区；物资区和食材区均分别设置有消毒区，通过消毒区将物资区和食材区分别隔离为未消毒区和已消毒区。本实用新型的有益效果为：本方案的消毒房充分结合养殖场运输场景和现场条件，所以匹配性较高，极大程度地解决了使用人员物资传统运输方式带来的生物安全问题，而且节省人力和时间成本，具有明显的优势。

61. 一种基于目标检测的畜禽动物体温评估方法

本发明提供了一种基于目标检测的畜禽动物体温评估方法，属于畜禽保温评估技术领域。其技术方案为：一种基于目标检测的畜禽动物体温评估方法，包括在畜禽舍内预设测点；设置巡栏测温路线将各测点串联；智能巡栏机器人沿巡栏测温路线由起点至终点进行巡栏测温，利用智能巡栏机器人上的热成像相机在每一检测点对相对应的畜禽进行热成像图像拍摄，拍摄后的热成像图像传输至中控平台由算法程序进行实时处理，生成数据结果报表；对数据结果报表进行对比分析，判断出体温异常的畜禽。本发明的有益效果为：检测效率高，分析运行效率高，评估结果置信度高，省时省力。

62. 一种动物尾根微量血液采样器

本实用新型提供了一种动物尾根微量血液采样器，涉及养殖设备领域。其技术方案为：包括针体，针体远离针尖的一端设置手持部，针体上套设棉球，棉球位于手持部和针尖之间。本实用新型的有益效果为：本装置将动物采血的穿刺采血止血样品封装保存，以及后期样品处理的离心结合到了一起，极大地方便了动物微量血液采集工作；在采血环节引入本装置可显著提高采样效率，降低采样人员工作强度，减少对动物的应激以及造成的身体损伤；每个采集的样品均独立存放在其所对应的管体中，可避免采样过程对样品的污染；以及降低了采样过程扩散疾病的风险。

63. 一种畜牧业粪沟清洗机器人

本实用新型提供了一种粪沟清洗装置，属于畜牧业清洗设备技术领域。其技术方案为：一种粪沟清洗装置，其特征在于包括具有驱动机构的底盘，底盘内部设置有空腔，空腔内设置有为驱动机构供电的移动电源，空腔上方设置有防护盖，防护盖的上方设置有供水机构。本实用新型的有益效果为：具有移动电源的粪沟清洗装置。

64. 一种清洗机器人

本外观设计产品的名称：一种清洗机器人。

本外观设计产品的用途：本外观设计产品用于对地面的清洗，尤其适用与粪沟的清洗。

本外观设计的设计要点是：本产品的整体形状。

65. 一种仔猪被压死淘率和存活率的统计方法

本发明提供了一种仔猪被压死淘率和存活率的统计方法，涉及养殖技术领域。其技术方案为：在猪产房栏位内设置麦克风阵列，麦克风阵列设置在栏位壁的中部，麦克风阵列包括偶数个麦克风，麦克风对称分布朝向两个栏位；通过麦克风采集两个栏位的声源；根据采集到的声音，结合仔猪被压发生建模，判断声音来向，进而确定仔猪被压的栏位，进而统计出仔猪因被压导致的死淘率和存活率具体情况。本发明的有益效果为：本方法结合实际的生产性能指标，可以分析检测的被压事件是否被有效处理，并评估其应用可行性，即对成熟的商业产品和方案应用前的评估，提高数智化生产管理水平。

66. 一种风速在线测量设备

本实用新型提供了一种风速在线测量设备，属于风速测量技术领域。其技术方案为：一种风速在线测量设备，包括导风框，导风框的前端两侧设有立板，两个立板上设有向内弯折的挡板，立板之间滑动设有探头连接杆，挡板上且位于挡板及导风框之间均布有若干弹性支撑部，弹性支撑部能够承载探头连接杆；探头连接杆上均布有若干风速探头。本实用新型的有益效果为：本装置能够快速对风机口进行多点测量，提高风机口测风精度。

67. 无人机

本外观设计产品的名称：无人机。

本外观设计产品的用途：本外观设计产品用于非接触式物资转运。

本外观设计产品的设计要点：在于形状与结构。

68. 人员自助采样系统

本外观设计产品的名称：人员自助采样系统。

本外观设计产品的用途：本外观设计产品用于人员信息自助采集系统，是一种信息采集设备。

本外观设计产品的设计要点：在于形状与结构。

69. 一种基于 3D 视觉技术的猪体重预估方法

本发明提供了一种基于 3D 视觉技术的猪体重预估方法，涉及养殖设备领域。其技术方案为：以 RGBD 相机对猪只进行图像采集；对采集到的图像进行筛选；经过筛选后的图像进入 3D 点云建模通道，获取猪的三维空间模型；根据获取的三维模型，结合体重预估模型获得猪的体重；对所得的体重结果进行评估。本发明的有益效果为：本方案基于机器视觉技术的检测精度虽然在某些方面低于人工检测，但具有检测速度快，长期运行成本低，省时省力等优势；本方案提出的基于 3D 视觉技术估重的方法可以捕捉猪体全局信息，在提高猪体重预估的准确性方面具有很大潜力，有助于提高该技术的可行性和鲁棒性。

70. 一种基于机器视觉技术的猪只图像理想帧的筛选方法

本发明提供了一种基于机器视觉技术的猪只图像理想帧的筛选方法，涉及养殖设备领域。其技术方案为：使用 RGBD 相机进行猪只图像采集，对采集到的图像进行预处理，转换为图像帧进行保存；对获取图像帧中的猪体进行轮廓框选得到轮廓框；根据猪不同生长阶段的体尺，预设对应生长阶段的标准定位框参数；将图像帧中的轮廓框和标准定位框比较，进行一次筛选；根据一次筛选后的图像中轮廓框的倾斜度，进行二次筛选，作为理想帧。本发明的有益效果为：本方案提出的筛选方法有助于机器视觉估重技术在图像采集环节中的效率提升，精准筛选，通过降低图像冗余度，提高图像库的质量，以确保视觉估重技术输出结果的准确可靠。

71. 一种通用无人车底盘

本实用新型提供了一种通用无人车底盘，属于自动化设备技术领域。其技术方案为：一种通用无人车底盘，包括承载板，承载板设置于机壳上，机壳下方设有底盘；承载板上设有机械安装位及穿线孔，机械安装位能够安装上装装置，机壳上设有系列航空插座作为与上装装置连接的通用电气接口，上装装置的连接线能够穿过穿线孔通过航空插头与相应的航空插座连接。本实用新型的有益效果为：本装置能够承载多种功能的上装装置，组合成各种自动化机器人，适应不同的使用环境和作业操作；具有统一的机械安装位及电气接口，更换上装装置简单；设有检修门，便于后期的电路维修；防撞防水性能良好。

72. 一种猪场粪便转运装置

本实用新型提供了一种猪场粪便转运装置，属于畜牧养殖技术领域。其技术方案为：一种猪场粪便转运装置，其特征在于包括设置养殖区外侧的转运泵，转运泵的进粪口处通过管道连接有中转容器，中转容器的上端设置有转运通道，转运通道与养殖区内的绞龙输送装置连接，转运泵的出粪口通过管道与粪便处理区连接，靠近出粪口的管道上设置有压力阀，压力阀通过回流管与中转容器的上部连接，靠近粪便处理区的管道上设置有手动控制阀。本实用新型的有益效果为：可以代替车辆将养殖区内的粪便转运到粪污处理区的，且安全性高。

73. 一种基于深度学习模型的仔猪被压检测方法

本发明提供了一种基于深度学习模型的仔猪被压检测方法，涉及养殖技术领域。其技术方案为：在猪产房栏位内设置麦克风阵列；通过麦克风采集两个栏位的声音；根据采集到的声音，判断声源位置；结合基于 CNN 和声谱图所建立的仔猪被压发声模型，判断仔猪是否出现被压情况。本发明的有益效果为：本方案提供的方法将解放产房对劳动力的需求，降低企业的生产管理成本，提高畜牧养殖的数智化水平；此外，从生产角度来看，通过本方案的方法，可以降低仔猪死淘率，解救更多的仔猪，使其可转化为继续培育的生长育肥猪，对进一步提高畜牧企业的经营利润有很大帮助。

74. 一种用于畜禽场关口的多功能消毒房

本实用新型提供了一种用于畜禽场关口的多功能消毒房，属于畜禽物资生物安全防控技术领域。其技术方案为：一种用于畜禽场关口的多功能消毒房，包括一端侧壁上设有污区门另一端侧壁上设有净区门的箱体，以及设置在第一通道与第二通道之间的浸泡组件货架和消毒箱；箱体内设与污区门相连通的第一通道以及与净区门相连通的第二通道；第一通道与第二通道相互隔断。本实用新型的有益效果为：将不同功能的消毒作业集成，占地面积及建设成本低，管控效果好，作业顺畅且安全，生物安全防护效果好。

75. 一种用于畜禽场关口的高温消毒房

本实用新型提供了一种用于畜禽场关口的高温消毒房，属于畜禽物资生物安全防控技术领域。其技术方案为：一种用于畜禽场关口的高温消毒房，其中包括箱体，设置在箱体一端部的净区门，相对应设置在箱体另一端部的污区门，相对应设置在箱体内两端的两个电热风机，以及首尾依次靠接设置在箱体内的若干个货架；箱体内设有与污区门相连通的第一通道，以及与净区门相连通的第二通道，第一通道与第二通道相互隔断，且第一通道第二通道位于若干个货架两侧。本实用新型的有益效果为：污区净区之间无交叉，确保进场物资的生物安全，结构简单，布置合理，建造及维护成本低，消毒过程实时监测且消毒效果好。

76. 一种用于畜禽场关口人员进场的洗消管控洗澡间

本实用新型提供了一种用于畜禽场关口人员进场的洗消管控洗澡间，属于畜禽饲养安全防控技术领域。其技术方案为：一种用于畜禽场关口人员进场的洗消管控洗澡间，包括洗澡模块和换衣模块，换衣模块包括输入端位于污区的污区换衣模块，以及输出端位于净区的净区换衣模块，洗澡模块包括与污区换衣模块的输出端相连通的第一走廊，与净区换衣模块的输入端相连通的第二走廊，以及分别设有 AB 门的若干个洗澡间；AB 门中其中一个门与第一走廊相连通，另一个门与第二走廊相连通。本实用新型的有益效果为：方便对人员流动进行监督与控制，满足生物安全要求，模块化设计方便拆装。

77. 一种可便携拆装的畜用微创采样器

本实用新型提供了一种可便携拆装的畜用微创采样器，包括手柄与针头，针头上设置有取样孔，手柄一端设置有弹性夹头，针头尾端位于弹性夹头内部，手柄柄体与弹性夹头之间设置有连接块，连接块外部螺纹连接有连接件，手柄外部还设置有防护套，防护套一端通过弹簧与手柄尾端连接，防护套外壁设置有限位结构。本实用新型的有益效果为：设计合理，结构简单，安全可靠，可便携操作，方便作业人员使用，并可提高工作效率。

78. 一种可快速拆装的畜用微创采样器

本实用新型提供了一种可快速拆装的畜用微创采样器，包括手柄与针头，针头设置有取样孔，尾端设置有固定块，手柄的一端设置有外螺纹，手柄与针头之间通过连接件固定，手柄的尾端设置有限位管，限位管内固定连接有弹簧，弹簧的另一端固定连接有套环，套环转动连接有防护套，防护套包裹与手柄外壁，防护套靠近套环一端设置有限位板，限位管靠近防护套一侧设置有限位壁，限位壁上设置有与限位板配合的通槽。本实用新型的有益效果为：设计合理，结构简单，安全可靠，可便携操作，方便作业人员使用，并可提高工作效率。

79. 一种动物口腔液采样用装置

本实用新型提供了一种动物口腔液采样用装置，包括手柄与取样结构，取样结构包括与手柄固定连接的储液容器，储液容器连接于手柄的一端，储液容器顶部活动连接有取液板，取液板为弧形板，板面上均匀分布有若干滤液孔，取液板通过限位结构与储液容器活动连接，限位结构包括卡扣与卡板，卡板为条状的弧形板，位于储液容器的内壁顶部，卡扣位于取液板底边，且卡扣与卡板配合卡接。本实用新型的有益效果为：设计合理，结构简单，安全可靠，可解决动物口腔采样难度大、时间久、成本代价高的问题。

80. 一种鼠洞取样装置

本实用新型提供了一种鼠洞取样装置，包括固定架，固定架顶部连接有固定杆，固定架铰接有旋转架，旋转架底部固定连接有吸水件，固定架内侧设置有两个挤水结构，两个挤水结构的间距小于吸水件的宽度，挤水结构内部中空，且外壁设置有若干滤水孔，旋转架上设置有与固定架铰接的旋转轴，旋转轴的一端连接有电机。本实用新型的有益效果为：设计合理，结构简单，安全可靠，可有效并准确采集老鼠携带的病毒。

81. 一种用于猪喉部液体快速采样装置

本实用新型提供了一种用于猪喉部液体快速采样装置，包括辅助筒，辅助筒的一端为封闭，另一端为开口，辅助筒内固定连接有弹簧，辅助筒的外壁活动连接有若干控制件，每个控制件的顶端均设置有限位块，每个限位块均贯穿辅助筒外壁的通孔延伸至辅助筒内部，本装置还包括取样棒，取样棒的一端设置有吸水棉。本实用新型的有益效果为：设计合理，结构简单，安全可靠，可快速对猪只喉部液体取样且可有效避免交叉感染。

82. 一种畜用组织微创采样器

本发明提供了一种畜用组织微创采样器，属于畜牧业专用器械技术领域。其技术方案为：一种畜用组织微创采样器，其中包括针筒，贯穿针筒两端设置的推杆，设置在针筒尾端，活动连接在推杆相对应端部的封头，设置在针筒头端，套接在推杆相对应端部，且与推杆沿其中心线方向滑动配合的套针。本发明的有益效果为：采样高效，采样时间短暂，活体组织流血少，结构及操作简单，设计巧妙，适应性好，省时省力，应用价值高。

83. 一种畜用组织医疗采样器

本发明提供了一种畜用组织医疗采样器，属于畜牧业医疗器械领域。其技术方案为：一种畜用组织医疗采样器，其中包括针筒，穿设于针筒尾端，轴线与针筒的轴线平行的推杆，轴线与针筒的轴线平行，一端设置在针筒头端的套针，以及一端设置在推杆内端，另一端穿设于套针内顶杆。本发明的有益效果为：结构简单，可靠性高，适应性好，可维护性强，使得对淋巴组织的采样工作更加高效便捷。

84. 一种畜用组织微创采样器

本实用新型提供了一种畜用组织微创采样器，属于畜牧业专用器械技术领域。其技术方案为：一种畜用组织微创采样器，其中包括针筒，贯穿针筒两端设置的推杆，设置在针筒尾端，活动连接在推杆相对应端部的封头，设置在针筒头端，套接在推杆相对应端部，且与推杆沿其中心线方向滑动配合的套针。本实用新型的有益效果为：采样高效，采样时间短，活体组织流血少，结构及操作简单，设计巧妙，适应性好，省时省力，应用价值高。

85. 一种畜用组织医疗采样器

本实用新型提供了一种畜用组织医疗采样器，畜牧业医疗器械领域。其技术方案为：一种畜用组织医疗采样器，其中包括针筒，穿设于针筒尾端，轴线与针筒的轴线平行的推杆，轴线与针筒的轴线平行，一端设置在针筒头端的套针，以及一端设置在推杆内端，另一端穿设于套针内顶杆。本实用新型的有益效果为：结构简单，可靠性高，适应性好，可维护性强，使得对淋巴组织的采样工作更加高效便捷。

86. 一种可便捷操作的畜用取样装置

本实用新型提供了一种可便捷操作的畜用取样装置，属于畜牧业专用设备技术领域。其技术方案为：一种可便捷操作的畜用取样装置，包括针筒套针及推杆，套针通过其尾端的针冠设置在针筒头端，套针头端设有针尖取样孔及倒刺，推杆贯穿针筒，其头端穿设于套针内。本实用新型的有益效果为：节省耗材，避免活体之间交叉感染，操作及结构简单，方便携带及批量运输，节省人力和时间成本。

87. 一种猪用咽拭子采样工具生物安全系统的生产车间

本实用新型提供了一种猪用咽拭子采样工具生物安全系统的生产车间，涉及猪用生产设备技术领域。其技术方案为：包括设置在生产车间中部的隔板，隔板一侧依次设置有沥水间、烘干间、储物间一、隔板另一侧设置有臭氧消毒间，臭氧消毒间一侧设置有储物间二。

本实用新型的有益效果为：自动分拣机一侧设置有两个输送带，通过设定分拣可浸泡物与不可浸泡物和物体的密度，大于水的密度的同时又使可浸泡的原料如采样棒或者不锈钢凳子在输送带上传送，一次进入沥水间、烘干间、储物间，最后进入生产间，对不可浸泡的原料如标签或者植绒进入臭氧消毒间。间隔屋的设计可以避免原料或设备与外界接触，在提高生产效率的同时，在提高了生产的安全性。

88. 一种指导非洲猪瘟净化的电子罗盘及其使用方法

本发明提供了一种指导非洲猪瘟净化的电子罗盘及其使用方法，包括罗盘盘体，盘体背面设置有连接件，连接于墙体或栏杆，盘面中心设置有第一指针与第二指针，第一指针手动控制旋转，第二指针通过电机转动，盘体盘面为圆形的液晶屏幕，且盘面被分为六个模块，包括以下步骤：非洲猪瘟疫情发生后，手动调节第一指针指向内一环的其一阶段；第二指针从"0"值开始自动进行旋转，一天行进一个间隔；内三环、内四环、内五环与外环均根据第二指针的位置显示文字与颜色；每天通过互动模块触屏点击是否完成操作；疫情稳定至20 d，屏幕熄灭，疫情结束。本发明的有益效果为：设计合理，结构简单，安全可靠，有效对非洲猪瘟疫情进行把控与防治。

89. 一种可便捷操作的畜用取样装置

本实用新型提供了一种可便捷操作的畜用取样装置，属于畜牧业专用设备技术领域。其技术方案为：一种可便捷操作的畜用取样装置，包括针筒、套针及推杆，套针通过其尾端的针冠设置在针筒头端，套针头端设有针尖、取样孔及倒刺，推杆贯穿针筒，其头端穿设于套针内。本实用新型的有益效果为：节省耗材，避免活体之间交叉感染，操作及结构简单，方便携带及批量运输，节省人力和时间成本。

90. 一种猪舍用便携式保湿空气采样器

本发明提供了一种猪舍用便携式保湿空气采样器，包括筒体，筒体内设置有采样材料，采样材料通过固定机构固定设置于筒体内，固定机构包括压板，压板包括上压板与下压板，采样材料位于上压板与下压板之间，固定机构滑动连接于筒体内，筒体一端设置有锁定机构，锁定机构包括若干设置于筒体端部的爪钩，爪钩均通过驱动机构驱动向内弯曲，筒体端部开设有注水孔，注水孔连通筒体内的凹槽，当采样材料固定于筒体内时，采样材料与凹槽连通。本发明的有益效果为：设计合理，结构简单，安全可靠，可长时间保持舍内风机口处采样材料的湿润，以提高猪舍内 PCV、PRV 等 DNA 病原的采样效率。

91. 一种猪用大栏病原微生物采集装置

本实用新型提供了一种猪用大栏病原微生物采集装置，包括基体，基体为球状，一端连接有杆体；基体的表面开设有若干镶嵌槽，镶嵌槽内连接有采集件，每个采集件上均设置有若干刷毛，基体内部设置有超声清洁器。本实用新型的有益效果为：设计合理，结构简单，安全可靠，可吸引猪只撕咬从而收集口腔液碎片，以便进行抗原、抗体检测。

图书在版编目（CIP）数据

现代养猪应用研究进展 / 闫之春主编 . —北京：
中国农业出版社，2022.9
（现代养猪前沿科技与实践应用丛书）
ISBN 978-7-109-29992-4

Ⅰ.①现…　Ⅱ.①闫…　Ⅲ.①养猪学－研究　Ⅳ.
①S828

中国版本图书馆 CIP 数据核字（2022）第 170063 号

现代养猪应用研究进展
XIANDAI YANGZHU YINGYONG YANJIU JINZHAN

中国农业出版社出版
地址：北京市朝阳区麦子店街 18 号楼
邮编：100125
责任编辑：王森鹤　周晓艳
版式设计：杜　然　责任校对：刘丽香
印刷：中农印务有限公司
版次：2022 年 9 月第 1 版
印次：2022 年 9 月北京第 1 次印刷
发行：新华书店北京发行所
开本：787mm×1092mm　1/16
印张：22.75
字数：510 千字
定价：168.00 元